Smart Antennas for Wireless Communications

IS-95 and Third Generation CDMA Applications

Prentice Hall Communications Engineering and Emerging Technologies Series

Theodore S. Rappaport, Series Editor

LIBERTI & RAPPAPORT *Smart Antennas for Wireless Communications: IS-95 and Third Generation CDMA Applications*

GARG & WILKES *Principles and Applications of GSM*

RAPPAPORT *Wireless Communications: Principles & Practice*

RAZAVI *RF Microelectronics*

STARR, CIOFFI & SILVERMAN *Understanding Digital Subscriber Line Technology*

FORTHCOMING

CIMINI & LI *Orthogonal Frequency Division Multiplexing for Wireless Communication*

GARG & HALPERN *CDMA Networks and CDMA Technologies*

POOR & WANG *Wireless Communication Systems: Advanced Techniques for Signal Reception*

REED & WOERNER *Software Radio: A Modern Approach to Radio Engineering*

TRANTER, KOSBAR, RAPPAPORT & SHANMUGAN *Simulation of Modern Communications Systems with Wireless Applications*

Smart Antennas for Wireless Communications

IS-95 and Third Generation CDMA Applications

Joseph C. Liberti, Jr.
Bellcore
Theodore S. Rappaport
Mobile & Portable Radio Research Group
Virginia Tech

Prentice Hall PTR
Upper Saddle River, NJ 07458
http://www.phptr.com

Library of Congress Cataloging-in-Publication Data

Liberti, Joseph C., Jr.
Smart antennas for wireless communications: IS-95 and third generation CDMA applications/ Joseph C. Liberti, Jr., Theodore S. Rappaport.
p. cm. -- (Prentice Hall communications engineering and emerging technologies series)
Includes bibliographical references and index.
ISBN 0-13-719287-8
1. Adaptive antennas. 2. Wireless communication systems. 3. Code division multiple access. I. Rappaport, Theodore S. II. Title. III. Series.
TK7871.6.L53 1999
621.382'4--dc21 99-11800
CIP

Editorial/production supervision: *Joan L. McNamara*
Cover designer: *Talar Agasyan*
Cover illustrator: *John Christiana*
Cover design director: *Jerry Votta*
Manufacturing manager: *Alan Fischer*
Marketing manager: *Lisa Konzelmann*
Acquisitions editor: *Bernard Goodwin*
Editorial assistant: *Diane Spina*
Book design and layout: *Aurelia Scharnhorst*

Prentice Hall books are widely used by corporations and government agencies for training, marketing, and resale.
The publisher offers discounts on this book when ordered in bulk quantities. For more information, contact Corporate Sales Department. Phone: 800-382-3419; Fax: 201-236-7141; E-mail: corpsales@prenhall.com; Or write: Prentice Hall PTR, Corp. Sales Dept., One Lake Street, Upper Saddle River, NJ 07458

In Chapter 6, Figures 6–7, 6–8, and 6–9 were prepared through collaborative participation in the Advanced Telecommunication Information Distribution Research Program (ATIRP) Consortium sponsored by the U.S. Army Research Laboratory under the Federated Laboratory Program, Cooperative Agreement DAAL01-96-2-0002. The U.S. government is authorized to reproduce and distribute reprints for Government purposes notwithstanding any copyright notation thereon.

Printed in the United States of America
10 9 8 7 6 5 4 3 2 1

ISBN 0-13-719287-8

Prentice-Hall International (UK) Limited, *London*
Prentice-Hall of Australia Pty. Limited, *Sydney*
Prentice-Hall Canada Inc., *Toronto*
Prentice-Hall Hispanoamericana, S.A., *Mexico*
Prentice-Hall of India Private Limited, *New Delhi*
Prentice-Hall of Japan, Inc., *Tokyo*
Prentice-Hall (Singapore) Pte. Ltd., *Singapore*
Editora Prentice-Hall do Brasil, Ltda., *Rio de Janeiro*

To my family - J. C. L., Jr.
To Brenda - T. S. R.

Table of Contents

6 Characterization of Spatio-Temporal Radio Channels 161

9 Direction-Of-Arrival Estimation Algorithms 253

10 RF Position Location Systems 285

Preface

This text has been created to satisfy the growing demand for knowledge in two emerging areas: adaptive antennas (also known as *smart antennas*) and Code Division Multiple Access. CDMA was commercialized in the early 1990s by Qualcomm, Inc., a San Diego, California, company that pioneered the use of a classic military concept for the burgeoning cellular telephone industry. Adaptive arrays, first conceptualized in the 1960s with the birth of digital signal processing, only recently have become practical for deployment; the intense growth rates of wireless services around the world are beckoning for their commercial use.

This text has been developed through years of research by the authors and their colleagues at the Mobile and Portable Radio Research Group of Virginia Tech and at Bell Communications Research. Our goal in creating this text is to provide fundamental and practical information for practicing engineers, students, and researchers in industry as well as in academia. To complement the book, the second author was asked by the Institute of Electrical and Electronics Engineers (IEEE) to provide a compendium of selected readings of key journal papers dedicated to the topic of smart antennas. The compendium, when used in conjunction with this text, provides a convenient single source of literature for use in classrooms or industry short courses.

The material and organization of this book stemmed from the first author's 1995 Ph.D. dissertation on the subject of CDMA and smart antennas. Since then, a great deal of work has transpired in the field, including the adoption of the IS-95 J-STD-008 CDMA standard, the new 14,400 bps voice coder for Rate Set 2 channels, new methods and models for implementation and modeling of smart antennas in CDMA, and the stringent wireless E-911 position location requirement [125m, 67% of the time] imposed by the Federal Communications Commission. We have worked diligently to include up-to-the-minute information in this text.

The text is arranged into 10 chapters. Chapter 1 provides an overview of CDMA and smart antennas; it includes a glossary of terms and a fundamental treatment of synchronous and asynchronous CDMA. Antenna and propagation fundamentals, as they relate to CDMA systems, are also presented. Chapter 2 provides valuable practical information on the IS-95 J-STD-008 standard, and it provides in-depth descriptions of all of the CDMA channels. Also included is an actual link budget design for a PCS CDMA system. Chapter 3 provides fundamental material on adaptive antenna arrays and array theory. The concepts of beamforming, weighting vectors, and fixed-beam vs. adaptive beam antennas are covered. Chapter 4 applies this material to specific CDMA implementations that may be used for today's IS-95 and future CDMA systems. Chapter 5 combines the concepts of CDMA and adaptive antennas to derive analytical expressions that allow wireless system designers to predict the coverage and capacity gains that adaptive antennas provide in a multi-cell CDMA system. This chapter derives classic results that have led to system capacity predictions using CDMA with and without adaptive antennas.

Chapter 6 provides an overview of multipath and Direction-Of-Arrival models for wireless channels. A host of propagation models which are useful for analysis and simulation of adaptive array algorithms are presented. Chapter 7 then describes complete details of one multipath propagation model, the Geometrically Based Single Bounce Elliptical Model, which provides complete characterization of a multipath environment in microcell/picocell applications.

Chapter 8 describes optimal spatial filtering approaches that use both adaptive arrays and characteristics of the CDMA signals. Building on the fundamentals provided in Chapter 3, this chapter presents optimal methods that null interference while maximizing the carrier-to-noise ratio of a desired user. Chapter 9 describes the algorithmic techniques for determining the Direction-Of-Arrival (DOA) of a signal in a multi-user interference environment. Such capabilities will be required for position location techniques. Chapter 10 concludes this text with a thorough treatment of position location algorithms and approaches. Appendix A covers the derivation of the Gaussian Approximation and its many derivatives for spread spectrum systems. Other appendices provide information that engineers and educators may find useful.

The authors wish to acknowledge the invaluable assistance, skill, and patience of Aurelia Scharnhorst, a research associate with Virginia Tech's Mobile and Portable Radio Research Group (MPRG), in formatting this text. The ingenuity and hard work of Zhigang Rong, Rias Muhamed, and George Mizusawa are represented in parts of Chapters 8, 9, and 10 of this book, as portions of their masters' theses have been used with their gracious permission. Other MPRG researchers who played an important role in building the knowledge base presented in this text are Rich Ertel, Kevin Krizman, Neal Patwari, Paulo Cardieri, and Tom Biedka. The authors would also like to thank Prof. M. Zoltowski of Purdue University, Prof. A. Paulraj of Stanford University, M. Feuerstein of Metawave, C. Thompson of Virginia Tech, and Prof. W. Tranter and Prof. B. Woerner of Virginia Tech's MPRG for their review of this text and encouragement to pursue this project. Kevin Sowerby of the University of Auckland, New Zealand also helped inspire this work during his 1997 sabbatical at MPRG.

The authors would also like to thank Joe Wilkes, Paul Zablocky, and Shimon Moshavi of Bellcore, for valuable discussions regarding IS-95. Daniel Devasirvathm, Scott Seidel, and John Koshy provided insight and assistance that allowed the book to become a reality.

This text is the product of funded research supported at Virginia Tech through the MPRG industrial affiliates program. It is our pleasure to bring this book to you, and we hope you find it useful.

J. C. L., Jr.
T. S. R.

CHAPTER 1

Introduction

This text provides fundamental and practical technical information on smart antennas and Code Division Multiple Access (CDMA), two technologies that will play a major role in the future of wireless telecommunications. The information presented here is targeted at engineering professionals and wireless practitioners and is also suitable for graduate level course work.

In the past, wireless systems were deployed using fixed antenna systems, with antenna patterns that were carefully engineered to achieve desired coverage characteristics, but that could not change to react dynamically to changing traffic requirements. Smart antennas are a new technology for wireless systems that use a fixed set of antenna elements in an *array*. The signals from these antenna elements are combined to form a movable beam pattern that can be steered, using either digital signal processing, or RF hardware, to a desired direction that tracks mobile units as they move. This allows the smart antenna system to focus *Radio Frequency* (RF) resources on a particular subscriber, while minimizing the impact of noise, interference, and other effects that can degrade signal quality.

Code Division Multiple Access (CDMA) is a new wireless technology that allows multiple radio subscribers to share the same frequency band at the same time by assigning each user a unique code. The technology makes very efficient use of limited spectral resources and allows robust communication over time-varying radio channels. Both smart antenna technology and CDMA promise to revolutionize the field of wireless communications.

To gain a historic perspective, it is worth noting that land mobile radio was introduced as early as the 1920s to provide two-way communications to automobiles. These radio systems evolved into a number of specialized services and offered paging, dispatch, and two-way voice and data communications to mobile users. In the early 1980s, analog cellular radio systems, including the Advanced Mobile Phone System (AMPS) in the United States, the Total Access

Communications System (TACS) in Europe, and the Japanese TACS System (JTACS) in Japan, were deployed, bringing untethered wireless voice access to the Public Switched Telephone Network (PSTN). In the mid-1990s, digital cellular and Personal Communications Services evolved to provide universal coverage to users, using lightweight, low power, portable subscriber handsets with the clarity of digital voice and enhanced features and services such as digital messaging and caller ID.

The recent explosion of Wireless Local Loop (WLL) in emerging countries has generated tremendous interest in adaptive antennas to provide rapid, inexpensive wireless infrastructure in emerging countries and to allow Competitive Local Exchange Carriers (CLECs) to compete in new markets. Emerging WLL systems which make use of CDMA and adaptive antenna technology have been developed to provide enhanced range, reliability, and capacity.

Recently deployed digital cellular and Personal Communications Systems will be complemented shortly by new systems that combine the clarity and capacity of digital voice systems with high rate data services, allowing video delivery, high speed Internet access, and broadband data services. Smart antennas and CDMA are critical enabling technologies that will allow the efficient use of the radio spectrum needed to provide high mobility in dense traffic areas, as well as broadband wireless access.

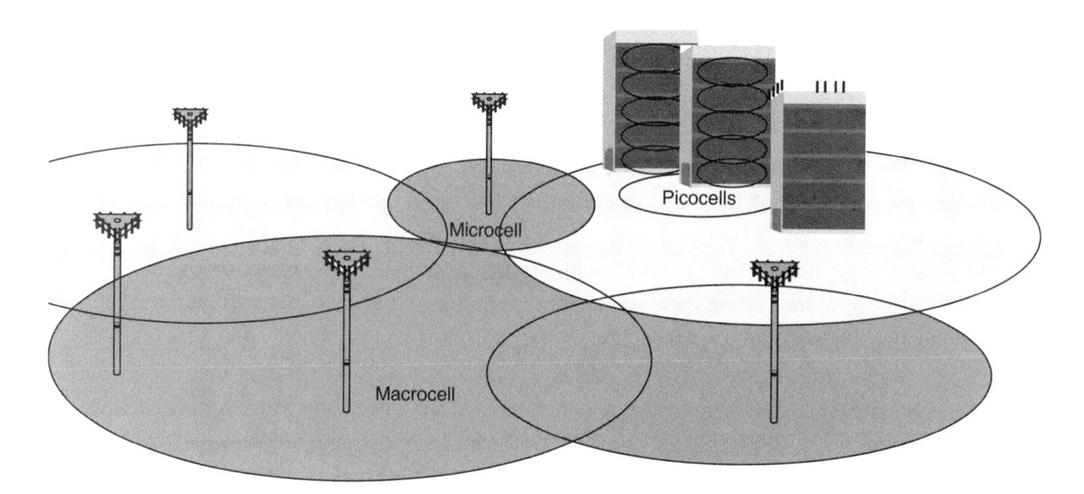

Figure 1–1 Cellular systems provide service by dividing a coverage region into cells. Each cell is served by a base station. Large Macrocells can have radii of several kilometers. Microcells, with ranges of hundreds of meters to a kilometer, provide service to dense, high-capacity areas. Picocells are used to provide spot coverage indoors and in high-traffic-density pedestrian areas.

1.1 The Cellular Radio Concept

The concept behind cellular radio is that a finite spectrum, or bandwidth, allocation is made available throughout a geographical area by dividing the region into a number of smaller cells, as illustrated in Figure 1–1. In traditional analog cellular systems, each cell uses a portion of the spectrum. Cells which are sufficiently far apart can reuse the same spectrum resources [Rap96a]. Modern CDMA systems, as we show later in this chapter, also reuse spectral resources from one cell to the next, but in a somewhat different manner.

The user communicates with the base station in each cell through a set of logical *channels* which are used for paging, access, and traffic. A channel is allocated to a subscriber when it is active in a cell and is released when the portable unit terminates the call or hands-off to another cell.

When a mobile unit is engaged in transmitting or receiving voice signals, data, or fax information, the unit is *active*. A mobile unit which is prepared to receive or place a call, but is not actively transmitting or receiving, is *idle*, as shown in Figure 1–2.

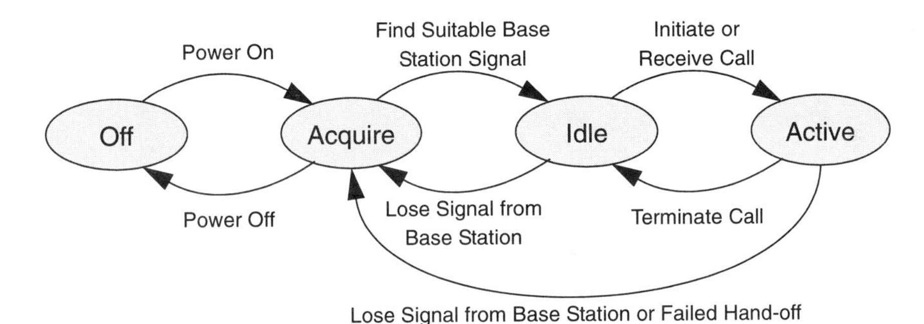

Figure 1–2 Typical state diagram for cellular and PCS subscriber units.

The first cellular radio systems had cell radii of several kilometers. These systems, which primarily served users in vehicles, are referred to as *macro-cellular* systems. Over time, as market penetration increases, the number of users in a given area which must be supported by the system increases. The number of subscribers that can be supported in an area is limited by the spectrum available and the air interface technology. To support more capacity, the power of each base station is lowered, and finite spectrum resources are reused more frequently over a geographic area. This is accomplished using very small cells, or *microcells*, with closely spaced base stations. In general, systems with smaller cells and densely spaced base stations have higher capacity. More recently, PCS systems have emerged with cells as small as a few hundred meters, or even small enough to cover a portion of a single floor of an office building using *picocells*.

While microcell systems offer enhanced capacity, they do so at the expense of high infrastructure costs, since many more base stations are required. Microcell systems can also present difficulties for high speed users that hand-off rapidly from one cell to another. To exploit the benefits of both types of systems, multi-tier systems use a combination of macrocells to serve high speed subscribers, with microcells to support high capacity areas [Rap96a].

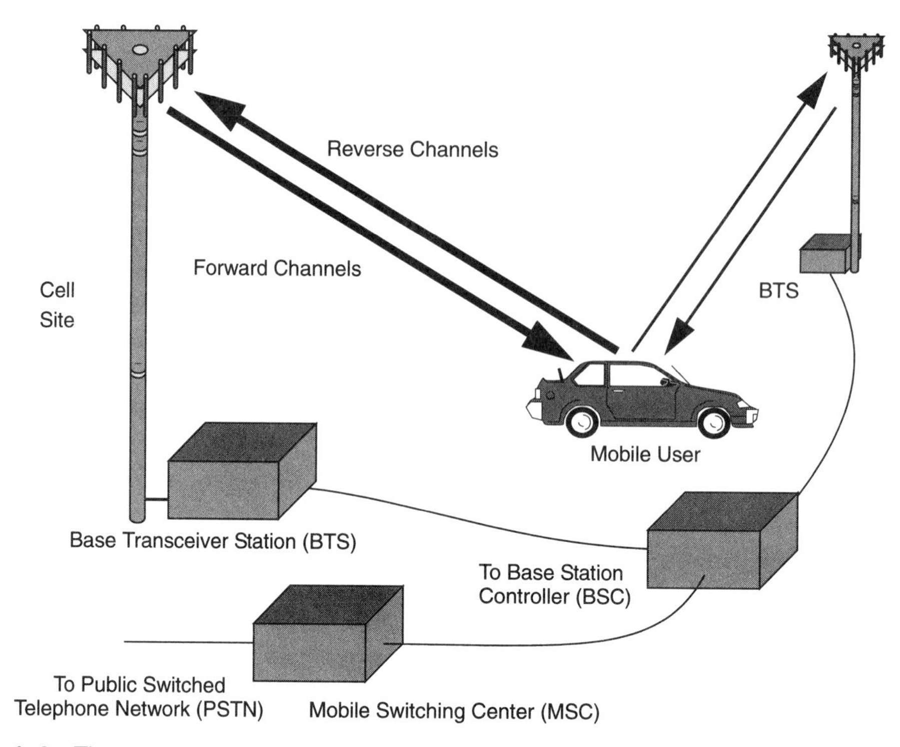

Figure 1–3 There are two main types of forward channels. Control and access channels are used to set up calls and provide security and management functions. Traffic channels are used to carry voice traffic. The reverse channels are also divided into access channels and traffic channels. In some systems, the Base Station Controller (BSC) may be integrated directly into the cell site. In other systems, as shown here, the Base Transceiver Stations (BTSs) are connected to a Base Station Controller.

In wireless communication systems, the radio link from the base station transmitter to the portable receiver is the *forward* link or *downlink*. The link in which the portable unit is the transmitter and the base station is the receiver is called the *reverse* link or *uplink*. Typically, the forward and reverse channels are divided into different types of channels. Control and access channels are used to set up calls and handle other control functions for idle units. The traffic channels are used to carry voice and data information. These channels are illustrated in Figure 1–3. In many systems, the uplink and downlink use channels which are separated in frequency. This technique is called Frequency Division Duplex (FDD). Other systems use the same frequency channel for both the uplink and the downlink, allowing the uplink to use the frequency slot during one time period and the downlink to use the frequency slot during the next time period. This method is called Time Division Duplex (TDD). These concepts are illustrated in Figure 1–4.

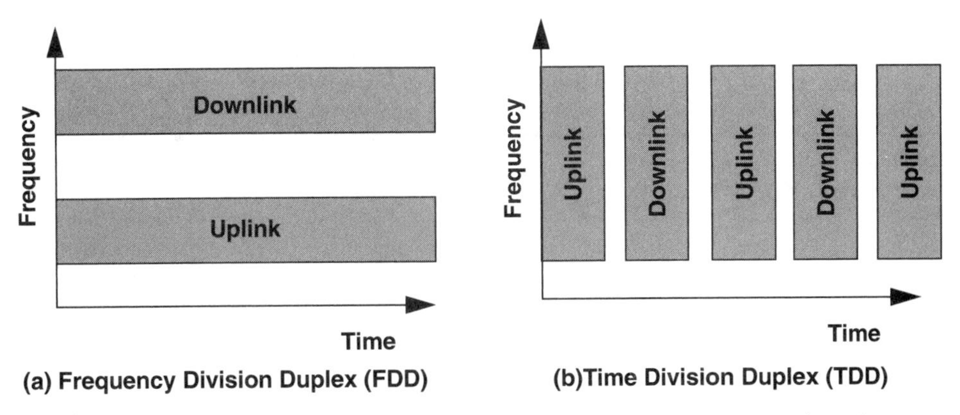

Figure 1–4 Frequency and Time Division Duplex strategies. In FDD systems, since the transmitter and receiver are simultaneously active, they must be widely separated in frequency or highly selective filters must be used to protect a subscriber unit's receiver from its own transmissions. When a limited bandwidth is available, TDD systems are attractive because the subscriber is not receiving while it is transmitting, eliminating the need for tight filtering. TDD is also useful in variable rate, asymmetric bandwidth systems. FDD systems eliminate stringent timing requirements needed by TDD technologies and are suitable for high power, long range systems where propagation delays are significant. TDD can also be complicated to implement in dense reuse schemes without strict inter-base station synchronization.

1.2 Evolution of Wireless Communications

Around the world, a large number of cellular and PCS systems have been proposed and deployed. The first analog services were widely deployed in the mid-1980s. These systems were Frequency Division Multiple Access (FDMA) systems where different channels were separated by giving each channel a unique frequency band. In the United States, the Advanced Mobile Phone System (AMPS) gained widespread acceptance, but the analog modulation scheme constrained capacity.

To overcome the limitations of these analog systems, second generation systems, as shown in Table 1–1, were developed using digital modulation. Digital systems offer many advantages over analog systems, including the ability to reliably recreate a virtually noise-free copy of the transmitted signal at the receiver provided that only a limited number of errors are made in receiving the digital signal. Advanced transmission and signal processing techniques can also be used with digital signals to combat the effects of noise and multipath encountered in mobile environments. Digital technologies also enable discontinuous transmission because users do not talk 100% of the time in both directions. These two features combine to provide higher capacity, better voice quality, and much longer battery life than was possible in analog systems. Second generation systems also incorporate integrated data transmission capabilities. Third generation systems, discussed in Section 1.2.5, add a range of broadband data capabilities.

In order to fully exploit the benefits of digital modulation, FDMA is not the best selection. Since FDMA systems allocate a frequency band to a channel for the duration of a connection, it

Table 1–1 Evolution of Wireless Networks

	First Generation Systems	Second Generation Systems	Third Generation Systems
Time Frame	1984-1996	1996-2000	2000-2010
Services	Analog Mobile Telephony Voice Band Data	Digital voice, messaging	High speed data Broadband video Multimedia
Architecture	Macrocellular	Microcellular, Picocellular Wireless Local Loop	
Radio Technology	Analog FM, FDD-FDMA	Digital modulation, CDMA, TDMA using TDD and FDD	CDMA, possibly combined with TDMA, with TDD and FDD variants
Frequency Band	800 MHz	800+1900 MHz	2 GHz+
Examples	AMPS TACS ETACS NMT450/900 NTT JTACS/NTACS	cdmaOne (IS-95) GSM/DCS-1900 US TDMA IS-136 PACS PHS	cdma2000 WCDMA

is not simple to share an FDMA channel among multiple users or accommodate variable bandwidth signals. A fundamental result from communications theory is that the signals from multiple users may share a transmission medium if their signals can be made orthogonal [Proa89]. In FDMA systems, the signals are made orthogonal by separating channels into distinct frequency bands. The channels may alternatively be separated by making them mutually orthogonal based on some other characteristic, such as the time slot occupied by the channel, an underlying signal property, or even spatial position.

In Time Division Multiple Access (TDMA), the channels are made orthogonal by separating them in time, with all users using the same frequency band. The channels may also be made orthogonal by using a different underlying code sequence for each channel which is orthogonal to the code sequences used by other channels. This is called Code Division Multiple Access (CDMA). These multiple access strategies are illustrated conceptually in Figure 1–5.

In TDMA a time slot, which recurs once per frame, is assigned to each user. The users each transmit at a very high data rate during the brief time slot. For instance, if there are N time slots, each of which is T_s seconds long, then each user, producing data at some data rate R_d, must store data for NT_s seconds and transmit at a data rate NR_d during the T_s second time slot.

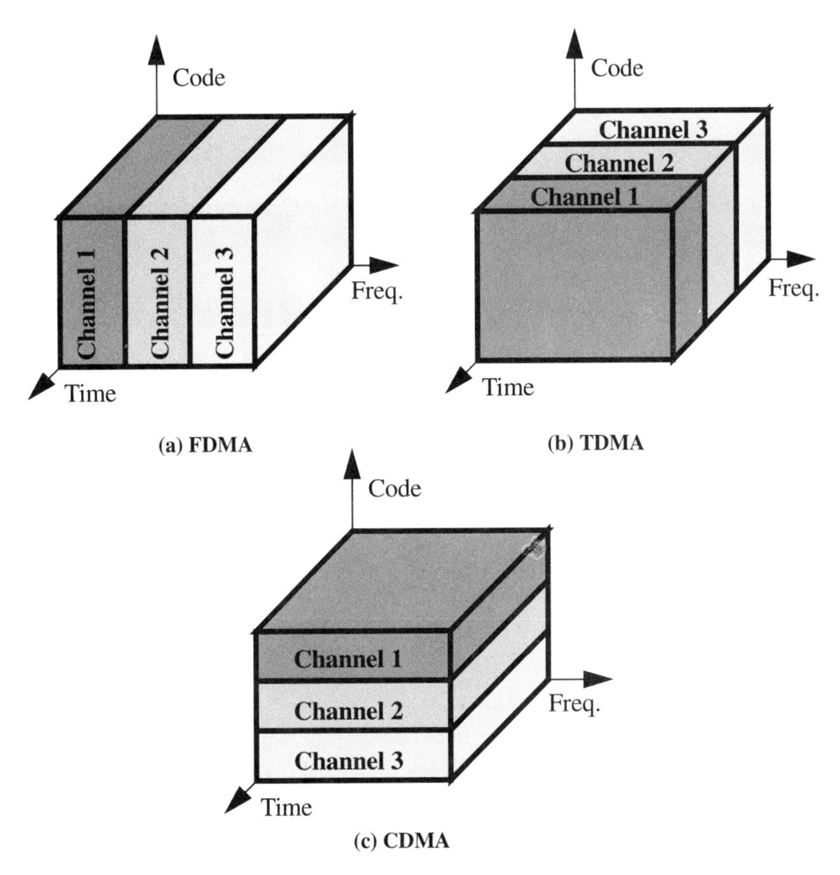

Figure 1–5 Three multiple access schemes: (a) Frequency Division Multiple Access (FDMA), in which different channels are assigned to different frequency bands; (b) Time Division Multiple Access (TDMA), where each channel occupies a cyclically repeating time slot; and (c) Code Division Multiple Access (CDMA), in which each channel is assigned a unique signature sequence code. A system could also combine all three of these multiple access strategies [Jer92].

High data rate users may be accommodated by assigning more than one time slot to a user. For instance, if M time slots are assigned to a user, then they can produce data at a rate MR_d bits per second at the expense of allowing fewer users on the system.

While TDMA users within a cell are separated by their separate time slots, different cells use different frequency channels. There may also be different frequency slots, or carriers within a cell, each of which may support several users. Examples of TDMA cellular systems are the Global System for Mobile Communications (GSM), Digital-AMPS (IS-136), the Personal Access Communications System (PACS), and the Personal Handyphone System (PHS).

Alternatively, in a Direct Sequence CDMA (DS-CDMA) system, all users transmit in the same frequency band at the same time. Different channels are distinguished by assigning a different underlying pseudo-noise (PN) sequence to each channel. The bandwidth of the PN-sequence is much larger than the bandwidth of the data sequence transmitted by the user. The method for despreading the signal at the receiver and the underlying concepts of direct sequence spread spectrum are discussed more fully in Section 1.3.

1.2.1 Key Terms and Concepts in Wireless Communications

To assist the reader, we now present a brief glossary of key terms which are the subject of this book.

Direct Sequence Code Division Multiple Access (DS-CDMA) - A wireless access technique, where signals from different users are assigned a unique *spreading code*. Signals from different users occupy the same spectrum at the same time. The receiver separates signals using their codewords, as shown in Figure 1–6.

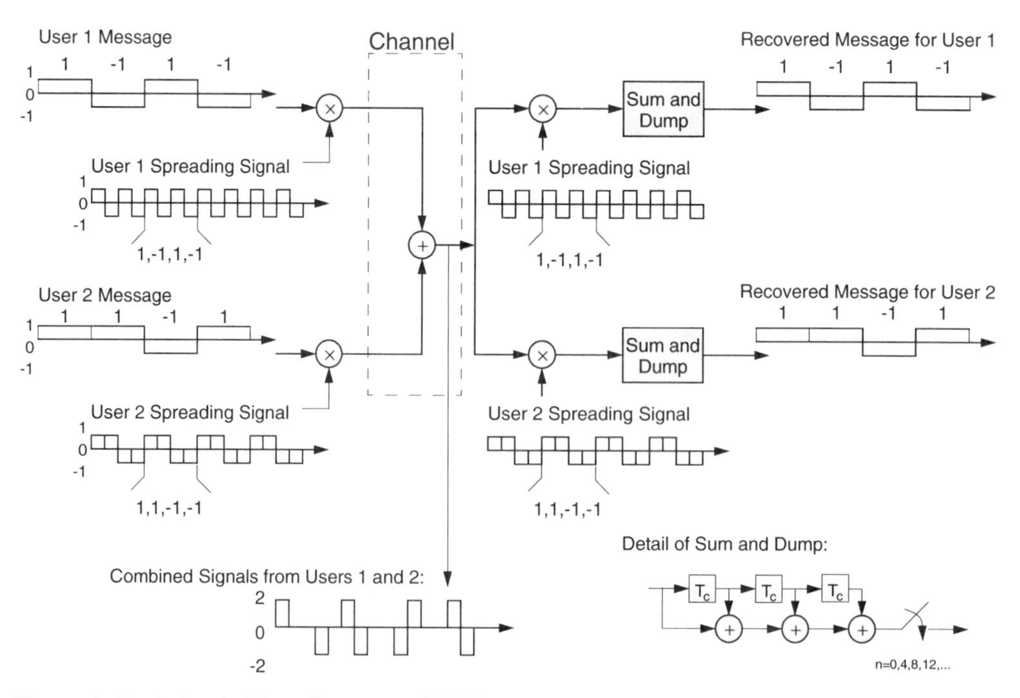

Figure 1–6 A simple Direct Sequence CDMA system. The signals from each user are spread by a spreading code, which is known at the transmitter and receiver. CDMA signals can occupy the same spectrum at the same time.

Adaptive Antenna - An array of antennas which is able to change its antenna pattern dynamically to adjust to noise, interference, and multipath. Adaptive array antennas can adjust their pattern to track portable users. Adaptive antennas are used to enhance received signals and may also be used to form beams for transmission.

Switched Beam Systems - Switched beam systems use a number of fixed beams at an antenna site. The receiver selects the beam that provides the greatest signal enhancement and interference reduction. Switched beam systems may not offer the degree of performance improvement offered by adaptive systems, but they are often much less complex and are easier to retro-fit to existing wireless technologies.

Smart Antennas - Smart antenna systems can include both adaptive antenna and switched beam technologies, as shown in Figure 1–7.

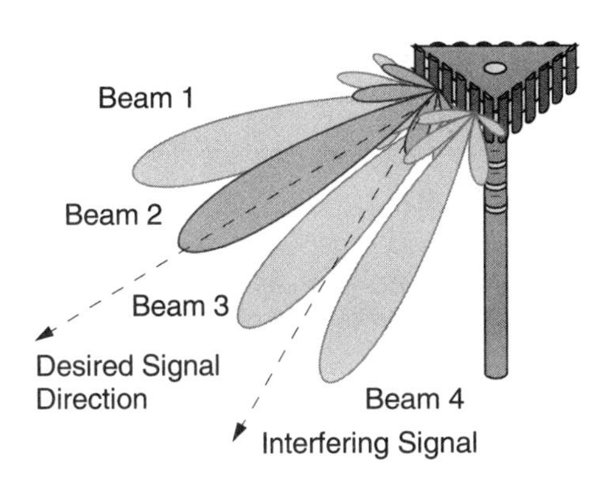

(a) Switched Beam Systems can select one of several beams to enhance receive signals. Beam 2 is selected here for the desired signal.

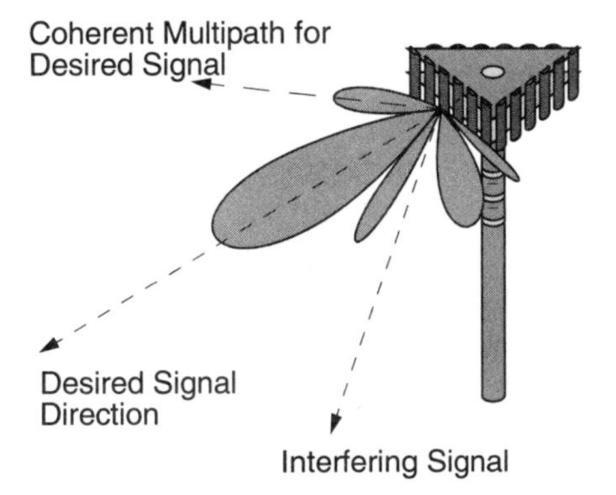

(b) An adaptive antenna can adjust its antenna pattern to enhance the desired signal, null or reduce interference, and collect correlated multipath power.

Figure 1–7 Two smart antenna technologies.

Macrocells - Large cells, covering several kilometers with each base station.

Microcells - Smaller cells, used to provide increased capacity with cell spacing of a few hundred meters to a kilometer.

Picocells - Very small cells, used to provide extremely high capacity indoors and in high-traffic pedestrian areas.

Low Tier Systems - Wireless systems using picocells and microcells to provide low power service to pedestrian users indoors and in pedestrian areas. Typically, low tier systems, such as Digital European Cordless Telephone (DECT) and Personal Access Communications System (PACS), are not as well suited for high vehicle speeds as High Tier systems, and are more susceptible to multipath. However, low tier systems are usually able to support higher voice quality. Because low tier systems do not have to contend with high time delay spread, base stations and subscriber units can be less expensive, and low power requirements allow very long battery life. Low Tier systems excel at covering high density areas, but are less effective in rural, low density, and high speed environments.

High Tier Systems - High Tier Systems use macrocells and microcells to provide wide area, universal coverage. High Tier Systems, such as IS-95 and GSM, use a variety of techniques to combat multipath.

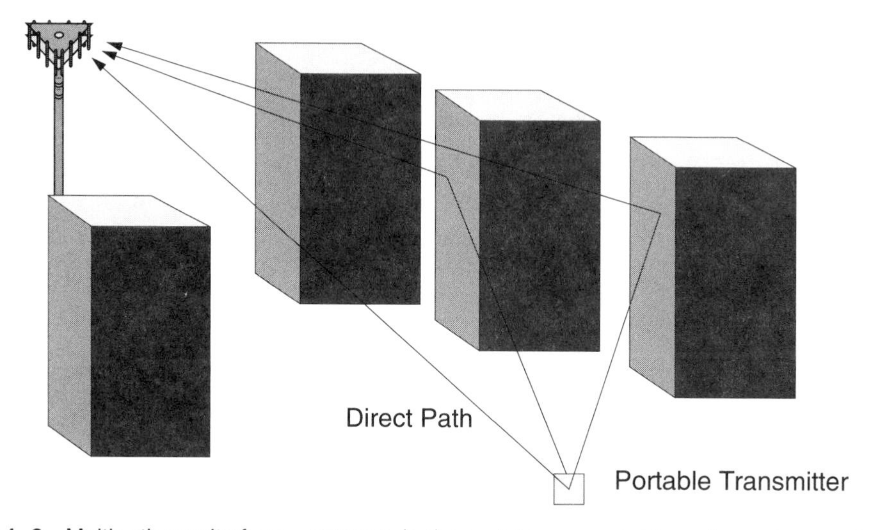

Figure 1–8 Multipath results from scatterers in the mobile and portable radio environment.

Multipath - In mobile and portable radio channels, as shown in Figure 1–8, there are multiple radio paths between the transmitter and receiver. The longer paths result in delayed versions of the desired signal arriving at the receiver. When the difference in delays between the different multipath components, quantified by the *Time Delay Spread*, is large, symbols spread into one another, leading to *InterSymbol Interference* (ISI) at the receiver. This can result in poor signal reception, even when the signal level is high, particularly in TDMA systems. Even when the time delay spread is small, if strong multipath is present, the phases of the multipath signal components can combine destructively over a narrow bandwidth, leading to *fading* of the received signal level [Rap96a].

Digital Modulation Techniques - In digital modulation, transmitted signals are comprised of symbols from a finite alphabet. The key advantage of digital modulation techniques is that if the receiver correctly determines the transmitted symbol based on the received signal, then a perfect, noise-free copy of the transmitted signal can be recreated at the receiver. Unlike analog signals, error control coding can be easily applied to digital signals, providing the ability to add controlled redundancy, which provides robustness to any errors made at the receiver.

One of the simplest digital modulation techniques is Binary Phase Shift Keying (BPSK). In BPSK, a "1" is assigned to a phase shift of $\theta_i = 0°$, and a "0" is assigned to a phase shift of $\theta_i = 180°$, so that the transmitted signal is

$$s(t) = A_c \cos(\omega_c t + \theta_i) \qquad iT_b \le t < (i+1)T_b \tag{1.1}$$

In Quadrature Phase Shift Keying (QPSK) systems, θ_i can take one of four values, such as 45°, 135°, -45°, or -135°. A variation of this approach, called $\pi/4$-shifted Differential QPSK or $\pi/4$-DQPSK, is used in IS-136 and PACS. Another technique used in GSM and DCS-1900, is *Gaussian Minimum Shift Keying* (GMSK).

1.2.2 Digital Cellular and PCS Technologies

In the mid 1990s, second generation Wireless Systems were widely deployed around the world. These systems replaced earlier analog technologies and provided improved voice quality, greater system capacity, and enhanced coverage and reliability. Table 1–2 summarizes the major Second Generation Wireless Systems deployed in Europe and the United States.

Table 1–2 Second Generation Wireless Systems

	cdmaOne, IS95, ANSI J-STD-008	**GSM, DCS-1900, ANSI J-STD-007**	**NADC, IS-54/IS-136, ANSI J-STD-011**	**PACS, ANSI J-STD-014**
Uplink Frequencies	824-849 MHz (US Cellular) 1850-1910 MHz (US PCS)	890-915 MHz (Europe) 1850-1910 MHz (US PCS)	824-849 MHz (US Cellular) 1850-1910 MHz (US PCS)	1850-1910 MHz (US PCS)
Downlink Frequencies	869-894 MHz (US Cellular) 1930-1990 MHz (US PCS)	935-960 MHz (Europe) 1930-1990 MHz (US PCS)	869-894 MHz (US Cellular) 1930-1990 MHz (US PCS)	1930-1990 MHz (US PCS)
Duplexing	FDD	FDD	FDD	FDD
Multiple Access Technology	CDMA	TDMA	TDMA	TDMA
Modulation	BPSK with Quadrature Spreading	GMSK with BT=0.3	$\pi/4$ DQPSK	$\pi/4$ DQPSK
Carrier Separation	1.25 MHz	200 kHz	30 kHz	300 kHz

Continued on next page

Table 1–2 Second Generation Wireless Systems (Continued)

	cdmaOne, IS95, ANSI J-STD-008	GSM, DCS-1900, ANSI J-STD-007	NADC, IS-54/IS-136, ANSI J-STD-011	PACS, ANSI J-STD-014
Channel Data Rate	1.2288 Mchips/sec	270.833 kbps	48.6 kbps	384 kbps
Voice and Control Channels per Carrier	64	8	3	8 (16 with 16 kbps vocoder)
Speech Coding	Code Excited Linear Prediction (CELP) @ 13 kbps, Enhanced Variable Rate Codec (EVRC) @ 8 kbps	Residual Pulse Excited-Long Term Prediction (RPE-LTP) @ 13 kbps	Vector Sum Excited Linear Predictive Coder (VSELP) @ 7.95 kbps	Adaptive Differential Pulse Code Modulation (ADPCM) @ 32 kbps

1.2.3 CDMA Wireless Local Loop

Wireless Local Loop (WLL) systems are used in lieu of wired access for fixed residential and business customers. In WLL deployments, base stations are deployed similar to a cellular service; however, instead of providing service to mobile subscribers, WLL services provide wireline-quality service to fixed users that have traditionally been served by wired links to the PSTN. At the fixed subscriber location, some systems use a wireless link directly to a handset or home base unit, whereas other systems use an antenna mounted outside of the house or business.

WLL systems have been deployed widely in emerging nations, where infrastructure is simply not available and providing access to remote populations would be prohibitively expensive. Whereas wired links to the PSTN can cost approximately $2000 per customer to install, WLL systems can be implemented at less than half that cost, making wireless access an attractive technology for Competitive Local Exchange Carriers (CLECs) hoping to compete with existing wired infrastructure.

Many of these systems use directional antennas mounted outside at the subscriber unit to minimize the impact of reverse link signals in adjacent cells. This technique, used in commercial CDMA Wireless Local Loop products, can help to maximize system capacity at the expense of wireless mobility. Other systems, particularly IS-95-based systems, use non-directional antennas at the subscriber unit. This sacrifices some capacity, but allows the Wireless Local Loop system to be completely compatible and inter-operable with mobile CDMA systems. A sampling of CDMA-based WLL systems is shown in Table 1–3.

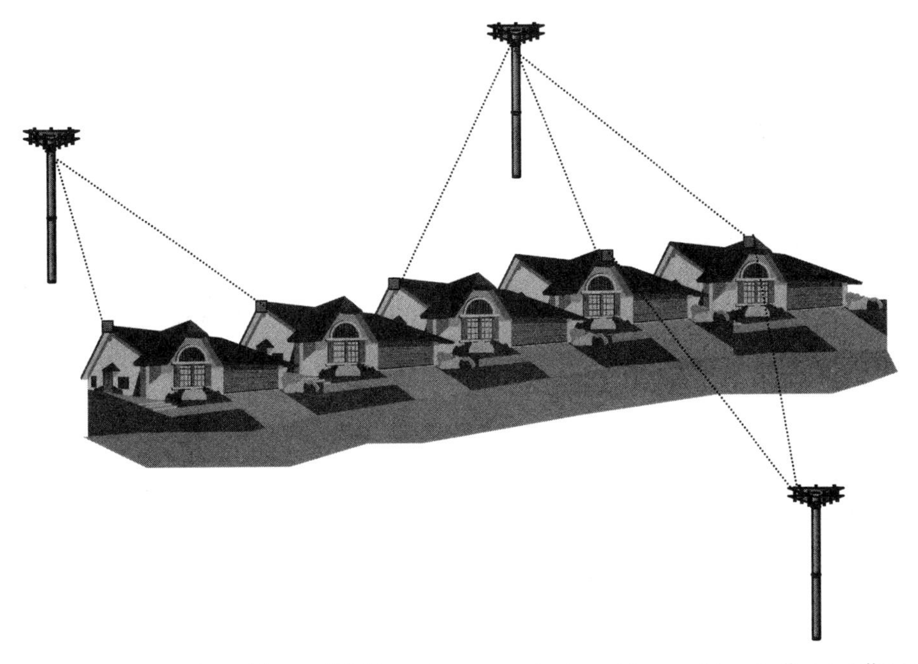

Figure 1–9 Wireless Local Loop (WLL) systems are designed to provide wireline-quality voice and data services to residential and business customers. These systems may link to a Fixed Access Unit mounted on the outside of the subscriber building, or they may attempt to penetrate the structure to provide service directly to the subscriber handset.

Table 1–3 A sampling of typical (non-exhaustive) CDMA Wireless Local Loop Technologies.

Technology	Companies Offering Service	Base Station Antenna/ Subscriber Unit Antenna	Frequency Band	Service
IS-95-Based CDMA	Qualcomm, Sanyo, Nortel Proximity-C, Vodaphone	Sector/Omni	800 MHz, 1900 MHz	Voice, Fax, Low Rate Data
Multi-Gain Wireless™ (MGW™): TDMA/TDD with Frequency Hop Spread Spectrum	Tadiran, Raychem	Sector/ Directional	2.4 GHz	Voice, Fax, Low Rate Data
Airloop™: Orthogonal CDMA	Airspan, Lucent	Sector/ Directional	2.2, 2.4, 3.5 GHz	ISDN-based Voice and Data

These systems are deployed in North America, Europe, Asia, and South America in frequency bands allocated for cellular, PCS, and unlicensed operations from 800 MHz to 3.5 GHz.

1.2.4 MMDS and LMDS

In addition to voice services, other new wireless technologies can add broadband data, video, and high-speed Internet access for fixed users. The Multichannel Multipoint Distribution Service (MMDS) in 2150-2160 MHz, and 2500-2700 MHz provides "wireless cable television" service, with some limited uplink capability [Rap96a, Ch. 10]. Recently, Local Multipoint Distribution Service (LMDS) spectrum at 28 GHz has been allocated, promising 1300 MHz of spectrum to provide these services to fixed users, as shown in Figure 1–10 [Sei98].

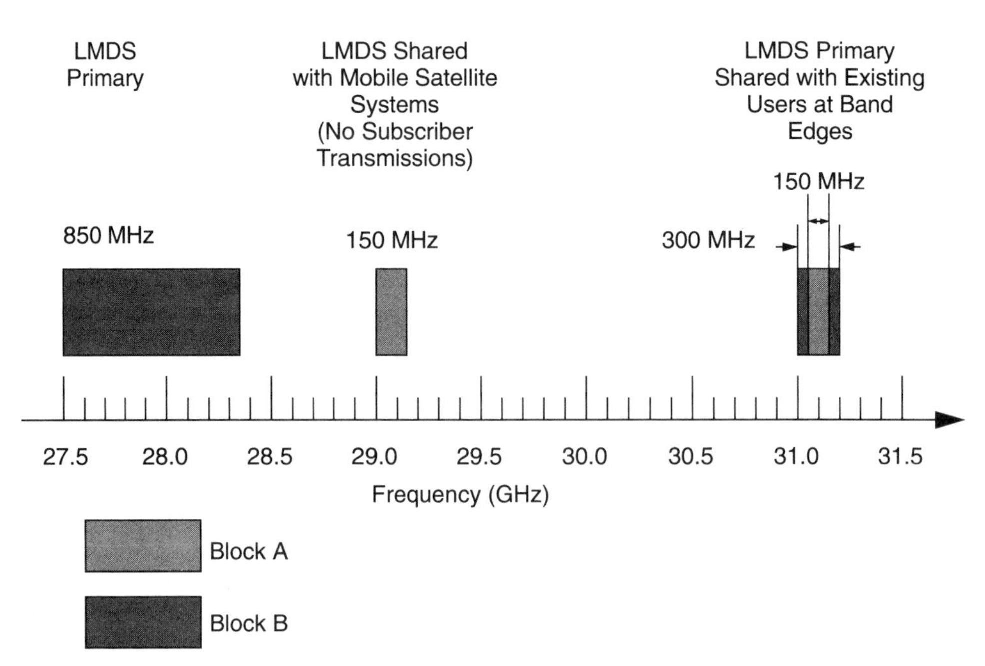

Figure 1–10 The Local Multipoint Distribution Service (LMDS) provides 1300 MHz in the 28-31 GHz band [Sei98] (©1998 IEEE).

1.2.5 Third Generation (3G) Wireless Systems

Just as second generation digital wireless networks have been widely deployed, work is well underway to develop third generation (3G) wireless networks. These networks will add broadband data to support video, Internet access, and other high speed data services for untethered devices.

As described in Chapter 1 of [Rap96a], in 1992 the World Administrative Radio Commission (WARC) of the International Telecommunications Union (ITU) formulated a plan to implement a global frequency band in the 2000 MHz range that would be common to all countries for universal wireless communication systems. This visionary concept, originally known as Future Public Land Mobile Telephone Systems (FPLMTS), was renamed International Mobile Telecommunications 2000 (IMT-2000) in 1995.

IMT-2000 was originally intended to define a single, ubiquitious wireless communication standard for all countries throughout the world, thereby promoting truly ubiquitious wireless communications systems and equipment for all citizens of the world through commonality in infrastructure and handset design. However, in the mid 1990s, it became clear that this vision of commonality would not be tenable due to the great commercial success of each of the various digital wireless technologies. Proponents of the leading European digital wireless technology, based upon TDMA and the GSM standard, and the leading North American digital technology, based upon IS-95 CDMA, have been unable, up until now, to come to agreement on a single united third generation (3G) wireless technology that could be standardized throughout the world. Thus, the ITU, in recent years, expressed a desire to determine a family of standards that would be implemented in a common frequency band throughout the globe.

In June 1998, ITU received a total of 15 comprehensive IMT-2000 proposals from the leading governmental and industrial standards bodies throughout the world. Five of these proposals dealt with satellite communication systems and ten of these dealt with terrestrial PCS-like wireless systems. It is remarkable to note that most of the satellite and terrestrial proposals include some form of CDMA in the air interface protocol. The Outdoor IMT-2000 systems will be deployed with a Mobile Transmit Frequency of 1920-1980 MHz and a base station transmit band of 2110-2170 MHz. Indoor systems will use *Time Division Duplexing* (TDD).

At the time of this writing, the ITU is scheduled to select a family of "winning" world standards within the next few months. The leading U.S. proposal, cdma2000, will be backward-compatible with IS-95 systems (discussed in Chapter 2), but will support chip rates up to 12 times the 1.2288 Mchip/second rate used in the current standard, enabling higher capacity and user data rates [ITU98]. The cdma2000 standard will also incorporate auxiliary pilots to support smart antenna technology, which are discussed further in Chapter 4. It is likely that some variation of cdma2000 will be developed for U.S. markets, regardless of ITU actions [TIA98].

The *European Telecommunications Standards Institute* (ETSI) has developed the *Universal Mobile Telecommunications System* (UMTS) as an evolutionary path for GSM. The *UMTS Terrestrial Radio Access* (UTRA) standard is a *Wideband CDMA* (W-CDMA) technology that includes features that ensure easy integration with existing GSM Technology. The UTRA proposal shares a number of features, including direct sequence spreading rates, frame rate, and slot structure with other proposals from North America (W-CDMA/NA) and Asia (W-CDMA/Japan and CDMA-II from Korea), as illustrated in Table 1–4.

The UWC-136 proposal is designed to provide an upgrade path for both the North American TDMA digital technology (IS-136) and GSM. Unlike most of the other standards, UWC-136 is a TDMA technology, although frequency hopping is supported.

Table 1–4 Ten IMT-2000 candidate standards proposed to ITU in 1998 [ITU98].

<table>
<tr><th>Air Interface</th><th>Mode of Operation</th><th>Duplexing Method</th><th>Key Features and Support of Smart Antenna Technology</th></tr>
<tr><td>cdma2000
US TIA
TR45.5</td><td>Multi-Carrier and Direct Spreading DS-CDMA at $N = 1.2288$ Mcps with N=1,3,6,9,12</td><td>FDD and TDD Modes</td><td>• Backward compatibility with IS-95A and IS-95B. Downlink can be implemented using either Multi-Carrier or Direct Spreading. Uplink can support a simultaneous combination of Multi-Carrier and Direct Spreading.
• Auxiliary carriers to help with downlink channel estimation in forward link beamforming.</td></tr>
<tr><td>UTRA
(UMTS Terrestrial Radio Access)
ETSI SMG2</td><td rowspan="5">DS-CDMA at Rates of $N \times 1.024$ Mcps with N=4,8,16</td><td rowspan="5">FDD and TDD Modes</td><td rowspan="5">• Wideband DS-CDMA System.
Backward compatibility with GSM/DCS-1900.
Up to 2.048 Mbps on Downlink in FDD Mode.
• The collection of proposed standards represented here each exhibit unique features, but support a common set of chip rates, 10 ms frame structure, with 16 slots per frame.
• Connection-dedicated pilot bits assist in downlink beamforming.</td></tr>
<tr><td>W-CDMA/NA
(Wideband CDMA/ North America)
USA T1P1-ATIS</td></tr>
<tr><td>W-CDMA/Japan
(Wideband CDMA)
Japan ARIB</td></tr>
<tr><td>CDMA II
South Korea TTA</td></tr>
<tr><td>WIMS/W-CDMA
USA
TIA TR46.1</td></tr>
<tr><td>CDMA I
South Korea TTA</td><td>DS-CDMA at $N \times 0.9216$ Mcps with N=1,4,16</td><td>FDD and TDD Modes</td><td>• Up to 512 kbps per spreading code, code aggregation up to 2.048 Mbps.</td></tr>
<tr><td>UWC-136
(Universal Wireless Communications Consortium)
USA TIA TR45.3</td><td>TDMA - Up to 722.2 kbps (Outdoor/Vehicular),
Up to 5.2 Mbps (indoor office)</td><td>FDD (Outdoor/ Vehicular)
TDD (Indoor Office)</td><td>• Backward compatibility and upgrade path for both IS-136 and GSM.
• Fits into existing IS-136 RF frequency plan.
• Explicit plans to support adaptive antenna technology.</td></tr>
<tr><td>TD-SCDMA
China Academy of Telecommunication Technology (CATT)</td><td>DS-CDMA 1.1136 Mcps</td><td>TDD</td><td>• RF channel bit rate up to 2.227 Mbps
• Use of smart antenna technology is fundamental (but not strictly required) in TD-SCDMA</td></tr>
<tr><td>DECT
ETSI Project (EP)
DECT</td><td>1150-3456 kbps TDMA</td><td>TDD</td><td>• Enhanced version of 2G DECT technology.</td></tr>
</table>

As government and industry bodies struggle over the complex technical and market issues surrounding these 3G proposals, the IMT-2000 vision is likely to be delayed. Nevertheless, the IMT-2000 proposals illustrate the flexibility of CDMA for providing high capacity, wideband user data, and multimedia transmission.

1.3 Spread Spectrum and Code Division Multiple Access

Traditionally, in radio communication systems, the carrier was modulated with user data using techniques that minimize the transmitted bandwidth to conserve spectrum resources. This was because radio systems were designed so that only a single channel occupied a given frequency band, as illustrated in Figure 1–5(a). If signals are transmitted in multiple non-overlapping frequency bands, they do not interfere with each other and the signals may each be recovered, provided that the power levels are high enough relative to the noise which is always present in the channel. In a spread spectrum system, rather than trying to minimize the bandwidth of the modulated signal, the goal is to create a modulated signal that uses a large amount of bandwidth. There are two main types of spread spectrum systems: direct sequence (DS) and frequency hop (FH).

1.3.1 Direct-Sequence Spread Spectrum

In DS systems, a narrowband signal, containing a message with bandwidth B_1, is multiplied by a signal with a much larger bandwidth B_2, which is called the spreading signal. Essential features of Direct Sequence Spread Spectrum are that

- The bandwidth of the spreading signal, B_2, is much larger than the bandwidth of the message signal, B_1, and
- The spreading signal is independent of the message signal.

Thus, assuming that the bandwidth of the spreading signal is much larger than the bandwidth of the narrowband information signal, the transmitted signal will have a bandwidth which is essentially equal to the bandwidth of the spreading signal.

The spreading signal is comprised of symbols that are defined by a pseudo random sequence which is known to both the transmitter and receiver. These symbols are called *chips*. Typically, the rate of chips in the spreading code, R_c, is much greater than the symbol rate, R_d, of the original data sequence. The pseudo random chip sequence is also called a *Pseudo Noise* (PN) sequence because the power spectral density of the PN-sequence looks approximately like white noise filtered to have the same spectral envelope as a chip. The spreading factor, or processing gain, is defined as the ratio of the chip rate to the data symbol rate, or $N = R_c/R_d$.

To illustrate the operation of a DS spread spectrum system, consider an information signal $b(t)$. This signal may be a voice or data signal. We will assume that $b(t)$ is a digital signal composed of a sequence of symbols, b_j, each of duration T_s. The data signal is given by

$$b(t) = \sum_{j=-\infty}^{\infty} b_j \Psi\left(\frac{t - jT_s}{T_s}\right) \tag{1.2}$$

where $\Psi(t/T)$ is the unit pulse function:

$$\Psi(t/T) = \begin{cases} 1 & 0 \le t < T \\ 0 & \text{otherwise} \end{cases} \tag{1.3}$$

This signal is multiplied by a spreading sequence $a(t)$ which is composed of a sequence of chips:

$$a(t) = \sum_{j=-\infty}^{\infty} \sum_{i=0}^{M-1} a_i \Psi\left(\frac{t-(i+jM)T_c}{T_c}\right) \qquad |a_i| = 1 \tag{1.4}$$

where T_c is the chip period, and M is the number of PN symbols in the sequence before the sequence repeats. It should be noted that the chips serve to spread and identify the signal, whereas the data symbols convey information. In the IS-95 standard, spreading symbols are represented by quadrature chips having a carrier initial phase from the symbol set $\{e^{j\pi/4}, e^{-j\pi/4}, e^{j3\pi/4}, e^{-j3\pi/4}\}$ [EIA93].

The multiplied signal, $a(t)b(t)$, is upconverted to a carrier frequency f_c by multiplying $a(t)b(t)$ by the carrier,

$$s(t) = a(t)b(t)\cos(\omega_c t) \tag{1.5}$$

Let us assume, for the moment, that the channel does not distort the signal in any way, so that the received signal, $r(t)$, consists of a weighted version of the transmitted signal with added white Gaussian noise, $n(t)$:

$$r(t) = As(t) + n(t) = Aa(t)b(t)\cos(\omega_c t) + n(t) \tag{1.6}$$

At the receiver, a local replica of the chip sequence, $a(t-\tau_o)$, is generated, where τ_o is a random time offset between 0 and MT_c. To despread the signal at the receiver, the local chip sequence must be "delay locked" with the received signal. This means that the timing offset, τ_o, must be set to zero. One means of doing this is through the use of a *Delay Locked Loop* (DLL). Similarly, a *Phase Locked Loop* (PLL) may be used to create a replica of the carrier, $\cos(\omega_c t)$. Using these two quantities, a decision statistic is formed by multiplying the received signal by the local PN sequence and the local oscillator and integrating the result over one data symbol:

$$\begin{aligned} Z_j &= \int_{jT_s}^{(j+1)T_s} r(t)a^*(t)\cos(\omega_c t)dt \qquad (1.7) \\ &= \int_{jT_s}^{(j+1)T_s} [Ab(t)a(t)\cos(\omega_c t) + n(t)]a^*(t)\cos(\omega_c t)dt \end{aligned}$$

Let us assume that $a(t)a^*(t) = 1$, then

$$Z_j = \int_{jT_s}^{(j+1)T_s} [Ab(t)\cos^2(\omega_c t) + n(t)a^*(t)\cos(\omega_c t)]dt \tag{1.8}$$

$$= \int_{jT_s}^{(j+1)T_s} \left[A\left(\sum_{i=-\infty}^{\infty} b_i \Psi\left(\frac{t-iT_s}{T_s}\right)\right)\cos^2(\omega_c t) + n(t)a^*(t)\cos(\omega_c t) \right] dt$$

$$= A\int_{jT_s}^{(j+1)T_s} \sum_{i=-\infty}^{\infty} b_i \Psi\left(\frac{t-iT_s}{T_s}\right)\cos^2(\omega_c t)dt + \int_{jT_s}^{(j+1)T_s} n(t)a^*(t)\cos(\omega_c t)dt$$

$$= Ab_j\left(\frac{T_s}{2} + \frac{1}{4\omega_c}(\sin(2\omega_c(j+1)T_s) - \sin(2\omega_c jT_s))\right) + \eta$$

where η represents the influence of channel noise on the decision statistic. Assuming that the carrier frequency is large relative to the reciprocal of the bit period, then

$$Z_j = \frac{Ab_jT_s}{2} + \eta \tag{1.9}$$

Therefore, the decision statistic, Z_j, is an estimate, $\hat{b}_j$, of the transmitted data symbol b_j.

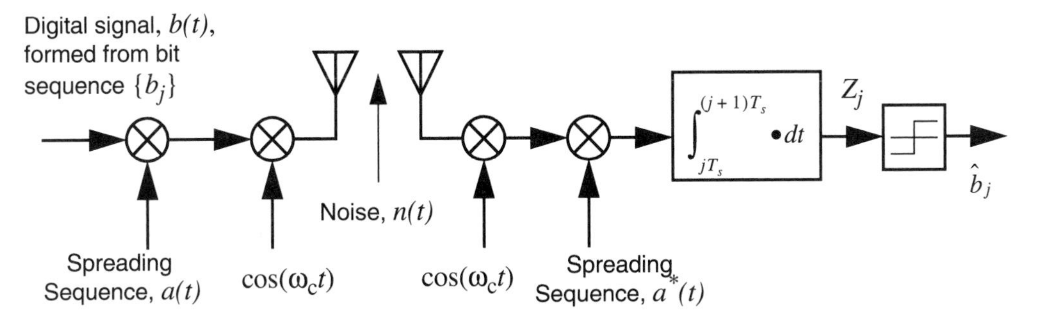

Figure 1–11 A direct sequence spread spectrum transmitter and receiver.

If the channel noise is additive white Gaussian noise with two-sided power spectral density $N_o/2$, then the variable η is a zero mean Gaussian random variable $u_\eta = 0$ with variance

$$\begin{aligned}
\sigma_\eta^2 &= E[|\eta - \mu_\eta|^2] = E[\eta\eta^*] \\
&= E\left[\int_{\lambda=jT_s}^{(j+1)T_s}\int_{t=jT_s}^{(j+1)T_s} n(t)n^*(\lambda)a_0^*(t)a_0(\lambda)\cos(\omega_c t)\cos(\omega_c\lambda)dtd\lambda\right] \\
&= \int_{\lambda=jT_s}^{(j+1)T_s}\int_{t=jT_s}^{(j+1)T_s}\frac{N_0}{2}\delta(t-\lambda)a_0^*(t)a_0(\lambda)\cos(\omega_c t)\cos(\omega_c\lambda)dtd\lambda \\
&= \frac{N_0}{2}\int_{t=jT_s}^{(j+1)T_s}\cos^2(\omega_c t)dt = \frac{N_0}{4}\int_{t=jT_s}^{(j+1)T_s}(1+\cos(2\omega_c t))dt \\
&= \frac{N_0}{4}\left(T_s + \frac{1}{\omega_c}\sin(\omega_c T_s)\cos(\omega_c T_s(2j+1))\right) \\
&= \frac{N_0T_s}{4} \qquad \text{For } \omega_c \gg \frac{1}{T_s}
\end{aligned} \tag{1.10}$$

The energy per symbol of the signal component of the decision statistic, Z_j, is

$$E_s = \frac{A^2T_s^2}{4} \tag{1.11}$$

Let us assume that the symbols, b_j, being transmitted are from a binary symbol set such that $b_j \in \{-1, 1\}$. Then the energy per symbol, E_s, in (1.11) is the same as the energy per bit, E_b. The bit error rate at the output of the matched filter receiver is given by

$$P_e = Q\left(\sqrt{\frac{2E_b}{N_0}}\right) \tag{1.12}$$

where $Q(x)$ is the standard Q-function which is given in Appendix B. The Q-function is a monotonically decreasing function, so that a larger ratio of symbol energy to noise power produces a smaller bit error rate.

The bit error rate given by (1.12) is the same as if we had used no spreading sequence. However, there are several differences between this case and the case where no spreading sequence is used. First of all, since the signal is spread over a large bandwidth with the same total power, the spectral density of the signal can be very small. This makes it difficult for a potential eavesdropper to detect the presence of the signal. This property is known as *Low Probability of Detection* (LPD). Additionally, if the eavesdropper does not know the PN-sequence used, then it will be very difficult to extract the information signal portion of $r(t)$. This property is called *Low Probability of Interception* (LPI) [Proa89].

In mobile and portable radio channels, the fact that the DS signal is spread over a large bandwidth can significantly mitigate the effects of fading which can degrade the performance of narrowband systems. The coherence bandwidth of the channel quantifies the region of spectrum over which the amplitudes of all signal components at the receiver undergo approximately the same fading level. Typical values for coherence bandwidth are 3 MHz in indoor channels and 0.1 MHz for outdoor channels [Rap96a]. If the signal bandwidth is smaller than the coherence bandwidth of the channel and one part of the spectrum of the signal experiences a fade, then, in all likelihood, the entire signal spectrum experiences a fade and it is not possible to recover the lost data symbols except through error control coding. *Using DS techniques, the signal can be made to cover a large enough bandwidth so that if one part of the signal spectrum is in a fade, other parts of the signal spectrum are not faded.* One result of this is that the short-time variation of the total received power for a wide band DS signal is much less than that of the narrowband signal [Ste92]. Various techniques can be used to recover the signal as long as some portion of the received signal is not in a fade.

Finally, one property of the DS signal alluded to earlier is that many DS signals can be overlaid on top of each other in the same frequency band. Let us assume that there are two users in the system using the same frequency band at the same time. We assume that the PN sequence from user 0 is $a_0(t)$ and the PN sequence from user 1 is $a_1(t)$ where

$$a_k(t) = \sum_{j=-\infty}^{\infty} \sum_{i=0}^{M-1} a_{k,i} \Psi\left(\frac{t-(i+jM)T_c}{T_c}\right) \qquad |a_{k,i}| = 1 \tag{1.13}$$

The signals are said to be orthogonal over a symbol period if

$$\int_{jT_s}^{(j+1)T_s} a_0(t) a_1^*(t) dt = 0 \tag{1.14}$$

Let us assume for the moment that the repetition rate of the PN-sequence is the same as the symbol period such that $MT_c = T_s$. When the length of the PN sequence, MT_c, is equal to the message symbol period, T_s, the system is called a *code-on-pulse* system. If the sequences are chip aligned, and bit aligned, then the orthogonality over a symbol period may be expressed as

$$\sum_{j=0}^{M-1} a_{0,j} a_{1,j}^* = 0 \tag{1.15}$$

If each user has a different orthogonal underlying code sequence, then many users can share the same medium without interfering with each other. This property may be used as a multiple access technique, namely Code Division Multiple Access. The case in which the user's sequences are chip and bit aligned is called *synchronous* CDMA.

The received signal at user 0's receiver consists of the desired signal, multiple access interference from user 1, and noise

$$r_0(t) = A_0 b_0(t) a_0(t) \cos(\omega_c t) + A_1 b_1(t) a_1(t) \cos(\omega_c t + \alpha_1) + n(t) \tag{1.16}$$

where

$$b_k(t) = \sum_{j=-\infty}^{\infty} b_{k,j}\Psi\left(\frac{t-jT_s}{T_s}\right) \tag{1.17}$$

The decision statistic for user 0 is obtained using Equation (1.7). The component of the decision statistic due to the desired signal is $A_0T_sb_{0,j}/2$, just as before. The component of the decision statistic due to the noise component is again a zero mean Gaussian random variable with variance $N_0T_s/4$. In synchronous CDMA, the component due to the interference is

$$\int_{jT_s}^{(j+1)T_s} A_1b_1(t)a_1(t)a_0^*(t)\cos(\omega_ct)\cos(\omega_ct+\alpha_1)dt \tag{1.18}$$

$$= A_1b_{1,j}\left(\sum_{j=0}^{M-1} a_{1,j}a_{0,j}^*\right)\int_{jT_s}^{(j+1)T_s} \cos(\omega_ct)\cos(\omega_ct+\alpha_1)dt$$

$$\approx A_1b_{1,j}\left(\sum_{j=0}^{M-1} a_{1,j}a_{0,j}^*\right)\frac{T_s}{2}\cos\alpha_1$$

Thus if the spreading sequences of user 0 and user 1 are orthogonal, so that they satisfy (1.15), then the component of the decision statistic due to interference will be zero. In this case, the addition of the users in the same time and frequency slots will not affect the performance of other users. In most cases, it is possible to operate only in a synchronous CDMA mode on the forward link since the base station can simultaneously modulate and spread the signals for all users. On the reverse link, since it is difficult to synchronize spatially separated mobile users on the bit and chip level, and since the signals travel different path lengths from the transmitter to the base station, it is not feasible to operate in a synchronous mode.

In the asynchronous case, where usually the sequence for user 1 is delayed by τ_1 seconds relative to the chip sequence for user 0, the expression describing the interaction between the signals from two users at the receiver for user 0 becomes considerably more complicated. We may express the delay as an integer number of chip periods γ_1 such that $\tau_1 = \gamma_1T_c + \Delta_1$ where $0 \le \Delta_1 < T_c$. We will further assume that for values of $i \ge M$ and $i < 0$ that $a_{1,i} = a_{1,i-mM}$, so that $0 \le i - mM < M - 1$. Then the signals from user 0 and 1 will not interfere with each other if

$$\left(b_{1,j-1}\sum_{i=0}^{\gamma_1-1} a_{1,i-\gamma_1}a_{0,i}{}^* + b_{1,j}\sum_{i=\gamma_1}^{M-1} a_{1,i-\gamma_1}a_{0,i}{}^*\right)(T_c-\Delta_1) \tag{1.19}$$

$$+\left(b_{1,j-1}\sum_{i=-1}^{\gamma_1-1} a_{1,i-\gamma_1}a_{0,i+1}{}^* + b_{1,j}\sum_{i=\gamma_1}^{M-2} a_{1,i-\gamma_1}a_{0,i+1}{}^*\right)\Delta_1 = 0$$

It is not possible to select useful spreading codes which satisfy (1.19) over all possible values of $b_{1,j}$, Δ_1 and γ_1. Thus, in an asynchronous CDMA system, the signals from different users interfere with each other, resulting in higher bit error rates as compared with orthogonal CDMA. Therefore, in asynchronous CDMA systems, every user contributes interference, called *Multiple Access Interference* (MAI), to the decision statistic for other users.

1.3.2 Multiple Access Interference in DS-CDMA Systems

Consider a CDMA system in which K users occupy the same frequency band at the same time. These users may share the same cell, or some of the K users may communicate with other base stations.

The signal received at the base station from user k is given by [Pur77]

$$s_k(t-\tau_k) = \sqrt{2P_k}a_k(t-\tau_k)b_k(t-\tau_k)\cos(\omega_c t+\phi_k) \tag{1.20}$$

where $b_k(t)$ is the data sequence for user k, $a_k(t)$ is the spreading (or chip) sequence for user k, τ_k is the delay of user k relative to a reference user 0, P_k is the received power of user k, and φ_k is the phase offset of user k relative to a reference user 0. Since τ_k and φ_k are relative terms, we can define $\tau_0 = 0$ and $\varphi_0 = 0$.

Let us assume that both $a_k(t)$ and $b_k(t)$ are binary sequences having values of -1 or +1. As noted earlier, the IS-95 CDMA system, which we will discuss in the next section, uses a somewhat different waveform, with quadrature spreading. The binary values used here, however, are useful for demonstrating some fundamental concepts in DS-CDMA.

The chip sequence $a_k(t)$ is of the form,

$$a_k(t) = \sum_{j=-\infty}^{\infty}\sum_{i=0}^{M-1} a_{k,i}\Psi\left(\frac{t-(i+jM)T_c}{T_c}\right) \qquad a_{k,i}\in\{-1,1\} \tag{1.21}$$

where M is the number of chips sent in a PN sequence period and T_c is the chip period. MT_c is the repetition period of the PN sequence.

For the data sequence, $b_k(t)$, T_b is the bit period. As noted in the introduction to this chapter, it is assumed that the bit period is an integer multiple of the chip period such that $T_b = NT_c$. Note that M and N do not need to be the same. The binary data sequence $b_k(t)$ is given by

$$b_k(t) = \sum_{j=-\infty}^{\infty} b_{k,j}\Psi\left(\frac{t-jT_b}{T_b}\right) \qquad b_{k,j}\in\{-1,1\} \tag{1.22}$$

At the receiver, illustrated in Figure 1–12, the signal available at the input to the correlator is given by

$$r_0(t) = \sum_{k=0}^{K-1} s_k(t-\tau_k)+n(t) \tag{1.23}$$

where $n(t)$ is additive Gaussian noise with two-sided power spectral density $N_0/2$. It is assumed in (1.23) that there is no multipath in the channel with the possible exception of multipath that leads to flat fading such that the coherence time of the channel is considerably larger than a symbol period.

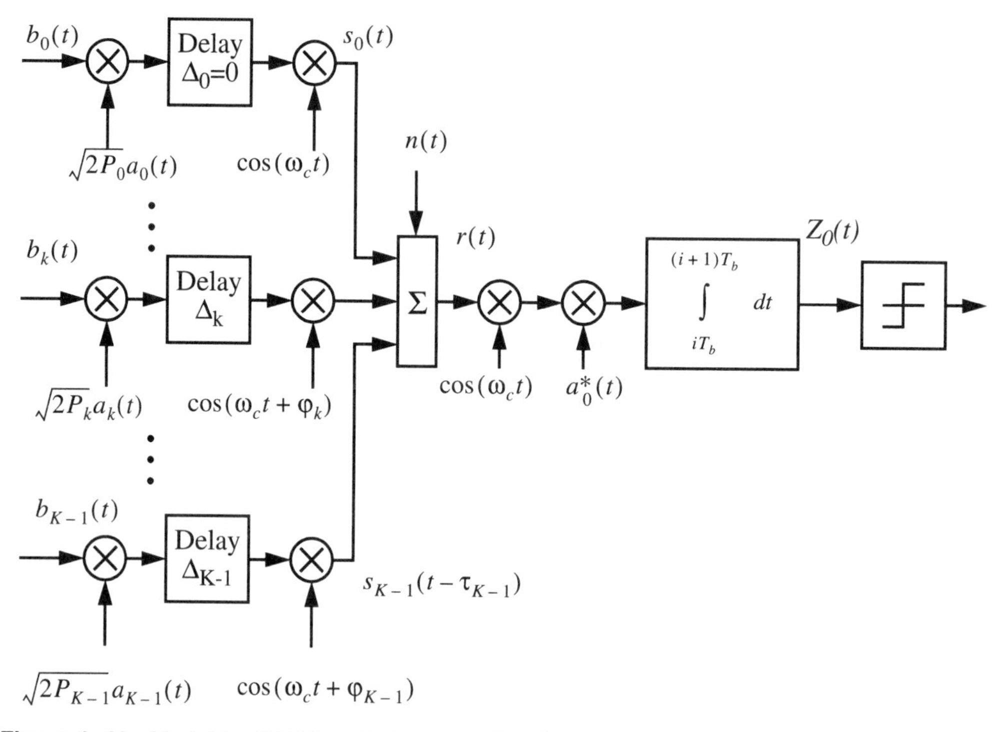

Figure 1–12 Model for CDMA multiple access interference [Pur77] (©1997 IEEE).

At the receiver, the received signal is mixed down to baseband, multiplied by the PN sequence of the desired user (user 0 for example) and integrated over one bit period. Thus, assuming that the receiver is delay and phase synchronized with user 0, the decision statistic for user 0 is given by

$$Z_0 = \int_{jT_b}^{(j+1)T_b} r_0(t)\, a_0^*(t) \cos(\omega_c t) dt \tag{1.24}$$

For convenience and simplicity of notation, the remainder of the analysis will be presented for the case of bit 0 ($j = 0$ in (1.24)). Substituting (1.20) and (1.23) into (1.24),

$$Z_0 = \int_{t=0}^{T_b} \left[\left(\sum_{k=0}^{K-1} \sqrt{2P_k} a_k(t-\tau_k) b_k(t-\tau_k) \cos(\omega_c t + \phi_k) \right) + n(t) \right] a_0^*(t) \cos(\omega_c t) dt \tag{1.25}$$

which may be expressed as

$$Z_0 = I_0 + \eta + \zeta \tag{1.26}$$

where I_0 is the contribution to the decision statistic from the desired user ($k = 0$), ζ is the multiple access interference, and η is the noise contribution. The contribution from the desired user is given by

$$\begin{aligned} I_0 &= \sqrt{2P_0}\int_{t=0}^{T_b} |a_0(t)|^2 b_0(t)\cos^2(\omega_c t)dt \\ &= \sqrt{\frac{P_0}{2}}\int_{t=0}^{T_b}\left(\sum_{i=-\infty}^{\infty} b_{0,i}\Psi\left(\frac{t-iT_b}{T_b}\right)\right)(1+\cos(2\omega_c t))dt \\ &= \sqrt{\frac{P_0}{2}}b_{0,0}\int_{t=0}^{T_b}(1+\cos(2\omega_c t))dt \\ &\approx \sqrt{\frac{P_0}{2}}b_{0,0}T_b \end{aligned} \tag{1.27}$$

As shown in Appendix A, the noise term, η is given by

$$\eta = \int_{t=0}^{T_b} n(t)\ a_0^*(t)\cos(\omega_c t)dt \tag{1.28}$$

where it is assumed that n(t) is white Gaussian noise with two-sided power spectral density, $N_0/2$. The mean of η is

$$\mu_\eta = E[\eta] = \int_{t=0}^{T_b} E[n(t)]\ a_0^*(t)\cos(\omega_c t)dt = 0 \tag{1.29}$$

From (1.10), the variance of η is

$$\sigma_\eta^2 = \frac{N_0}{4}\left(T_b + \frac{1}{2\omega_c}\sin(2\omega_c T_b)\right) \approx \frac{N_0 T_b}{4} \qquad \text{For } \omega_c \gg \frac{2}{T_b} \tag{1.30}$$

The third component in (1.26), ζ, represents the contribution of multiple access interference to the decision statistic. ζ is the summation of $K-1$ terms, I_k,

$$\zeta = \sum_{k=1}^{K-1} I_k \tag{1.31}$$

each of which is given by

$$I_k = \int_{t=0}^{T_b} \sqrt{2P_k} b_k(t-\tau_k) a_k(t-\tau_k) a_0^*(t) \cos(\omega_c t + \phi_k) \cos(\omega_c t) dt \tag{1.32}$$

If the spreading sequences are not orthogonal, then the cross correlation properties of the sequences determine the degree to which one DS signal interferes with another DS signal. For a given length of PN-sequence, there is generally a much larger number of different sequences with low cross-correlation properties than mutually orthogonal sequences. From (A.49), it follows that

$$\sigma_\xi^2 = \sigma_\zeta^2 + \sigma_\eta^2 \tag{1.33}$$

$$= \frac{NT_c^2}{6} \sum_{k=1}^{K-1} P_k + \frac{N_0 T_b}{4}$$

where N is the spreading factor,

$$N = \frac{R_c}{R_b} = \frac{T_b}{T_c} \tag{1.34}$$

where R_c is the chip rate of the PN sequence and R_b is the message bit rate.

In Appendix A it is shown that, using the *Gaussian Approximation*, the Multiple Access Interference can be approximated as a Gaussian random variable. Then we approximate the Bit Error Rate using (A.51) as

$$P_e = Q\left(\frac{\sqrt{\frac{P_0}{2}} T_b}{\sqrt{\frac{NT_c^2}{6} \sum_{k=1}^{K-1} P_k + \frac{N_0 T_b}{4}}}\right) \tag{1.35}$$

$$= Q\left(\sqrt{\frac{P_0}{\frac{1}{3N} \sum_{k=1}^{K-1} P_k + \frac{N_0}{2T_b}}}\right)$$

in the interference limited case, where $\frac{N_0}{2T_b} \ll \frac{1}{3N} \sum_{k=1}^{K-1} P_k$, this becomes

$$P_e = Q\left(\sqrt{\frac{3N}{\sum_{k=1}^{K-1} P_k/P_0}}\right) \tag{1.36}$$

1.3.3 Power Control and the Near-Far Problem in CDMA

Consider the CDMA uplink where multiple subscribers are transmitting signals that are received at a single base station receiver. Generalizing (1.36), the bit error rate for user p can be expressed as

$$P_{e,\,\text{User p}} = Q(\sqrt{3N\gamma_p}) \tag{1.37}$$

where N is the spreading factor and γ_p represents the *Carrier-to-Interference-Ratio* (CIR) for subscriber p. Letting P_p represent the received power from user p, γ_p is

$$\gamma_p = \frac{P_p}{\sum_{\substack{k=0 \\ k \neq p}}^{K-1} P_k} \tag{1.38}$$

As the CIR increases, the bit error rate for a subscriber decreases.

Recall that P_k represents the power received at the base station from a particular subscriber. If every user transmits at the same power level, then the received power from users closer to the base station will tend to be higher than the received power from subscribers that are farther away from the base station. This leads to a different performance for subscriber links, depending on where the subscribers are located in the cell. More importantly, a few subscribers closest to the base station may contribute so much Multiple Access Interference that they prevent other uplink signals from being successfully received at the base station. This is the *near-far* effect [Rap96a].

To solve this problem, *power control* is used in CDMA systems. Power control forces all users to transmit the minimum amount of power needed to achieve acceptable signal quality at the base station. Power control typically reduces the power transmitted by subscribers closest to the base station, while increasing the power of subscribers farthest away from the base station. Compared with the case where all users transmit at the same power level, power control reduces the denominator of (1.38), increasing the CIR, and lowering the error rate for all subscribers.

If all subscribers in a system have the same bandwidth, data rate, and other signal characteristics, then one reasonable approach to power control is to set all of the received power levels, P_k, to a constant value, P_c. Then (1.38) can be rewritten as

$$\gamma_p = \frac{1}{(K-1)} \tag{1.39}$$

This is called *perfect power control*. In practice, this sort of perfect power control is not achievable and under many circumstances, despite its name, may not even be desirable.

Perfect power control requires exact knowledge of the loss in the radio propagation channel between the subscriber transmitter and the base station receiver, commonly referred to as *path loss*. As shown later in this chapter, there are a number of factors that impact the path loss between the transmitter and receiver, causing signal levels to fluctuate with time. In practical systems, the path loss between the uplink path is estimated through a combination of techniques. These include *Open Loop Power Control*, where the path loss on the downlink is measured, and it is assumed that the path loss is approximately the same on the uplink, so the subscriber unit adjusts its power accordingly. To account for the time varying nature of the channel, and the fact that path loss in the forward and reverse directions may not be the same, practical systems also use *Closed Loop Power Control*, where the base station instructs the subscriber unit to raise or lower its transmit power to meet some targeted received power level. This allows the mobile to adjust its transmit power to track the time varying nature of the channel.

Rather than adjust the subscriber transmitter power to meet a desired received power level at the base station, practical CDMA systems use other criteria for adjusting power levels. These include adjusting the power level to meet specific error rate or *Carrier-to-Interference-and-Noise-Ratio* (CINR) requirements. Chapter 2 illustrates that the IS-95 CDMA standard uses a complex power control mechanism that adjusts the subscriber transmitter power 800 times per second.

Assuming perfect power control, (1.36) can be used to illustrate how the addition of subscribers to a CDMA system increases the bit error rate seen by the receiver for any particular subscriber.

In the non-interference limited case, for perfect power control where $P_k = P_0$ for all $k = 0 \ldots K-1$, we have

$$P_e = Q\left(\sqrt{\frac{1}{\frac{K-1}{3N} + \frac{N_0}{2T_b P_0}}}\right) \tag{1.40}$$

In the interference limited case with perfect power control, equation (1.40) may be approximated by

$$P_e = Q\left(\sqrt{\frac{3N}{K-1}}\right) \tag{1.41}$$

Equation (1.41) is plotted in Figure 1–13. As the number of users increases, the raw bit error rate seen by the uplink receiver for any particular receiver increases. CDMA systems use error control coding so they are tolerant of bit errors introduced by Multiple Access Interference. Note that the agreement is excellent for average bit error rates larger than 10^{-3}, where most voice systems operate. However, due to the tail of the Gaussian distribution, the approximation slightly over estimates capacity for bit error rates less than 10^{-3}.

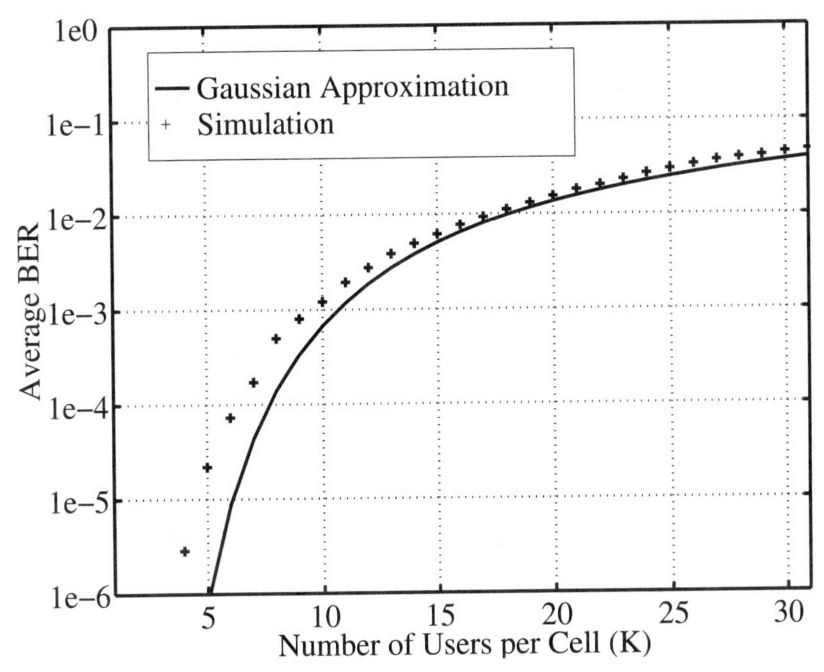

Figure 1–13 The bit error rate for an asynchronous reverse link obtained through computer simulation and through application of the Gaussian Approximation given in (1.41). The spreading factor is $N = 31$ and perfect power control is applied such that all users have the same power level. Omni-directional antennas are used at the base station.

1.3.4 Frequency Hop Spread Spectrum

An alternative to Direct Sequence (DS) is *Frequency Hop Spread Spectrum* (FH-SS). In many ways, these systems are similar to an FDMA system, in which the bandwidth available for multiple users is divided into N channels. At any point in time, a frequency hopping signal for a particular user occupies only a single frequency channel. The difference between this system and a traditional FDMA system is that the frequency hopping signal changes frequency carriers, or hops, at rapid periodic intervals. If the signal hops at, or close to, the symbol rate, the system is termed a fast frequency hopping system. If hopping occurs at a lower rate, it is called slow frequency hopping. The sequence of frequency slots occupied by the FH signal is a pseudo random sequence.

The pseudo random sequence is recreated at the receiver which retunes to the proper channel to match the transmitted signal. It is assumed that not all of the channels are continuously occupied. If two multiple access users are using the same channel set, each using a different pseudo random channel sequence, the signals will occasionally be transmitted in the same frequency slot. A bit error may occur in this case, but the effects of this can be minimized through the use of error control coding [Lin83][Rap96a].

A frequency hopping system provides a level of security, especially when a large number of channels are used, since an unintended receiver that does not know the pseudo random sequence of frequency slots must constantly retune to search for the signal that it is attempting to intercept. FH systems also have the effect of randomizing interference which can be beneficial [Par89]. In today's emerging wireless CDMA technology, DS spread spectrum has been widely accepted, and the remainder of this text focuses on Direct Sequence rather than Frequency Hop Spread Spectrum.

1.4 Antenna Systems

In any wireless system, antennas are used at each end of the link. The antenna is a means of coupling radio frequency power from a transmission line into free space, allowing a transmitter to radiate, and a receiver to capture incident electromagnetic power. Antennas can be as simple as a piece of wire, or they can be complex systems with active electronics. Despite the range of technologies comprising antenna systems, there are a number of concepts which are common to all antenna systems.

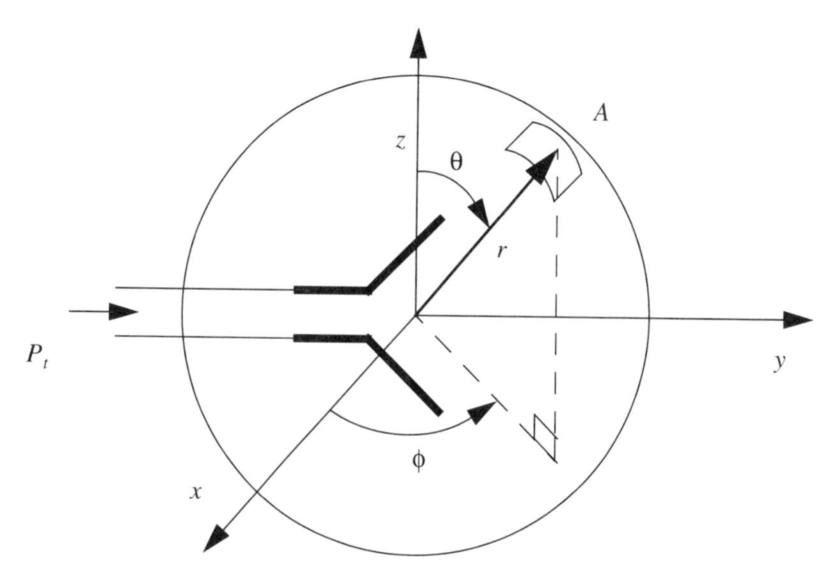

Figure 1–14 Geometry for analyzing antenna systems.

An antenna is illustrated in Figure 1–13. Here, a signal with time-average power of P_t is fed into the antenna. The antenna radiates this power in all directions. We define a direction using polar coordinates, θ and ϕ. If the vector $\boldsymbol{r}$ points in a particular direction, then ϕ is the angle between the x-axis and the projection of $\boldsymbol{r}$ into the x-y plane and θ is the angle between the z-axis and $\boldsymbol{r}$.

The power density, having units of W/(rad)^2, for a particular direction (θ, ϕ), is given by $U(\theta, \phi)$. If the antenna is *lossless*, and perfectly matched, then all of the power, P_t, flowing into the antenna will be radiated. In this case, the total power transmitted in all directions is

$$P_t = \int_{\phi=0}^{2\pi} \int_{\theta=0}^{\pi} U(\theta, \phi) \sin\theta d\theta d\phi \tag{1.42}$$

The average power density, U_{ave}, is

$$U_{ave} = \frac{P_t}{4\pi} = \frac{1}{4\pi} \int_{\phi=0}^{2\pi} \int_{\theta=0}^{\pi} U(\theta, \phi) \sin\theta d\theta d\phi \tag{1.43}$$

If the antenna transmits power equally in all directions, then $U(\theta, \phi)$ will be equal to U_{ave}, and the antenna is said to be *isotropic*. Isotropic antennas are useful for analytical purposes; however, most real antennas exhibit directionality, meaning that they transmit more power in one direction than in other directions. The maximum power density for the antenna is

$$U_m = \max_{\theta, \phi} \{U(\theta, \phi)\} \tag{1.44}$$

We define the *gain* of an antenna, with respect to isotropic, as

$$G(\theta, \phi) = \eta \frac{U(\theta, \phi)}{U_{ave}} \tag{1.45}$$

where η is the efficiency of the antenna, which accounts for losses. $G(\theta, \phi)$ is also called the *antenna power pattern*. The peak gain of the antenna is

$$G = \eta \frac{U_m}{U_{ave}} = \eta \frac{4\pi}{\displaystyle\int_{\phi=0}^{2\pi} \int_{\theta=0}^{\pi} \left[\frac{U(\theta, \phi)}{U_m}\right] \sin\theta d\theta d\phi} \tag{1.46}$$

When a single value is given for the gain of an antenna, it refers to the peak gain. The *Effective Isotropic Radiated Power* (EIRP) is defined as

$$P_{\text{EIRP}} = P_t G = 4\pi\eta U_m \tag{1.47}$$

The EIRP is the amount of power that would be required, using an isotropic antenna, to produce the same power density that is achieved in the boresight direction of the directional antenna. At a distance r from the antenna, in a direction (θ, ϕ), the total power available in an area, A is

$$P_r(\theta, \phi) = \frac{G(\theta, \phi) P_t A}{4\pi r^2} \tag{1.48}$$

In the direction of peak gain, total power available in an area A at a distance of r from the antenna is

$$P_r = \frac{G P_t A}{4\pi r^2} = \frac{A}{4\pi r^2} P_{EIRP} \tag{1.49}$$

Another important result is that the maximum gain of an antenna can be expressed in terms of its size or *maximum aperture*, A_m,

$$G = \eta \frac{4\pi}{\lambda^2} A_m \tag{1.50}$$

where $\lambda = c/f$ is the wavelength, with c, the speed of light, equal to $3 \times 10^8 m/s$, and f is the carrier frequency in Hz. This expression states that for an antenna of size, or maximum aperture, A_m, its gain will increase by the square of the frequency. Equivalently, the effective aperture, $A_e = \eta A_m$, is often used, where

$$A_e = \frac{\lambda^2 G}{4\pi} \tag{1.51}$$

1.5 Basic Concepts in Radiowave Propagation

Consider the wireless link illustrated in Figure 1–14. Here, a transmitter produces an RF signal with time-average power P_t, using an antenna with gain G_t. At the receiver, which is a distance d from the transmitter, an antenna with gain G_r is used.

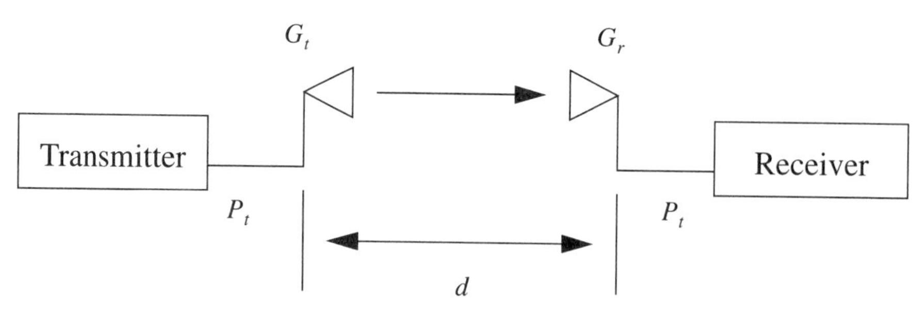

Figure 1–15 Basic model for a wireless transmitter-receiver pair.

Let us assume that the transmitter and receiver antennas are pointing toward each other so that their boresights are aligned. Then using (1.48), the power available at the receiving antenna is

$$P_r = \frac{G_t P_t A_e}{4\pi d^2} \tag{1.52}$$

where A_e is the effective aperture of the receiving antenna. Using (1.51), the received power can be expressed as

$$P_r = \frac{\lambda^2 G_t G_r P_t}{(4\pi d)^2} \tag{1.53}$$

This expression is *Friis' Free Space Link Equation.* The transmit power is measured in watts, the wavelength and distance are measured in meters, and the received power is given in watts. It is usually much more convenient to work in decibel units where

$$P_{r,\,dBm} = P_{t,\,dBm} + G_{t,\,dBi} + G_{r,\,dBi} - 20\log\left(\frac{4\pi f}{c}\right) - 20\log(d) \tag{1.54}$$

where log(x) is the base 10 logarithm of x.

Here, P_r and P_t are measured in dBm, G_t and G_r are in dBi, d is measured in meters, f is the carrier frequency in Hz, and c is the speed of light, 3×10^8 m/s. The received power in dBm is related to the received power in watts, P_w, by

$$P_{dBm} = 10\log\left(\frac{P_W}{1\text{mW}}\right) = 10\log(P_W) + 30 \tag{1.55}$$

The term dBi refers to the gain of an antenna, in dB, relative to an isotropic antenna. From (1.46), since $U_m = U_{ave}$ for an isotropic antenna, if the antenna is lossless, then the gain of an isotropic antenna, $G_{iso} = 1$. Therefore, the gain of an antenna relative to an isotropic antenna is

$$G_{dBi} = 10\log\left(\frac{G}{G_{iso}}\right) = 10\log(G) \tag{1.56}$$

Throughout the remainder of this text, we will typically use the terms P_r, P_t, G_t, and G_r to refer to either power levels in dBm and gain in dBi or, alternatively, power levels in watts and gain as a unitless quantity.

Equation (1.54) has several important implications for wireless communication systems. For adequate performance of a radio link, there is a minimum received power level, P_r, that must be available at the receiver. Given a particular T-R separation, *d*, and a frequency of operation, *f*, there are three ways that we can increase P_r. One way is to increase the transmitter power P_t. If the transmitter is a portable unit, however, increased transmitter power can reduce battery life, and high-power transmitters can be expensive and bulky. Alternatively, we can increase the gain of the transmitting or receiving antennas.

At the subscriber end of a mobile link, antenna gain is limited by several factors. Usually, tiny portable subscriber antennas must be able to provide uniform performance, regardless of the orientation of the antenna or the subscriber. Therefore, relatively non-directional antennas with low gain are often used. Another important factor is shown by (1.51). Here we see that the gain of an antenna is limited by its size. In order to achieve substantial gain at the portable unit, at frequencies used for cellular and PCS systems, the small size of portable transceivers severely limits the maximum gain obtainable.

As an example, suppose that a gain of 6 dBi is required from an antenna operating at 1900 MHz. Using (1.51), we find that an aperture of at least 80 cm^2 (8 cm x 10 cm) is required. This is much more than the area of a typical modern subscriber transceiver. Note, however, that wireless systems using antennas on cars, or Wireless Local Loop systems, which use antennas mounted on a residence, are important exceptions where larger antennas are possible.

For base stations, using conventional antennas with fixed antenna patterns, gain is usually limited to approximately 16-20 dBi. This is because base stations using conventional antennas must be able to supply adequate power throughout a coverage area. Smart antenna systems can offer improvements by offering dynamic antenna patterns.

We should note some key limitations and assumptions of Friis' Free Space Link Equation. First of all, as the name implies, the expression applies only in free space. Equation (1.54) does not account for multipath, or obstructions between the transmitter and receiver. We can generalize Friis' link equation by writing (1.54) as

$$P_r = P_t + G_t + G_r - L_p \tag{1.57}$$

where L_p represents the path loss between the transmitter and receiver.

1.5.1 Path Loss in Real World Channels

The mechanisms behind electromagnetic wave propagation are diverse, but they can generally be attributed to reflection, diffraction, and scattering. Most cellular radio systems operate in urban areas where there is often no direct line-of-sight path between the transmitter and the receiver, and where the presence of high-rise buildings causes severe diffraction loss. Due to multiple reflections from various objects, the electromagnetic waves travel along different paths of varying lengths. The interaction among these waves causes multipath fading, and the strengths of the waves decrease as the distance between the transmitter and receiver increases.

Propagation models have traditionally focused on predicting the average received signal strength at a given distance from the transmitter, as well as the variability of the signal strength in close spatial proximity to a particular location. Propagation models that predict the mean signal strength for an arbitrary transmitter-receiver (T-R) separation distance are useful in estimating the radio coverage area of a transmitter and are called *large-scale* propagation models, since they characterize gross path loss characteristics that apply over scales that are large compared to a wavelength. On the other hand, propagation models that characterize the rapid fluctuations of the received signal strength over very short travel distances (a few wavelengths) or short time durations (on the order of seconds) are called *small-scale* or *fading* models.

1.6 Small Scale Fading

As a mobile moves over very small distances, the instantaneous received signal strength may fluctuate rapidly, giving rise to small-scale fading. The reason is that the received signal is a sum of many contributions coming from different directions. Since the phases are random, the sum of the contributions varies widely; for example, a CW or narrowband signal obeys a Rayleigh fading distribution. In small-scale fading, the received power of a narrowband signal may vary by as much as three or four orders of magnitude (30 or 40 dB) when the receiver is moved by only a fraction of a wavelength. However, the local average signal power will be constant over a distance of several meters. As the mobile moves away from the transmitter over much larger distances, the local average received signal power will gradually decrease; it is this local average signal level that is predicted by large-scale propagation models. Typically, the local average received power is computed by averaging signal measurements over a measurement track of 5λ to 40λ in a local area. For cellular and PCS frequencies in the 1 GHz to 2 GHz band, this corresponds to measuring the local average received power over movements of 1 m to 10 m.

In wideband CDMA systems where the RF bandwidth exceeds the coherence bandwidth of the channel (or alternatively, when the chip duration is smaller than the multipath channel delay spread), the received signal power fades much less than a narrowband signal, as shown in Figure 1–16.

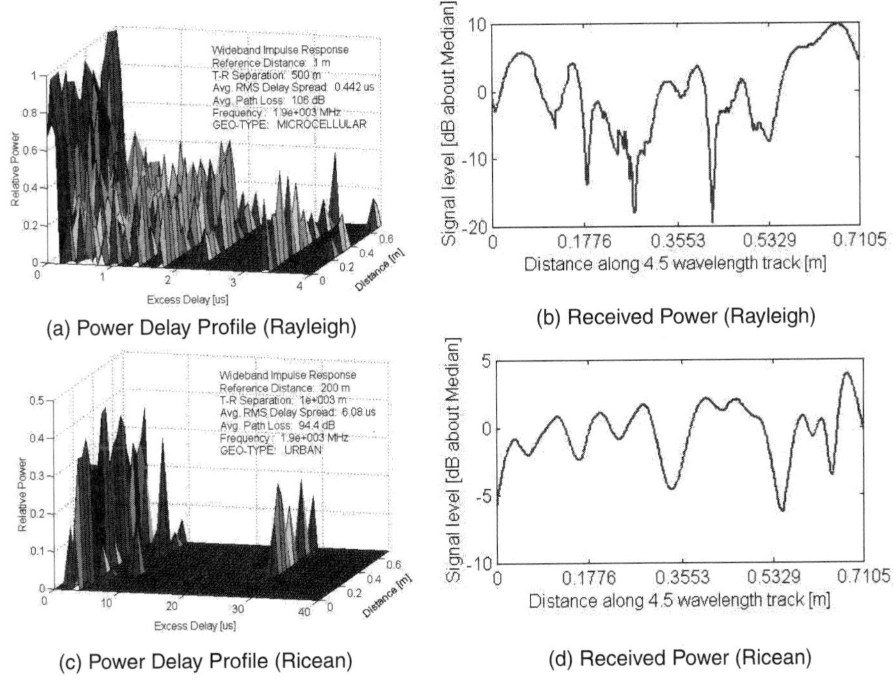

Figure 1–16 Plots showing the power delay profile and received power as a receiver moves over a 0.6 meter track. In the Rayleigh fading environment, shown in (a) and (b), many multipath components can combine destructively, resulting in deep fades in the total received power. In (c), the fact that very few later arriving components are weaker than earlier components leads to less deep fades, as illustrated in (d). These channels were simulated using *SMRCIM* 3.0 [©1998, Wireless Valley Communications, Inc., and Virginia Tech Intellectual Property, Inc., http://www.wvcomm.com].

1.7 Large Scale Path Loss

Most large scale radio propagation models are derived using a combination of analytical and empirical methods. The empirical approach is based on fitting curves or analytical expressions that recreate a set of measured data in the environment of interest. This approach has the advantage of implicitly taking into account all propagation factors, both known and unknown, through actual field measurements. However, the validity of an empirical model at transmission frequencies or environments other than those used to derive the model can be established only by additional measured data in the new environment at the required transmission frequency. Over time, some classical propagation models have emerged which are now used to predict large-scale coverage for wireless communication systems design. By using path loss models to estimate the received signal level as a function of distance, it becomes possible to predict the Carrier-to-Noise-Ratio for a mobile communication system.

1.7.1 Log-distance Path Loss Model

Both theoretical and measurement-based propagation models indicate that average received signal power decreases by the logarithm of the distance between the transmitter and the receiver, whether in outdoor or indoor radio channels. Such models have been used extensively in the literature. The average large-scale path loss for an arbitrary T-R separation is expressed as a function of distance by using a path loss exponent, n.

$$\overline{PL}(d) \propto \left(\frac{d}{d_0}\right)^n \tag{1.58}$$

or

$$\overline{PL}_{dB}(d) = \overline{PL}_{dB}(d_0) + 10n\log\left(\frac{d}{d_0}\right) \tag{1.59}$$

where n is the path loss exponent which indicates the rate at which the path loss increases with distance, d_0 is the close-in reference distance which is determined from measurements close to the transmitter, and d is the T-R separation distance. The bars in equations (1.58) and (1.59) denote the ensemble average of all possible path loss values for a given value of d. When plotted on a log-log scale, the modeled path loss is a straight line with a slope equal to $10n$ dB per decade. The value of n depends on the specific propagation environment. For example, in free space, n is equal to 2, and when obstructions are present, n will have a larger value.

It is important to select a free space reference distance that is appropriate for the propagation environment. In large coverage cellular systems, 1 km reference distances are commonly used [Lee89], whereas in microcellular systems, much smaller distances (such as 100 m or 1 m) are used. The reference path loss is calculated by using the free space path loss formula given by equation (1.54) or through field measurements at distance d_0. Table 1–5 lists typical path loss exponents and log normal shadowing standard deviations obtained in various mobile radio environments.

Table 1–5 Path Loss Exponents and Log-normal Shadowing Standard Deviation for Different Environments [Rap96a]

Environment	Path Loss Exponent, n	Standard Deviation, σ
Free space	2	0 dB
Urban area cellular radio	2.7 to 3.5	10 – 14 dB
Shadowed urban cellular radio	3 to 5	11 – 17 dB
In-building line-of-sight	1.6 to 1.8	4 – 7 dB
Obstructed in-building	4 to 6	5 – 12 dB
Obstructed in factories	2 to 3	6 – 9 dB

1.7.2 Log-normal Shadowing

The model in equation (1.59) does not consider the fact that the surrounding environmental clutter may be very different at two different locations having the same T-R separation. This leads to measured signals which are much different than the *average* value predicted by equation (1.59). Measurements have shown that at any value of d, the path loss $PL(d)$ at a particular location is random and distributed log-normally (normal in dB) about the mean distance-dependent value [Cox84], [Ber87]. We can therefore express the path loss as

$$PL(d) = \overline{PL}(d) + X_\sigma = \overline{PL}(d_0) + 10n\log\left(\frac{d}{d_0}\right) + X_\sigma \tag{1.60}$$

and

$$P_r(d) = P_t - \overline{PL}(d) - X_\sigma + G_t + G_r \tag{1.61}$$

where X_σ is a zero-mean Gaussian distributed random variable (in dB) with standard deviation σ (also in dB).

The log-normal distribution describes the random *shadowing* effects which occur over a large number of measurement locations which have the same T-R separation, but have different levels of clutter on the propagation path. This phenomenon is termed *log-normal shadowing*. Simply put, log-normal shadowing implies that measured signal levels at a specific T-R separation have a Gaussian (normal) distribution about the distance-dependent mean of (1.59), where the measured signal levels have values in dB units. The standard deviation of the Gaussian distribution that describes the shadowing also has units in dB. Thus, the random effects of shadowing are accounted for, using the Gaussian distribution which lends itself readily to evaluation.

The close-in reference distance d_0, the path loss exponent n, and the standard deviation σ, statistically describe the path loss model for an arbitrary location having a specific T-R separation; this model may be used in computer simulation to provide received power levels for random locations in communication system design and analysis. An example of an IS-95 link design using (1.60) is provided in Chapter 2.

In practice, the values of n and σ are computed from measured data, using linear regression where the difference between the measured and estimated path losses is minimized in a mean square error sense over a wide range of measurement locations and T-R separations. The value of $\overline{PL}(d_0)$ in (1.60) is based on either close-in measurements or on a free space assumption from the transmitter to d_0. An example of how the path loss exponent is determined from measured data follows. Figure 1–17 illustrates actual measured data in several cellular radio systems and demonstrates the random variations about the mean path loss (in dB) due to shadowing at specific T-R separations.

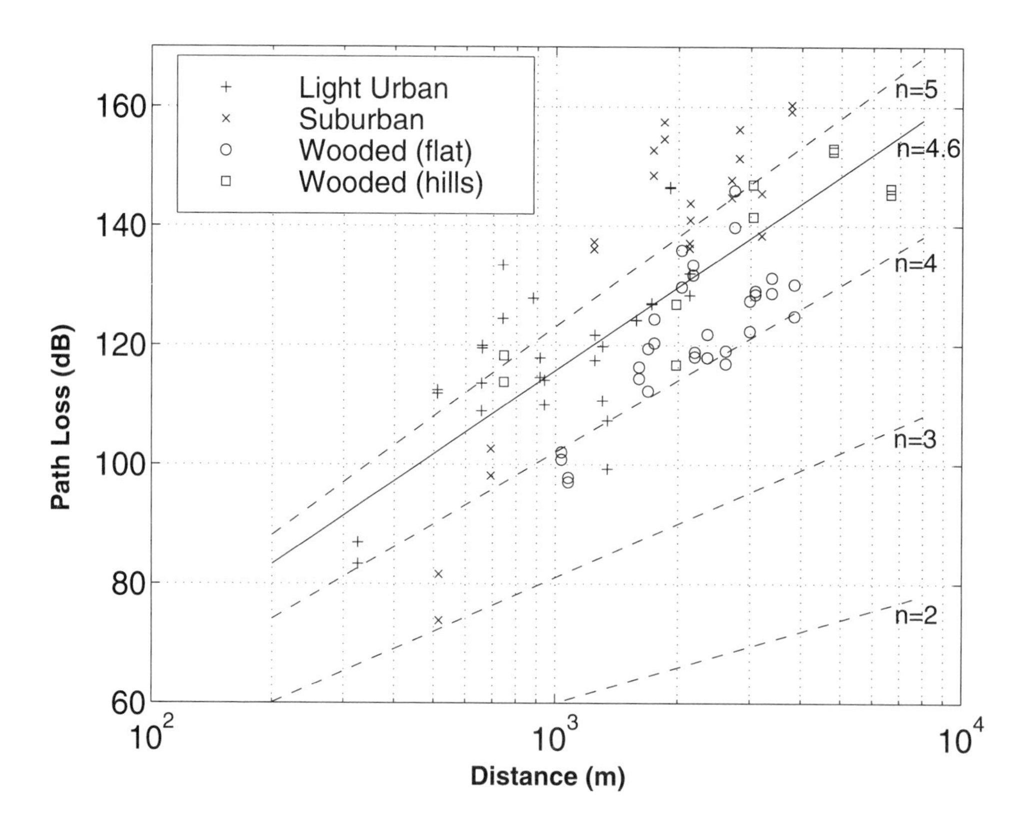

Figure 1–17 Plot of Path Loss vs. Distance at 1900 MHz for measurements performed with a receiver at microcell antenna heights (13 meters), and a mobile transmitter with an antenna height of 1.5 meters, in several different cluttered environments. The MMSE estimated path loss exponent for the collection of measurements is $n = 4.6$ with $\sigma = 12.5$ dB.

Since $PL(d)$ is a random variable with a normal distribution in dB about the distance-dependent mean, so is $P_r(d)$, and the Q-function or error function (*erf*) may be used to determine the probability that the received signal level will exceed (or fall below) a particular level. The Q-function is defined as

$$Q(z) = \frac{1}{\sqrt{2\pi}}\int_{z}^{\infty} \exp\left(-\frac{x^2}{2}\right)dx = \frac{1}{2}\left[1 - \mathrm{erf}\left(\frac{z}{\sqrt{2}}\right)\right] \quad (1.62)$$

where

$$Q(z) = 1 - Q(-z) \quad (1.63)$$

The probability that the received signal level will exceed a certain value γ can be calculated from the cumulative density function as

$$Pr[P_r(d) > \gamma] = Q\left(\frac{\gamma - \overline{P_r(d)}}{\sigma}\right) \quad (1.64)$$

Similarly, the probability that the received signal level will be below γ is given by

$$Pr[P_r(d) < \gamma] = Q\left(\frac{\overline{P_r(d)} - \gamma}{\sigma}\right) \quad (1.65)$$

Usually, shadowing is taken into account by determining a link margin that is required to achieve a certain coverage probability. For example, to achieve a 95% coverage probability at the cell edge with a shadowing standard deviation of $\sigma = 10\text{ dB}$, we find that we need $\overline{P_r}(d) - \gamma = 1.65\sigma$ or 16.5 dB. In other words, there is only a 5% probability that $Pr(d)$ will be less than γ, as long as the mean received power, $\overline{P_r}(d)$, is 16.5 dB above γ. We call this 16.5 dB value the *shadowing margin*. The link budget example in Chapter 2 uses shadowing margins to account for coverage probability due to shadowing.

1.8 Summary

In this chapter, we presented an overview of several state-of-art technologies in wireless technology, including several important emerging systems such as Wireless Local Loop, LMDS, and Third Generation Wireless Systems. We also developed a foundation in CDMA technology, and provided concepts in antennas and propagation. These tools will allow the reader to delve into the details of smart antennas for CDMA in Chapters 3 and 4. First, however, we take a detailed look at the most widely deployed CDMA technology, IS-95, in Chapter 2.

CHAPTER 2

IS-95 PCS & Cellular CDMA

Around the world, a number of different Code Division Multiple Access air interfaces are being deployed and proposed for mobile and portable wireless systems, Wireless Local Loop, satellite systems, and wireless Local Area Networks (LANs). However, by far the most widely deployed CDMA technology is the IS-95 based cellular and PCS CDMA technology deployed in North and South America, Asia, and elsewhere. This technology, which includes the IS-95A standard for 800 MHz cellular systems [EIA93], and the J-STD-008 standard for 1900 MHz PCS systems [Ans95b], will be collectively referred to as IS-95. As of late 1998, there are more than 12 million wireless subscribers using IS-95 or one of its variants.

The goal of this chapter is to present essential aspects of the IS-95 standard required to understand how smart antenna systems impact CDMA system performance. This requires a thorough understanding of uplink and downlink signal formats, multiple access mechanisms, and the interaction between signals from different users. In particular, in spatial processing it is necessary to understand aspects of underlying signal structure that can be exploited to extract and separate signals. Thus, we are primarily concerned with the *Layer 1*, or *Physical Layer*, characteristics of CDMA systems. High layer functions which help to ensure message delivery and important signaling features are very important aspects of wireless technology that are needed to ensure proper operation of a CDMA network. While this chapter provides a detailed overview of IS-95 technology, the reader will find that the two standards, TIA IS-95 and ANSI J-STD-008, provide even greater depth and therefore should be consulted for further reference.

2.1 Cellular and PCS Frequency Allocation

In the United States, CDMA (IS-95A) cellular service shares the existing AMPS spectrum allocation originally designed for the 30 kHz carrier spacing used by the analog system. One

CDMA channel uses 42 AMPS channels. In practice, 8 AMPS channels are used as guard bands on each side of the CDMA channel. The United States cellular frequency allocation is shown in Table 2–1.

Table 2–1 U.S. Cellular Frequency Allocation.

Band	Channel Numbers	Mobile Station Transmit Frequency (MHz)	Base Station Transmit Frequency (MHz)
A"	991-1023	824.0-825.0	869.0-870.0
A	1-333	825.0-835.0	870.0-880.0
B	334-666	835.0-845.0	880.0-890.0
A'	667-716	845.0-846.5	890.0-891.5
B'	717-799	846.5-849.0	891.5-894.0

2.1.1 PCS Frequency Allocation

PCS CDMA service does not use the same spectrum as the AMPS cellular service. PCS services use a total of 120 MHz in the 1850-1990 MHz band, allowing up to six providers (in addition to cellular providers) to serve an area. The PCS allocation shown in Table 2–2, uses a 50 kHz channel spacing, with center frequencies given by:

Mobile Station: $f_c = 1850.000 + 0.050*N$ MHz

Base Station: $f_c = 1930.000 + 0.050*N$ MHz

Table 2–2 U.S. PCS Frequency Allocation.

Band	Channel Numbers	Mobile Station Transmit Frequency (MHz)	Base Station Transmit Frequency (MHz)
A	0-299	1850-1865	1930-1945
D	300-399	1865-1870	1945-1950
B	400-699	1870-1885	1950-1965
E	700-799	1885-1890	1965-1970
F	800-899	1890-1895	1970-1975
C	900-1199	1895-1910	1975-1990

Each CDMA carrier requires 1.25 MHz, or 25 channels, each with a bandwidth of 50 kHz. The J-STD-008 suggests particular channels for CDMA operation. Operation near the edges (1.25 MHz) of the PCS band is prohibited, and operation near the block edges is permitted only if the adjacent block is allocated to the same operator or if different operators coordinate frequency planning. This fosters coordination between PCS service providers. Preferred channels for PCS CDMA are shown in Table 2–3.

Table 2–3 Preferred Channels for PCS CDMA Operation.

Band	Preferred Channels
A	25, 50, 75, 100, 125, 150, 175, 200, 225, 250, 275
B	425, 450, 475, 500, 525, 550, 575, 600, 625, 650, 675
C	925, 950, 975, 1000, 1025, 1050, 1075, 1100, 1125, 1150, 1175
D	325, 350, 375
E	725, 750, 775
F	825, 850, 875

2.2 How IS-95 CDMA PCS Systems Work

Two-way cellular communication relies on a variety of channels to ensure call connection and delivery. *Control Channels* include *Paging Channels* that are used to notify subscriber units of incoming calls and *Access Channels* that let the subscriber initiate outgoing calls. *Traffic Channels* are used to support customer voice calls and messages. The structures of the CDMA uplink and downlink are illustrated in Figure 2–1.

In a CDMA network, all base stations use the same frequency channel, or carrier. Spreading codes are used to separate signals. Each base station is synchronized to *CDMA system time*, which is derived from a precise time reference supplied by GPS satellites. In IS-95 CDMA, every base station transmits a *pilot* signal on the downlink using the same Pseudo-Noise (PN) sequence; however, each pilot is offset in time from the others, allowing the subscriber to differentiate the signals. We will see later that the pilot PN sequence is 32768 chips long, at a chip rate of 1.2288 Mchips/sec, so that each pilot PN sequence repeats every 26.67 ms. Each base station pilot is transmitted so that its PN sequence starts with an offset that is an integer multiple of 64 chips, or 52.08 μs, from other sequences. Pilot timing is described in further detail in Sections 2.4 and 2.5. Since there are 32768 chips in the pilot PN sequence, and PN offsets are integer multiples of 64 chips, there are 512 pilot offsets.

The IS-95 CDMA subscriber unit searches for pilot signals that are strong enough to detect. Each CDMA subscriber unit can combine at least 3 independent forward traffic signals, using a *Rake receiver* (discussed in Chapter 3), to improve reception. These signals can be mul-

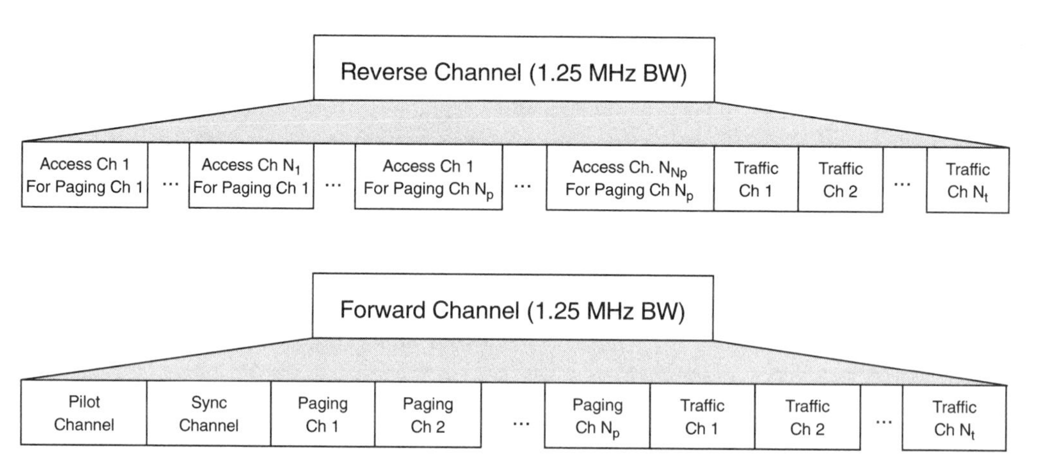

Figure 2–1 The channel structure of the CDMA uplink and downlink. In IS-95, $N_p \leq 7$, $N_t \leq 63$. Note that multiple channels exist simultaneously on a single RF carrier by using unique PN codes.

tipath replicas of the signal transmitted by a single base station. Alternatively, they may be different versions of the downlink signal, each transmitted by a different base station. A unique feature of CDMA systems is that adjacent base stations can transmit the same signal to a subscriber, enhancing downlink link quality. In this case, since the subscriber is not distinctly attached to one base station or the other, the subscriber is said to be in *soft handoff*. The uplink signals received from one subscriber can be combined in the CDMA network to enhance the reverse traffic channel, as well. As described in section 2.2.1, soft handoff can significantly improve system link performance, especially at the edge of the cell; however, because each forward link transmission takes up a traffic channel, soft handoff can reduce forward link capacity.

Each subscriber divides the pilot signals it receives into sets. The *active set* is comprised of the pilot channels (i.e., base stations or sectors) made available to the subscriber unit by the Mobile Switching Center (MSC). The active set can contain up to 6 pilot PN offsets (i.e., up to 6 sectors transmitting on the same carrier frequency). The pilot channels with sufficient signal strength to provide a suitably demodulated signal but which are not yet in the active set make up the *candidate set*. The *neighbor set* is a list of "nearby" pilots that are likely candidates for handoff, and is transmitted periodically to the subscriber by the active base stations. The *remaining set* is comprised of any pilot channels not in any of the above sets. This is shown geographically in Figure 2–2. The IS-95 subscriber unit monitors a subset of the 512 potential pilot channels, reporting the signal quality and Chip-Energy-to-Interference Ratio (E_c/I_o) to the Base Station and MSC. The channel list for the active set is sent to the subscriber unit from the BTS.

The mobile station continually searches for pilot signals on the current frequency assignment. When it detects a neighboring pilot with sufficient strength (greater than a threshold, *T_ADD*) that is not *active*, the mobile sends a *Pilot Strength Measurement Message* (PSMM) to the base station. The BTSs associated with the previous active set may then assign a new for-

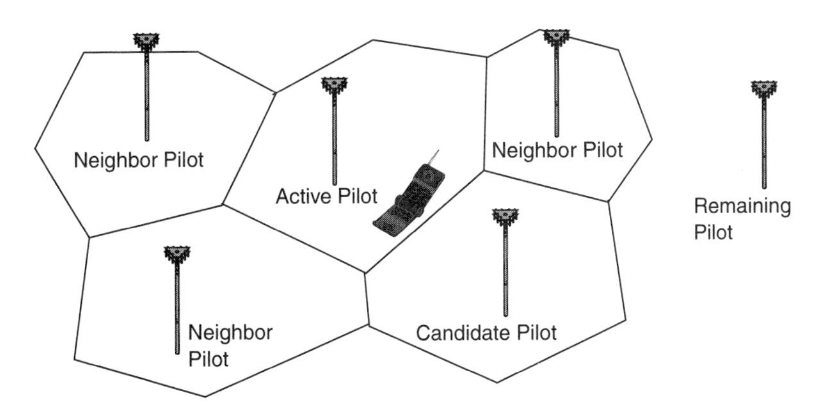

Figure 2–2 Geographic depiction of various pilot channels from different cells which share the same carrier frequency.

ward traffic channel to the subscriber unit by ordering the subscriber to add the new found pilot to its active set.

Subscriber units also can request the MSC to drop certain pilots from the active set. If a pilot in the active set drops below the threshold (T_DROP) and the *handoff drop timer* expires (T_TDROP), the subscriber unit sends a PSMM. The base station then determines whether a handoff is required and returns a *handoff direction message*, if necessary. If a handoff is needed, the mobile station then moves the pilot from the active set to the neighbor set.

Scanning across the entire code domain in a uniform manner would introduce intolerable delays, so the subscriber unit must scan pilots selectively. The subscriber unit searches around pilot offsets in the active, candidate, and remaining lists. A search window, typically between 4 and 130 chips, is specified for each pilot offset. The pilots in the active list are checked most often, candidate pilots and neighbor pilots are checked less often, and remaining pilots are checked infrequently. Active pilots typically have small search windows (~ 40 chips), while candidates and neighbors have large search windows (~120 chips).

A subscriber unit's time reference is offset from CDMA system time by the propagation delay between the base station and the subscribers. In Figure 2–3, a_1 is the known pilot offset of the active pilot and T_1 is the propagation delay. The pilot offset of the neighboring pilot, a_2, offers a smaller propagation delay, T_2, to the same subscriber. The time delay, τ, which determines the relative pilot offset between the two pilots, represents the pilot PN offset between the two downlink transmissions, $(a_2 - a_1)$.

The mobile will set a search window and look for the neighboring pilot around the time $\gamma = (a_1 + T_1 + \tau)$. If the propagation delays are significantly different between base stations, the pilot signal may fall outside of the search window designed for the original pilot. If this happens, the mobile will not be able to find the neighboring base station and a call may be dropped.

The mobile's search window size must be large enough to contain the apparent "wandering" of each neighbor pilot due to the propagation delay. At the same time, the search window

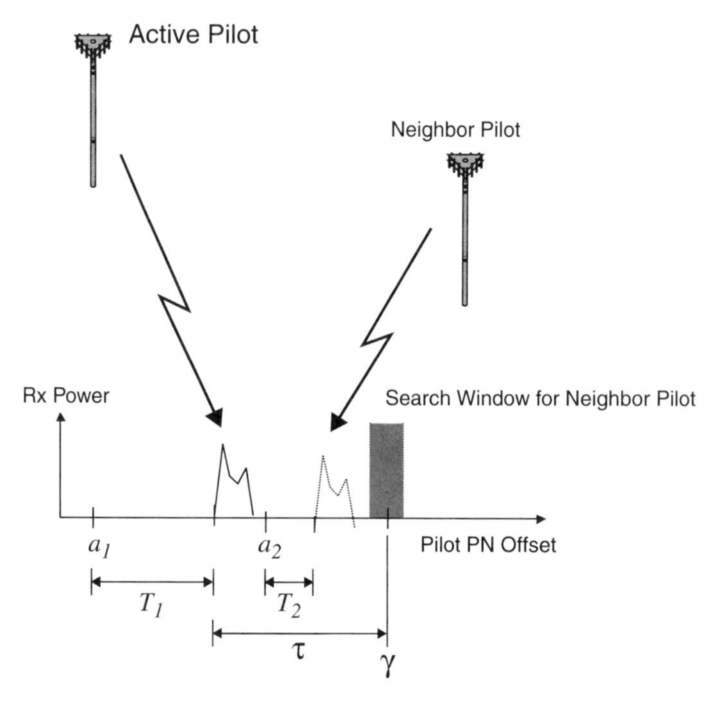

Figure 2–3 Various propagation delays for a mobile receiver, relative to two base stations.

must be set as small as possible so that the mobile can search through all of the neighbor pilots rapidly. Therefore, selection of a suitable search window size is critical to maximizing the performance of a CDMA system.

Each of the downlink signals generated by a single base station sector is orthogonal, which means that the downlink channels do not interfere with each other, provided that channel distortion is small. If downlink multipath introduces strong signal components that are delayed by more than one PN-chip, then downlink signals received at the subscriber unit from a single sector may lose orthogonality. In IS-95 there is a maximum of 64 downlink channels, including the pilot channel, the sync channel, paging channels, and traffic channels. If a system uses a pilot channel and a sync channel, with a single paging channel, there are potentially 61 traffic channels on each carrier. However, in Chapter 5, we show that a more realistic limit for IS-95 is 12-18 subscribers per carrier (i.e., per sector), due to *Multiple Access Interference* (MAI) from the other co-frequency base stations on the downlink, and both in-cell and out-of-cell subscribers on the uplink.

2.2.1 Soft and Softer Handoff in IS-95 CDMA Systems

In IS-95 all base stations may use the same frequency carrier. This makes it possible for a subscriber unit to simultaneously receive downlink signals from different co-channel base stations. Similarly, because uplink traffic signals are identified by their Long Code Mask (as we show in Section 2.4), any base station can monitor the uplink signal for a particular subscriber by dedicating a channel element (a Rake finger searcher and demodulator) to that subscriber. Uplink frames received by multiple Base Transceiver Stations (BTSs) are combined at the Base Station Controller (BSC), both of which are illustrated in Figure 1–3. When a CDMA mobile unit is simultaneously linked to two or more BTSs, the mobile is said to be in *soft handoff*.

On the uplink, soft handoff offers a major benefit for coverage, especially at the cell edge. In Section 1.7.2, it was shown that in log-normal shadowing to provide service with 95% probability at the cell edge, when using a single base station for coverage, 16.5 dB of additional margin is required to account for a 10 dB shadowing standard deviation. Now consider the case where subscribers at the cell edge can link to one of two cells. We write the received power, P_{r1}, for a subscriber at one base station as

$$P_{r1} = \overline{P_r(d_1)} - \chi_1 \tag{2.1}$$

The received power, P_{r2}, from the subscriber at the other base station is

$$P_{r2} = \overline{P_r(d_2)} - \chi_2 \tag{2.2}$$

where both χ_1 and χ_2 are log-normally distributed. For a boundary region which is approximately equidistant from the two cells, $d_1 \approx d_2$. Let us assume that the shadowing components between the two base stations are independent and identically distributed with zero mean and standard deviation σ. Then the probability that both signals fall below a required threshold γ, is

$$\begin{aligned} Pr(P_{r1} < \gamma, P_{r2} < \gamma) &= Pr(P_{r1} < \gamma) Pr(P_{r2} < \gamma) \\ &= Pr(\overline{P_r(d_1)} - \chi_1 < \gamma) Pr(\overline{P_r(d_2)} - \chi_2 < \gamma) \\ &= Pr(\chi_1 > \overline{P_r(d_1)} - \gamma) Pr(\chi_2 > \overline{P_r(d_2)} - \gamma) \\ &= [Pr(\chi_1 > \overline{P_r(d_1)} - \gamma)]^2 \\ &= \left[Q\left(\frac{\overline{P_r(d_1)} - \gamma}{\sigma}\right)\right]^2 \end{aligned} \tag{2.3}$$

For a coverage probability of 95%, the probability that both signals fall below the required threshold is 5%, and using a typical value of $\sigma = 10\text{ dB}$, the margin is only 7.6 dB. This is 8.9 dB less margin than required when a single base station is used to provide coverage.

In real world environments, the improvement is not quite this great. This is because the shadowing experienced from a given subscriber location is not independent from one base station location to another. As shown in [Vit95], the shadowing can be attributed to obstructions close to the mobile, χ_s, as well as obstructions which are distributed between the mobile and the base station — χ_{b1} for base station 1 and χ_{b2} for base station 2, then

$$\chi_1 = \alpha\chi_s + \beta\chi_{b1},\ \chi_2 = \alpha\chi_s + \beta\chi_{b2},\ \alpha^2 + \beta^2 = 1 \tag{2.4}$$

For $\alpha = \beta = 1/\sqrt{2}$, for $\sigma = 10$ dB, the required margin is 11.0 dB, indicating a 2.4 dB margin loss relative to the case of completely independent shadowing.

Just as a mobile can be linked to more than one BTS, a base station can receive uplink signals for a mobile subscriber from more than one sector. This allows the base station to receive multipath signals that arrive from widely different directions. In this case, which is called *softer handoff*, unlike soft handoff, signals are combined at the BTS.

2.3 Typical Link Budgets for IS-95 PCS

While the details of the forward and reverse channels of IS-95 are presented later in this chapter, an overall system link budget is useful for understanding the power allocation required in an IS-95 system design. Engineers who are tasked with the installation of CDMA systems must make decisions about power levels assigned to the various channels in the forward link and must design cell coverage based on the reverse link of weaker users.

IS-95 base stations typically use a 10 to 20-watt (40 to 43 dBm) RF amplifier to provide composite power for the pilot, sync, paging, and traffic channels. High-tier cellular CDMA systems may transmit as much as 50W, while low power PCS microcell systems may use as little as 8W. The base station antenna usually provides well over 10 dB of gain on the horizon, so that the effective isotropic radiated power (EIRP) of an IS-95 base station usually exceeds 50 dBm. The J-STD-008 specification limits the maximum EIRP of any IS-95 base station to 1640 watts EIRP, which is equivalent to 20W fed into a 19 dBi antenna.

Typically, 15-20% of the total transmitted power, at maximum traffic load, is reserved for the pilot. This is because the pilot is used for coherent demodulation of other downlink channels. Without adequate pilot power, the sync, paging, and traffic channels cannot be recovered.

Table 2–4 Percentage of composite RF power assigned to different IS-95 channels for a typical IS-95 forward link with 10 users.

Channel	% RF Power
Pilot	20%
Sync	2%
Paging	14%
All Traffic Channels	64%

The sync channel typically has 10% of the pilot channel power. The total number of paging channels in use is usually assigned a composite power of 75% of the pilot power. Each forward traffic channel is allocated between 4% and 80% of the pilot power. The percentage of composite RF power allocated to the forward traffic channels varies according to loading on the paging and traffic channels, which causes the maximum transmission range of the forward channels to vary based on the number of channels in use. When only one forward traffic channel is active, its power may be 80% of the pilot. Table 2–4 illustrates the typical percentage of power assigned to each of the forward link channels, with 10 to 15 active users.

2.3.1 IS-95 Forward Link Budget for 1900 MHz

When a path loss propagation model, as shown in Section 1.7.1, is used with transmit and receive power levels, it is possible to develop link budgets that predict RF coverage range. Under typical conditions, a signal to noise ratio (E_b/N_0) of 7.5 dB provides good link performance (better than 1% Frame Error Rate), and this is a typical target for practical deployment. Note, however, that the E_b/N_0 required for 1% FER can range from 4 dB to as much as 21 dB, depending on vehicle speed, hand-off state, rate set, and channel conditions. For back-of-the-envelope calculations, a path loss exponent of n = 3.5 (35 dB/decade) is often used, with a shadow fading standard deviation of 8 dB. Because IS-95 uses a relatively wide bandwidth, transmissions do not usually undergo deep fades, and a fade margin of only a few dB is typically used. Table 2–5 shows an actual IS-95 1900 MHz link budget design. Notice that the sensitivity of the subscriber unit is included in the link budget, as are effects such as building penetration losses and losses through automobiles for portable users operating from their cars. A fade margin of 5.5 dB is used, in addition to the shadow fading of 8 dB to account for random obstacles in the channel.

2.3.2 IS-95 Reverse Link Budget at 1900 MHz

For designing the reverse link, it is essential that the base station receive adequate signals from the farthest subscribers, while still providing power control. Proper reverse link design requires that the cable losses and noise figure from the base station components be included in order to determine the power received at the base station receiver antenna terminals (rather than just the power incident on the antenna). Table 2–6 illustrates a typical IS-95 PCS reverse link budget, which factors in approximately 2.6 dB of uplink interference that occurs from the asynchronous reverse channel provided to multiple users in the cell.

2.4 Reverse Traffic Channel Transmission for IS-95

In the next several sections, we consider each of the major functional blocks that comprise the reverse traffic channel transmitter in IS-95. We first present details of the various voice coders used in CDMA systems, along with the variations of the reverse traffic channel transmitter which are needed to support different data rates. Next, we discuss each of the functional blocks associated with the reverse traffic channel transmitter, including the error control coding systems, the symbol repetition and interleaving blocks, and the direct sequence spreading systems.

Table 2–5 Forward Link Budget for IS-95 Using 10W Transmitter.

Line #	Forward Link Budget	Value	Notes:
1	Base Tx Power (dBm)	40.0	
2	Base Tx Cable Loss	2.90	
3	Base Tx Antenna Gain (dBi)	17.20	
4	Base EIRP (dBm)	54.30	(#1+#3-#2)
5	Maximum Number of Users	13	This is the maximum number of users
6	Percent Power Allocated to each Forward Voice Channel	4.85%	This is the percentage of power allocated to each forward voice channel
7	Base EIRP per Mobile Channel (dBm)	41.15	(#4+10*log10(#6))
8	Portable Antenna Gain (dBi)	0.00	
9	Portable Polarization Loss (dB)	2.00	
10	Portable Diversity Gain (dB)	0.00	
11	Desired Rx Eb/No (dB)	7.50	About a 1% Frame Error Rate
12	Data Rate (bps)	14400.00	
13	Portable Receiver NF (dB)	10.00	
14	Signal Bandwidth (MHz)	1.22880	
15	Portable Receiver Sensitivity (dBm)	-114.92	(#11+10*log10(#12)+#13-174)
16	Power Required Incident on Antenna (dBm)	-112.92	(#15-#8-#10+#9)
17	Fade Margin (dB)	5.50	75.5% Edge Reliability and 90.1% area coverage for n=3.5 sigma=8
18	Interference Margin (dB)	0.00	Noise Added from Outside Interference
19	Modified Required Power Incident on Antenna (dBm)	-107.42	Minimum receive level at the base station antenna (#16+#17+#18)
20	Automobile Loss (dB)	6.00	Loss Through Vehicle
21	Building Loss (dB)	14.00	Loss Through Outside Walls (should get everybody on outside perimeter)
22	Maximum Downlink Loss Possible (dB) - Outside	148.57	For person outside (#7-#19)
23	Maximum Downlink Loss Possible (dB) - In Car	142.57	For person in car (#7-#19-#20)
24	Maximum Downlink Loss Possible (dB) - In Building	134.57	For person in building (#7-#19-#21)

2.4.1 Variable Rate Vocoders

In CDMA, a variable rate voice coder or *vocoder* is used to convert voice signals into a digital format that can be efficiently transmitted over the air. These vocoders take advantage of the fact that typically, subscribers talk only 40-50% of the time. During quiet periods, such as breaks in the conversation, or when the other party is talking, the vocoder drops to a very low bit rate. During speech periods, the vocoder operates at a higher rate, which is needed to maximize voice fidelity.

Table 2–6 Reverse Link Budget for IS-95 Using 10W Transmitter.

Line #	Reverse Link Budget	Value	Notes:
1	Portable Tx Power (dBm)	23.01	
2	Portable Antenna Gain (dBi)	0.00	
3	Polarization Loss and Body Loss (dB)	2.00	
4	Portable EIRP (dBm)	21.01	(#1+#2+#3)
5	Base Rx Antenna Gain (dBi)	17.20	
6	Base Rx Cable Loss (dB)	2.90	
7	Base Rx Diversity Gain (dB)	0.00	
8	Desired Rx Eb/No (dB)	7.50	About a 1% Frame Error Rate
9	Data Rate (bps)	14400	
10	Base Receiver NF (dB)	6.00	
11	Signal Bandwidth (MHz)	1.22880	
12	Base Receiver Sensitivity (dBm)	-118.92	(#8+10*log10(#9)+#10-174)
13	Power Required Incident on Antenna (dBm)	-133.22	(#12-#5-#7+#6)
14	Fade Margin (dB)	5.50	75.5% Edge Reliability and 90.1% Area Coverage for n=3.5, sigma=8
15	Cell Loading Margin (dB)	2.83	Noise Added for 50% Loading
16	Interference Margin (dB)	0.00	Noise Added from Outside Interference
17	Modified Required Power Incident on Antenna (dBm)	-124.89	Minimum receive level at the base station antenna (#13+#14+#15+#16)
18	Automobile Loss (dB)	6.00	Loss Through Vehicle
19	Building Loss (dB)	14.00	Loss Through Outside Walls (Should get everybody on outside perimeter)
20	Maximum Uplink Loss Possible (dB) - Outside	145.90	#4-#17
21	Maximum Uplink Loss Possible (dB) - In Car	139.90	For Person in Car (#4-#17-#18)
22	Maximum Uplink Loss Possible (dB) - In Building	121.90	For Person in Building (#4-#17-#19)

Variable rate vocoding presents two key advantages:

- The reduction in transmission results in a longer battery life.
- Since the level of interference directly determines capacity, variable rate voice coding increases capacity by a factor of approximately two [Ros96].

IS-95 was originally developed using the IS-96A vocoder, an 8.6 kbps speech coder. Table 2–7 shows the *Mean Opinion Score* (MOS) for each vocoder as a function of *Frame Error Rate* (FER). A MOS of 4.00-4.30 is equivalent to wireline quality voice, whereas a MOS of 5 is perfect intelligibility. The IS-96A vocoder is capable of achieving only a MOS score of 3.29, even under 0% FER conditions. The IS-96A vocoder degrades rapidly in performance as the FER increases.

Table 2–7 Speech Coder Performance (MOS) as a function of frame error rate. (V. K. Garg, K. Smolik, and J. E. Wilkes, *Applications of CDMA in Wireless Personal Communications*, Prentice Hall, NJ, ©1997. Reprinted by permission of Prentice Hall.)

Frame Error Rate, %	IS-96A (Rate Set 1)	CDG 13 kbps (Rate Set 2)	EVRC (Rate Set 1)
0	3.29	4.00	3.95
1	3.17	3.95	3.83
2	2.77	3.88	3.66
3	2.55	3.67	3.50

Because of the poor perceived link quality using the 8.6 kbps IS-96A vocoder, CDMA systems were initially deployed in the United States using the CDG-13 kbps vocoder (developed by the *CDMA Development Group*, now known as *cdmaOne*). The CDG-13 vocoder provides much better qualitative performance, as shown in Table 2–7. However, the IS-95 standard was originally developed to support a maximum vocoder rate of only 8.6 kbps. In order to support the new CDG-13 kbps vocoder, a revised version of the IS-95 standard was developed for the new *Rate Set 2*, which supports a 13.3 kbps vocoder. The original system that supports the 8.6 kbps maximum vocoder rate is called *Rate Set 1,* and is still supported in the standard, both for backward compatibility with the IS-96A vocoder and for a new vocoder that will be discussed shortly. Table 2–8 summarizes the data rates supported by each rate set.

Table 2–8 Rate Set 1 and Rate Set 2 vocoder rates. Each frame produced by the vocoder is 20 ms long.

Rate Set 1		Rate Set 2	
Bits per Frame (from Vocoder)	Bit Rate (at Input to Convolutional Coder)	Bits per Frame (from Vocoder)	Bit Rate (at Input to Convolutional Coder)
172 bits	9600 bps	267 bits	14400 bps
80 bits	4800 bps	125 bits	7200 bps
40 bits	2400 bps	55 bits	3600 bps
16 bits	1200 bps	21 bits	1800 bps

The reverse traffic channel transmitter for Rate Sets 1 and 2 are shown in Figures 2–4 and 2–5, respectively. In order to support the Rate Set 2 vocoder, the rate 1/3 convolutional coder

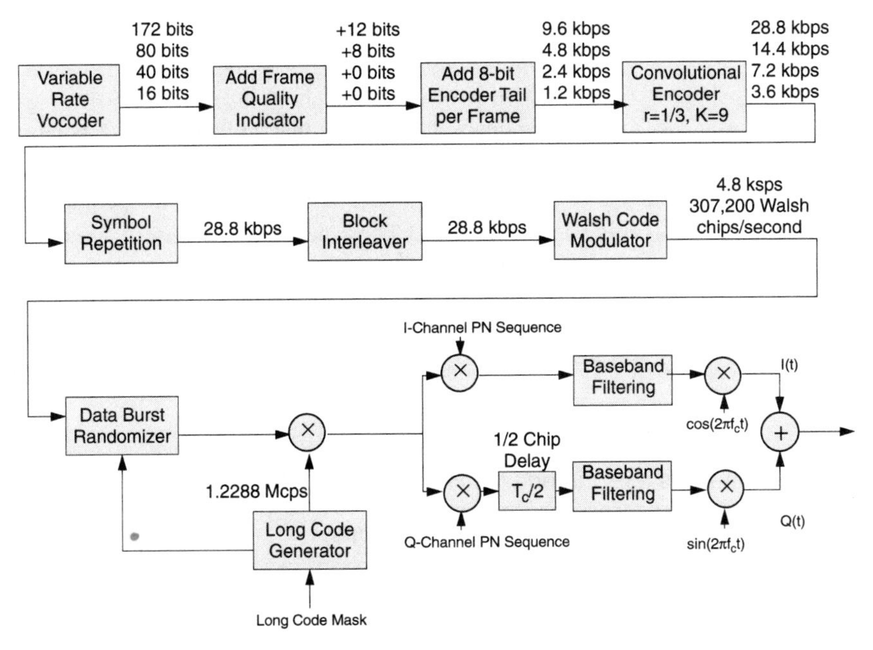

Figure 2–4 The IS-95 reverse channel for Rate Set 1. The number of bits, shown next to several of the blocks, refers to the number of bits contained in a 20 ms vocoder frame.

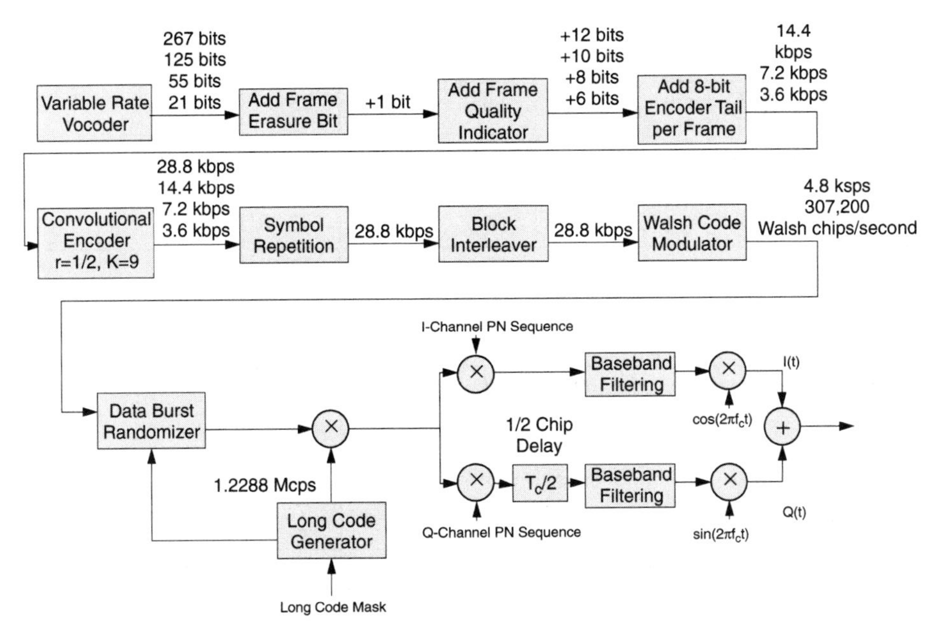

Figure 2–5 The IS-95 reverse channel for Rate Set 2. The number of bits, shown next to several of the blocks, refers to the number of bits contained in a 20 ms vocoder frame.

(for every 1 bit into the convolutional encoder, 3 symbols are produced) was replaced with a rate 1/2 convolutional coder. Rate 1/2 codes, discussed in Section 2.4.2, provide less error control protection than rate 1/3 codes. Since both Rate Set 1 and Rate Set 2 systems are designed to operate at an FER near 1%, the lower coding rate indicates that a higher E_b/N_0 is required for the Rate Set 2 vocoder.

More recently, the *Enhanced Variable Rate Coder* (EVRC) was developed. The 8.6 kbps EVRC gives performance similar to the CDG-13 kbps; however, because it uses Rate Set 1 rather than Rate Set 2, it can be operated at a lower signal-to-noise ratio. In CDMA systems, this means that each link can tolerate more Multiple Access Interference (MAI), thus providing increased capacity. Therefore, at this time, CDMA service providers are working to make use of the EVRC vocoder to increase CDMA system capacity while maintaining link quality. Details of the Rate Set 1 and Rate Set 2 transmitters are provided in Tables 2–9 and 2–10.

Table 2–9 Parameters associated with Rate Set 1.

Parameter	Data Rate (bps)				Units
	9600	**4800**	**2400**	**1200**	
Bits per 20 ms Frame out of Vocoder	172	80	40	16	Bits/20 ms Frame
Frame Quality Indicator Bits	12	8	0	0	Bits/20 ms Frame
Total Bits per Frame with 8-bit Encoder Tail	192	96	48	24	Bits/20 ms Frame
Bit Rate into 1/3 Rate Coder in each 20 ms Frame	9600	4800	2400	1200	bps
Rate out of 1/3 Rate Coder	28.8	14.4	7.2	3.6	sps
Repeated Code Symbol Rate	28.8	28.8	28.8	28.8	ksps
Transmit Duty Cycle	100.0	50.0	25.0	12.5	%
Walsh Code Modulation	6	6	6	6	Repeated Code Symbols/ Walsh Code
Walsh Symbol (64 chips) Rate	4800	4800	4800	4800	Symbols/sec
Walsh Chip Rate	307.2	307.2	307.2	307.2	kcps
Walsh Symbol Duration	208.33	208.33	208.33	208.33	μs
PN Chip Rate	1.2288	1.2288	1.2288	1.2288	Mchips/sec
PN Chips/Repeated Code Symbol	42.67	42.67	42.67	42.67	PN Chips/Repeated Code Symbol

Table 2–10 Parameters Associated with Rate Set 2.

Parameter	Data Rate (bps)				Units
	14400	7200	3600	1800	
Bits per 20 ms Frame out of Vocoder	267	125	55	21	Bits/20 ms Frame
Frame Erasure Bit	1	1	1	1	Bits/20 ms Frame
Frame Quality Indicator Bits	12	10	8	6	Bits/20 ms Frame
Total Bits per Frame with 8-bit Encoder Tail and Erasure Bit	288	144	72	36	Bits/20 ms Frame
Bit Rate into 1/2 Rate Coder in each 20 ms Frame	14400	7200	3600	1800	bps
Rate out of 1/2 Rate Coder	28.8	14.4	7.2	3.6	ksps
Repeated Code Symbol Rate	28.8	28.8	28.8	28.8	ksps
Transmit Duty Cycle	100.0	50.0	25.0	12.5	%
Walsh Code Modulation	6	6	6	6	Repeated Code Symbols/ Walsh Code
Walsh Symbol (64 chips) Rate	4800	4800	4800	4800	Symbols/sec
Walsh Chip Rate	307.2	307.2	307.2	307.2	kcps
Walsh Symbol Duration	208.33	208.33	208.33	208.33	μs
PN Chip Rate	1.2288	1.2288	1.2288	1.2288	Mchips/sec
PN Chips/Repeated Code Symbol	42.67	42.67	42.67	42.67	PN Chips/Repeated Code Symbol

2.4.2 Error Control - The Frame Quality Indicator and Convolutional Coding

The base station needs the ability to detect frame errors on the reverse link. The *Frame Quality Indicator* (FQI), illustrated in Figure 2–6, allows the base station receiver to determine whether or not a frame has been correctly received, by examining the parity-check bits at the end of the frame. For the reverse link, a parity check at the base station will determine if any bits in a frame are in error. If any are in error, the parity check fails, which is considered a *frame error.* The parity check bits are a *Cyclic Redundancy Code* (CRC), a class of linear error detecting codes which generate parity check bits by finding the remainder of a polynomial division. The FQI on the forward link operates in the same manner to give the mobile the ability to detect frame errors. The FQI does not change the frame data, it adds extra bits. In Rate Set 1, FQI adds twelve parity check bits at the 9600 bps data rate or 8 parity check bits at the 4800 bps data rate per frame. No error detection capability is provided at lower data rates for Rate Set 1. Table 2–10 shows the number of FQI parity bits associated with each data rate in Rate Set 2.

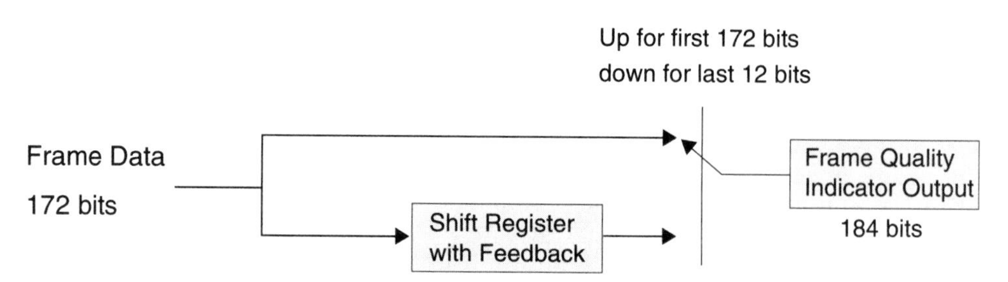

Figure 2–6 Operation of the 9600 bps FQI for Rate Set 1.

Convolutional coding is a type of error correcting code, as described in [Lin83]. A rate 1/3 encoder is specified in Rate Set 1; thus, 3 code bits are output for each information bit. In Rate Set 2, 2 code bits are produced for each information bit. Encoding involves the modulo-2 addition of selected time-delayed bits. Each of these codes has a constraint length of K=9; therefore, encoding requires a delay-line of length (K-1)=8, as shown in the block diagrams in Figures 2–7 and 2–8.

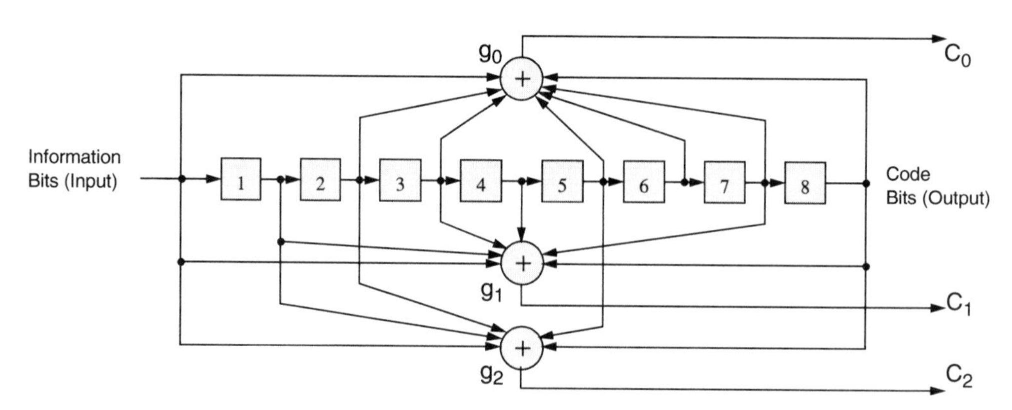

Figure 2–7 Rate 1/3 convolutional encoder for Rate Set 1.

Convolutional encoding greatly increases the number of bits and symbols per frame. The last 8 bits of each reverse traffic channel frame, which are all set to "0", are called the *encoder tail bits,* as shown in Figure 2–9. This tail is necessary to flush out the receiver decoder to a reset state. Convolutional encoding is highly effective against random channel errors, but not as effective against burst errors.

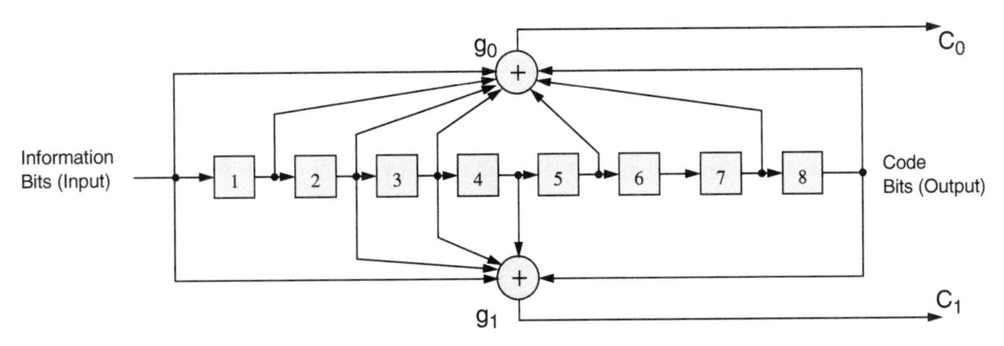

Figure 2–8 Rate 1/2 convolutional encoder for Rate Set 2.

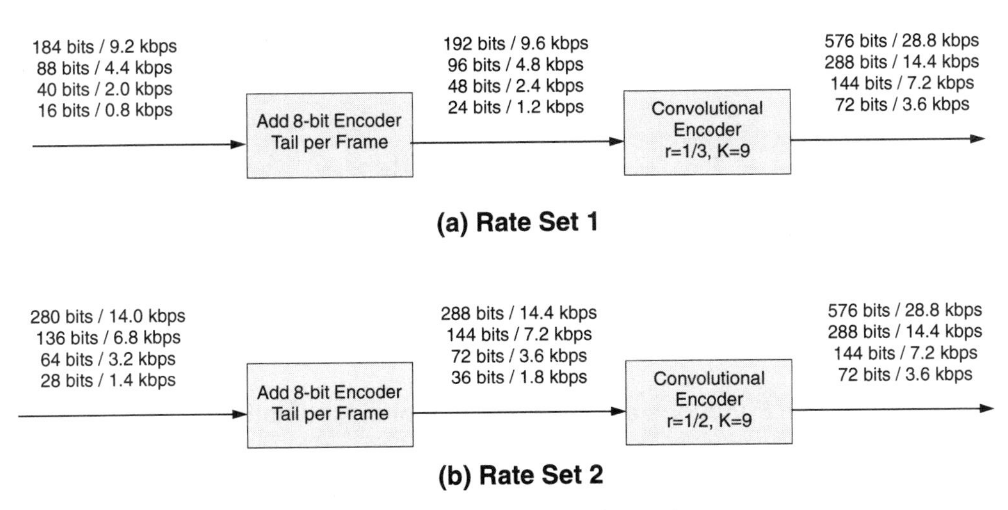

Figure 2–9 Error control coding, applied to each frame for each rate set.

2.4.3 Symbol Repetition and Block Interleaving

At data rates lower than the maximum rate, each symbol at the output of the convolutional encoder is repeated two, four, or eight times. At the output of the block interleaver, there will be two, four, or eight successive and identical groups of data. Redundant data will not be transmitted. This is ensured by the data burst randomizer. The symbol rate out of the Symbol Repetition block is 28.8 ksps.

The block interleaver "shuffles" the order of the symbols within a frame. Symbol errors in the radio channel tend to be bursty, meaning that several symbol errors often occur one after another. Without an interleaver, this defeats the convolutional coding techniques used in CDMA, which can correct a certain number of errors as long as there are not too many in a short period. With the interleaving and the subsequent de-interleaving process, burst errors are randomized in the symbol stream, improving the probability that convolutional coding will be successful in correcting channel errors.

A block interleaver in IS-95 is formed by a 32 x 18 matrix. A total of 576 code symbols (the number of symbols in a frame) is written column by column to the block interleaver. The output of the interleaver is read row by row for the 9600 and 14,400 bps data rates so that the output of the block interleaver in Figure 2–10 is: 1, 33, 65, ..., 545, 2, 34, 66, ..., 546, ...

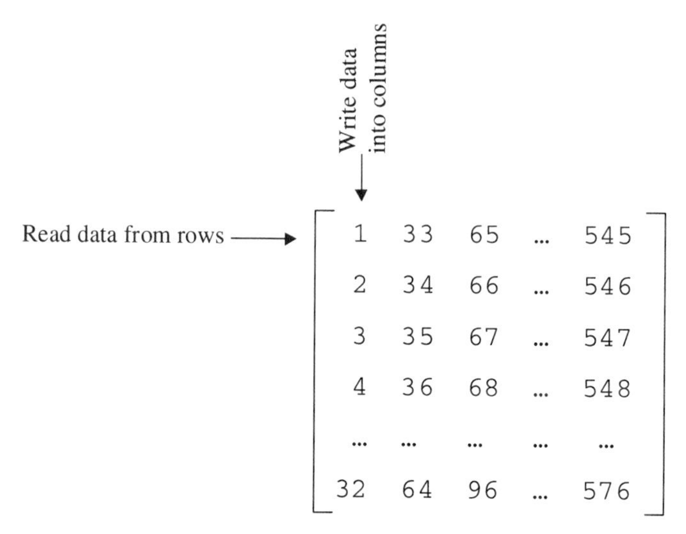

Figure 2–10 Reverse link interleaver operation.

2.4.4 Walsh Functions and 64-ary Orthogonal Modulation

Each group of six modulation symbols is replaced with a single 64-chip *Walsh Function*. Walsh Functions provide basis functions for orthogonal modulation. Walsh Functions, which are sequences of bits, have the property that each function is exactly orthogonal to all of the others. For a sequence of length 64, there are 64 possible Walsh Functions, which are grouped into a Walsh matrix. Walsh Function N is chosen from the six symbols, c_0 through c_5: $N = c_0 + 4c_2 + 8c_3 + 16c_4 + 32c_5$.

Walsh functions form a set of 2^N orthogonal sequences of length 2^N. In general, a Walsh matrix of order 2^{N+1} is determined by the Walsh matrix of order 2^N:

$$\boldsymbol{H}_{2^{N+1}} = \begin{bmatrix} \boldsymbol{H}_{2^N} & \boldsymbol{H}_{2^N} \\ \boldsymbol{H}_{2^N} & \overline{\boldsymbol{H}_{2^N}} \end{bmatrix}, \quad (2.5)$$

Beginning with $H_1 = 0$, then, using (2.5)

$$H_2 = \begin{bmatrix} 0 & 0 \\ 0 & 1 \end{bmatrix}, \text{ and } H_4 = \begin{bmatrix} 0 & 0 & 0 & 0 \\ 0 & 1 & 0 & 1 \\ 0 & 0 & 1 & 1 \\ 0 & 1 & 1 & 0 \end{bmatrix} \tag{2.6}$$

In IS-95 CDMA, H_{64} is used to provide orthogonal functions in the 64-ary modulation scheme. The IS-95 Walsh matrix is illustrated in Figure 2–11. In the 64-ary modulation scheme, if a bit pattern of [101100] is input into the modulator, then the 64 bits from row 44 of the Walsh Matrix in Figure 2–11 are output.

In 64-ary orthogonal modulation, each set of six bits, at a rate of 28.8 kbps, into the Walsh Code Modulator block, produces 64 Walsh chips at the output at a rate of 307.2 kcps. Each set of 64 Walsh chips is treated as a symbol from an alphabet of 64 characters. Therefore, the symbol rate at the output of the Walsh Modulator is 4.8 ksps.

2.4.5 Data Burst Randomization and Gating

Regardless of the vocoder data rate, the rate of symbols out of the Walsh Symbol Modulator is the same due to symbol repetition. During periods when low data rates are produced by the vocoder, the average transmitted power from the subscriber unit should be reduced to maximize system capacity. There are two ways to approach this problem. On the reverse link, data is grouped, depending on the vocoder rate, and redundant Walsh symbols are eliminated by turning off the transmitter during these periods. As we will see later in the chapter, on the forward link, each bit is extended to N times the original period and is transmitted continuously at full power divided by N. Therefore, on the reverse link, transmitted power is gated on and off, while on the forward link, transmitted power levels are continuous but reduced. Both result in the same, reduced average transmitted power.

The task of turning on and off the reverse link transmitter, or *gating*, is accomplished by the *Data Burst Randomizer*. On the reverse channel, for each Rate Set, when transmitting at a rate lower than the maximum (9600 bps for Rate Set 1, and 14400 bps for Rate Set 2), symbols are repeated. The gating patterns used are coordinated with the interleaver and symbol repeat structures, so that only one member of each repeated set is transmitted. Each Rate Set uses a specific procedure to determine the precise gating pattern, as described in the IS-95 standards.

For example, using the 7200 bps rate in Rate Set 2, symbols at the output of the 1/2 rate coder are repeated once. These symbols are written into the block interleaver by columns and are read out in a prescribed order (rows 1,3,2,4,5,7,6,8... and so on). Groups of six symbols from the output of the interleaver are converted into Walsh Symbols, each of which contains 64 Walsh chips.

For the 7200 bps rate, we find that Walsh symbols at the output of the Walsh Modulator are redundant in a specific pattern: w0, w1, ..., w5, w0, w1, ..., w5, w6, w7, ..., w11, w6, w7, ..., w11, and so on. In other words, for the 7200 bps data rate in Rate Set 2, every twelve Walsh symbols contain six Walsh symbols followed by a repeat of the same six Walsh symbols.

	00	01	02	03	04	05	06	07	08	09	10	11	12	13	14	15	16	17	18	19	20	21	22	23	24	25	26	27	28	29	30	31	32	33	34	35	36	37	38	39	40	41	42	43	44	45	46	47	48	49	50	51	52	53	54	55	56	57	58	59	60	61	62	63
00	**0**	**0**	0	0	0	0	0	0	0	0	0	0	0	0	0	0	0	0	0	0	0	0	0	0	0	0	0	0	0	0	0	0	0	0	0	0	0	0	0	0	0	0	0	0	0	0	0	0	0	0	0	0	0	0	0	0	0	0	0	0	0	0	0	0
01	**0**	**1**	0	1	0	1	0	1	0	1	0	1	0	1	0	1	0	1	0	1	0	1	0	1	0	1	0	1	0	1	0	1	0	1	0	1	0	1	0	1	0	1	0	1	0	1	0	1	0	1	0	1	0	1	0	1	0	1	0	1	0	1	0	1
02	0	0	1	1	0	0	1	1	0	0	1	1	0	0	1	1	0	0	1	1	0	0	1	1	0	0	1	1	0	0	1	1	0	0	1	1	0	0	1	1	0	0	1	1	0	0	1	1	0	0	1	1	0	0	1	1	0	0	1	1	0	0	1	1
03	0	1	1	0	0	1	1	0	0	1	1	0	0	1	1	0	0	1	1	0	0	1	1	0	0	1	1	0	0	1	1	0	0	1	1	0	0	1	1	0	0	1	1	0	0	1	1	0	0	1	1	0	0	1	1	0	0	1	1	0	0	1	1	0
04	0	0	0	0	1	1	1	1	0	0	0	0	1	1	1	1	0	0	0	0	1	1	1	1	0	0	0	0	1	1	1	1	0	0	0	0	1	1	1	1	0	0	0	0	1	1	1	1	0	0	0	0	1	1	1	1	0	0	0	0	1	1	1	1
05	0	1	0	1	1	0	1	0	0	1	0	1	1	0	1	0	0	1	0	1	1	0	1	0	0	1	0	1	1	0	1	0	0	1	0	1	1	0	1	0	0	1	0	1	1	0	1	0	0	1	0	1	1	0	1	0	0	1	0	1	1	0	1	0
06	0	0	1	1	1	1	0	0	0	0	1	1	1	1	0	0	0	0	1	1	1	1	0	0	0	0	1	1	1	1	0	0	0	0	1	1	1	1	0	0	0	0	1	1	1	1	0	0	0	0	1	1	1	1	0	0	0	0	1	1	1	1	0	0
07	0	1	1	0	1	0	0	1	0	1	1	0	1	0	0	1	0	1	1	0	1	0	0	1	0	1	1	0	1	0	0	1	0	1	1	0	1	0	0	1	0	1	1	0	1	0	0	1	0	1	1	0	1	0	0	1	0	1	1	0	1	0	0	1
08	0	0	0	0	0	0	0	0	1	1	1	1	1	1	1	1	0	0	0	0	0	0	0	0	1	1	1	1	1	1	1	1	0	0	0	0	0	0	0	0	1	1	1	1	1	1	1	1	0	0	0	0	0	0	0	0	1	1	1	1	1	1	1	1
09	0	1	0	1	0	1	0	1	1	0	1	0	1	0	1	0	0	1	0	1	0	1	0	1	1	0	1	0	1	0	1	0	0	1	0	1	0	1	0	1	1	0	1	0	1	0	1	0	0	1	0	1	0	1	0	1	1	0	1	0	1	0	1	0
10	0	0	1	1	0	0	1	1	1	1	0	0	1	1	0	0	0	0	1	1	0	0	1	1	1	1	0	0	1	1	0	0	0	0	1	1	0	0	1	1	1	1	0	0	1	1	0	0	0	0	1	1	0	0	1	1	1	1	0	0	1	1	0	0
11	0	1	1	0	0	1	1	0	1	0	0	1	1	0	0	1	0	1	1	0	0	1	1	0	1	0	0	1	1	0	0	1	0	1	1	0	0	1	1	0	1	0	0	1	1	0	0	1	0	1	1	0	0	1	1	0	1	0	0	1	1	0	0	1
12	0	0	0	0	1	1	1	1	1	1	1	1	0	0	0	0	0	0	0	0	1	1	1	1	1	1	1	1	0	0	0	0	0	0	0	0	1	1	1	1	1	1	1	1	0	0	0	0	0	0	0	0	1	1	1	1	1	1	1	1	0	0	0	0
13	0	1	0	1	1	0	1	0	1	0	1	0	0	1	0	1	0	1	0	1	1	0	1	0	1	0	1	0	0	1	0	1	0	1	0	1	1	0	1	0	1	0	1	0	0	1	0	1	0	1	0	1	1	0	1	0	1	0	1	0	0	1	0	1
14	0	0	1	1	1	1	0	0	1	1	0	0	0	0	1	1	0	0	1	1	1	1	0	0	1	1	0	0	0	0	1	1	0	0	1	1	1	1	0	0	1	1	0	0	0	0	1	1	0	0	1	1	1	1	0	0	1	1	0	0	0	0	1	1
15	0	1	1	0	1	0	0	1	1	0	0	1	0	1	1	0	0	1	1	0	1	0	0	1	1	0	0	1	0	1	1	0	0	1	1	0	1	0	0	1	1	0	0	1	0	1	1	0	0	1	1	0	1	0	0	1	1	0	0	1	0	1	1	0
16	0	0	0	0	0	0	0	0	0	0	0	0	0	0	0	0	1	1	1	1	1	1	1	1	1	1	1	1	1	1	1	1	0	0	0	0	0	0	0	0	0	0	0	0	0	0	0	0	1	1	1	1	1	1	1	1	1	1	1	1	1	1	1	1
17	0	1	0	1	0	1	0	1	0	1	0	1	0	1	0	1	1	0	1	0	1	0	1	0	1	0	1	0	1	0	1	0	0	1	0	1	0	1	0	1	0	1	0	1	0	1	0	1	1	0	1	0	1	0	1	0	1	0	1	0	1	0	1	0
18	0	0	1	1	0	0	1	1	0	0	1	1	0	0	1	1	1	1	0	0	1	1	0	0	1	1	0	0	1	1	0	0	0	0	1	1	0	0	1	1	0	0	1	1	0	0	1	1	1	1	0	0	1	1	0	0	1	1	0	0	1	1	0	0
19	0	1	1	0	0	1	1	0	0	1	1	0	0	1	1	0	1	0	0	1	1	0	0	1	1	0	0	1	1	0	0	1	0	1	1	0	0	1	1	0	0	1	1	0	0	1	1	0	1	0	0	1	1	0	0	1	1	0	0	1	1	0	0	1
20	0	0	0	0	1	1	1	1	0	0	0	0	1	1	1	1	1	1	1	1	0	0	0	0	1	1	1	1	0	0	0	0	0	0	0	0	1	1	1	1	0	0	0	0	1	1	1	1	1	1	1	1	0	0	0	0	1	1	1	1	0	0	0	0
21	0	1	0	1	1	0	1	0	0	1	0	1	1	0	1	0	1	0	1	0	0	1	0	1	1	0	1	0	0	1	0	1	0	1	0	1	1	0	1	0	0	1	0	1	1	0	1	0	1	0	1	0	0	1	0	1	1	0	1	0	0	1	0	1
22	0	0	1	1	1	1	0	0	0	0	1	1	1	1	0	0	1	1	0	0	0	0	1	1	1	1	0	0	0	0	1	1	0	0	1	1	1	1	0	0	0	0	1	1	1	1	0	0	1	1	0	0	0	0	1	1	1	1	0	0	0	0	1	1
23	0	1	1	0	1	0	0	1	0	1	1	0	1	0	0	1	1	0	0	1	0	1	1	0	1	0	0	1	0	1	1	0	0	1	1	0	1	0	0	1	0	1	1	0	1	0	0	1	1	0	0	1	0	1	1	0	1	0	0	1	0	1	1	0
24	0	0	0	0	0	0	0	0	1	1	1	1	1	1	1	1	1	1	1	1	1	1	1	1	0	0	0	0	0	0	0	0	0	0	0	0	0	0	0	0	1	1	1	1	1	1	1	1	1	1	1	1	1	1	1	1	0	0	0	0	0	0	0	0
25	0	1	0	1	0	1	0	1	1	0	1	0	1	0	1	0	1	0	1	0	1	0	1	0	0	1	0	1	0	1	0	1	0	1	0	1	0	1	0	1	1	0	1	0	1	0	1	0	1	0	1	0	1	0	1	0	0	1	0	1	0	1	0	1
26	0	0	1	1	0	0	1	1	1	1	0	0	1	1	0	0	1	1	0	0	1	1	0	0	0	0	1	1	0	0	1	1	0	0	1	1	0	0	1	1	1	1	0	0	1	1	0	0	1	1	0	0	1	1	0	0	0	0	1	1	0	0	1	1
27	0	1	1	0	0	1	1	0	1	0	0	1	1	0	0	1	1	0	0	1	1	0	0	1	0	1	1	0	0	1	1	0	0	1	1	0	0	1	1	0	1	0	0	1	1	0	0	1	1	0	0	1	1	0	0	1	0	1	1	0	0	1	1	0
28	0	0	0	0	1	1	1	1	1	1	1	1	0	0	0	0	1	1	1	1	0	0	0	0	0	0	0	0	1	1	1	1	0	0	0	0	1	1	1	1	1	1	1	1	0	0	0	0	1	1	1	1	0	0	0	0	0	0	0	0	1	1	1	1
29	0	1	0	1	1	0	1	0	1	0	1	0	0	1	0	1	1	0	1	0	0	1	0	1	0	1	0	1	1	0	1	0	0	1	0	1	1	0	1	0	1	0	1	0	0	1	0	1	1	0	1	0	0	1	0	1	0	1	0	1	1	0	1	0
30	0	0	1	1	1	1	0	0	1	1	0	0	0	0	1	1	1	1	0	0	0	0	1	1	0	0	1	1	1	1	0	0	0	0	1	1	1	1	0	0	1	1	0	0	0	0	1	1	1	1	0	0	0	0	1	1	0	0	1	1	1	1	0	0
31	0	1	1	0	1	0	0	1	1	0	0	1	0	1	1	0	1	0	0	1	0	1	1	0	0	1	1	0	1	0	0	1	0	1	1	0	1	0	0	1	1	0	0	1	0	1	1	0	1	0	0	1	0	1	1	0	0	1	1	0	1	0	0	1
32	0	0	0	0	0	0	0	0	0	0	0	0	0	0	0	0	0	0	0	0	0	0	0	0	0	0	0	0	0	0	0	0	1	1	1	1	1	1	1	1	1	1	1	1	1	1	1	1	1	1	1	1	1	1	1	1	1	1	1	1	1	1	1	1
33	0	1	0	1	0	1	0	1	0	1	0	1	0	1	0	1	0	1	0	1	0	1	0	1	0	1	0	1	0	1	0	1	1	0	1	0	1	0	1	0	1	0	1	0	1	0	1	0	1	0	1	0	1	0	1	0	1	0	1	0	1	0	1	0
34	0	0	1	1	0	0	1	1	0	0	1	1	0	0	1	1	0	0	1	1	0	0	1	1	0	0	1	1	0	0	1	1	1	1	0	0	1	1	0	0	1	1	0	0	1	1	0	0	1	1	0	0	1	1	0	0	1	1	0	0	1	1	0	0
35	0	1	1	0	0	1	1	0	0	1	1	0	0	1	1	0	0	1	1	0	0	1	1	0	0	1	1	0	0	1	1	0	1	0	0	1	1	0	0	1	1	0	0	1	1	0	0	1	1	0	0	1	1	0	0	1	1	0	0	1	1	0	0	1
36	0	0	0	0	1	1	1	1	0	0	0	0	1	1	1	1	0	0	0	0	1	1	1	1	0	0	0	0	1	1	1	1	1	1	1	1	0	0	0	0	1	1	1	1	0	0	0	0	1	1	1	1	0	0	0	0	1	1	1	1	0	0	0	0
37	0	1	0	1	1	0	1	0	0	1	0	1	1	0	1	0	0	1	0	1	1	0	1	0	0	1	0	1	1	0	1	0	1	0	1	0	0	1	0	1	1	0	1	0	0	1	0	1	1	0	1	0	0	1	0	1	1	0	1	0	0	1	0	1
38	0	0	1	1	1	1	0	0	0	0	1	1	1	1	0	0	0	0	1	1	1	1	0	0	0	0	1	1	1	1	0	0	1	1	0	0	0	0	1	1	1	1	0	0	0	0	1	1	1	1	0	0	0	0	1	1	1	1	0	0	0	0	1	1
39	0	1	1	0	1	0	0	1	0	1	1	0	1	0	0	1	0	1	1	0	1	0	0	1	0	1	1	0	1	0	0	1	1	0	0	1	0	1	1	0	1	0	0	1	0	1	1	0	1	0	0	1	0	1	1	0	1	0	0	1	0	1	1	0
40	0	0	0	0	0	0	0	0	1	1	1	1	1	1	1	1	0	0	0	0	0	0	0	0	1	1	1	1	1	1	1	1	1	1	1	1	1	1	1	1	0	0	0	0	0	0	0	0	1	1	1	1	1	1	1	1	0	0	0	0	0	0	0	0
41	0	1	0	1	0	1	0	1	1	0	1	0	1	0	1	0	0	1	0	1	0	1	0	1	1	0	1	0	1	0	1	0	1	0	1	0	1	0	1	0	0	1	0	1	0	1	0	1	1	0	1	0	1	0	1	0	0	1	0	1	0	1	0	1
42	0	0	1	1	0	0	1	1	1	1	0	0	1	1	0	0	0	0	1	1	0	0	1	1	1	1	0	0	1	1	0	0	1	1	0	0	1	1	0	0	0	0	1	1	0	0	1	1	1	1	0	0	1	1	0	0	0	0	1	1	0	0	1	1
43	0	1	1	0	0	1	1	0	1	0	0	1	1	0	0	1	0	1	1	0	0	1	1	0	1	0	0	1	1	0	0	1	1	0	0	1	1	0	0	1	0	1	1	0	0	1	1	0	1	0	0	1	1	0	0	1	0	1	1	0	0	1	1	0
44	0	0	0	0	1	1	1	1	1	1	1	1	0	0	0	0	0	0	0	0	1	1	1	1	1	1	1	1	0	0	0	0	1	1	1	1	0	0	0	0	0	0	0	0	1	1	1	1	1	1	1	1	0	0	0	0	0	0	0	0	1	1	1	1
45	0	1	0	1	1	0	1	0	1	0	1	0	0	1	0	1	0	1	0	1	1	0	1	0	1	0	1	0	0	1	0	1	1	0	1	0	0	1	0	1	0	1	0	1	1	0	1	0	1	0	1	0	0	1	0	1	0	1	0	1	1	0	1	0
46	0	0	1	1	1	1	0	0	1	1	0	0	0	0	1	1	0	0	1	1	1	1	0	0	1	1	0	0	0	0	1	1	1	1	0	0	0	0	1	1	0	0	1	1	1	1	0	0	1	1	0	0	0	0	1	1	0	0	1	1	1	1	0	0
47	0	1	1	0	1	0	0	1	1	0	0	1	0	1	1	0	0	1	1	0	1	0	0	1	1	0	0	1	0	1	1	0	1	0	0	1	0	1	1	0	0	1	1	0	1	0	0	1	1	0	0	1	0	1	1	0	0	1	1	0	1	0	0	1
48	0	0	0	0	0	0	0	0	0	0	0	0	0	0	0	0	1	1	1	1	1	1	1	1	1	1	1	1	1	1	1	1	1	1	1	1	1	1	1	1	1	1	1	1	1	1	1	1	0	0	0	0	0	0	0	0	0	0	0	0	0	0	0	0
49	0	1	0	1	0	1	0	1	0	1	0	1	0	1	0	1	1	0	1	0	1	0	1	0	1	0	1	0	1	0	1	0	1	0	1	0	1	0	1	0	1	0	1	0	1	0	1	0	0	1	0	1	0	1	0	1	0	1	0	1	0	1	0	1
50	0	0	1	1	0	0	1	1	0	0	1	1	0	0	1	1	1	1	0	0	1	1	0	0	1	1	0	0	1	1	0	0	1	1	0	0	1	1	0	0	1	1	0	0	1	1	0	0	0	0	1	1	0	0	1	1	0	0	1	1	0	0	1	1
51	0	1	1	0	0	1	1	0	0	1	1	0	0	1	1	0	1	0	0	1	1	0	0	1	1	0	0	1	1	0	0	1	1	0	0	1	1	0	0	1	1	0	0	1	1	0	0	1	0	1	1	0	0	1	1	0	0	1	1	0	0	1	1	0
52	0	0	0	0	1	1	1	1	0	0	0	0	1	1	1	1	1	1	1	1	0	0	0	0	1	1	1	1	0	0	0	0	1	1	1	1	0	0	0	0	1	1	1	1	0	0	0	0	0	0	0	0	1	1	1	1	0	0	0	0	1	1	1	1
53	0	1	0	1	1	0	1	0	0	1	0	1	1	0	1	0	1	0	1	0	0	1	0	1	1	0	1	0	0	1	0	1	1	0	1	0	0	1	0	1	1	0	1	0	0	1	0	1	0	1	0	1	1	0	1	0	0	1	0	1	1	0	1	0
54	0	0	1	1	1	1	0	0	0	0	1	1	1	1	0	0	1	1	0	0	0	0	1	1	1	1	0	0	0	0	1	1	1	1	0	0	0	0	1	1	1	1	0	0	0	0	1	1	0	0	1	1	1	1	0	0	0	0	1	1	1	1	0	0
55	0	1	1	0	1	0	0	1	0	1	1	0	1	0	0	1	1	0	0	1	0	1	1	0	1	0	0	1	0	1	1	0	1	0	0	1	0	1	1	0	1	0	0	1	0	1	1	0	0	1	1	0	1	0	0	1	0	1	1	0	1	0	0	1
56	0	0	0	0	0	0	0	0	1	1	1	1	1	1	1	1	1	1	1	1	1	1	1	1	0	0	0	0	0	0	0	0	1	1	1	1	1	1	1	1	0	0	0	0	0	0	0	0	0	0	0	0	0	0	0	0	1	1	1	1	1	1	1	1
57	0	1	0	1	0	1	0	1	1	0	1	0	1	0	1	0	1	0	1	0	1	0	1	0	0	1	0	1	0	1	0	1	1	0	1	0	1	0	1	0	0	1	0	1	0	1	0	1	0	1	0	1	0	1	0	1	1	0	1	0	1	0	1	0
58	0	0	1	1	0	0	1	1	1	1	0	0	1	1	0	0	1	1	0	0	1	1	0	0	0	0	1	1	0	0	1	1	1	1	0	0	1	1	0	0	0	0	1	1	0	0	1	1	0	0	1	1	0	0	1	1	1	1	0	0	1	1	0	0
59	0	1	1	0	0	1	1	0	1	0	0	1	1	0	0	1	1	0	0	1	1	0	0	1	0	1	1	0	0	1	1	0	1	0	0	1	1	0	0	1	0	1	1	0	0	1	1	0	0	1	1	0	0	1	1	0	1	0	0	1	1	0	0	1
60	0	0	0	0	1	1	1	1	1	1	1	1	0	0	0	0	1	1	1	1	0	0	0	0	0	0	0	0	1	1	1	1	1	1	1	1	0	0	0	0	0	0	0	0	1	1	1	1	0	0	0	0	1	1	1	1	1	1	1	1	0	0	0	0
61	0	1	0	1	1	0	1	0	1	0	1	0	0	1	0	1	1	0	1	0	0	1	0	1	0	1	0	1	1	0	1	0	1	0	1	0	0	1	0	1	0	1	0	1	1	0	1	0	0	1	0	1	1	0	1	0	1	0	1	0	0	1	0	1
62	0	0	1	1	1	1	0	0	1	1	0	0	0	0	1	1	1	1	0	0	0	0	1	1	0	0	1	1	1	1	0	0	1	1	0	0	0	0	1	1	0	0	1	1	1	1	0	0	0	0	1	1	1	1	0	0	1	1	0	0	0	0	1	1
63	0	1	1	0	1	0	0	1	1	0	0	1	0	1	1	0	1	0	0	1	0	1	1	0	0	1	1	0	1	0	0	1	1	0	0	1	0	1	1	0	0	1	1	0	1	0	0	1	0	1	1	0	1	0	0	1	1	0	0	1	0	1	1	0

Figure 2–11 The Walsh Matrix used in CDMA.

The gating function uses *power control groups*, which are exactly six Walsh symbols (1.25 ms) long. There are 16 power control groups in each 20 ms frame. For the specific case of the 7200 bps rate, the gating function works by removing either the first six or the second six Walsh symbols out of every twelve Walsh symbols. In other words, the gating function transmits one out of every pair of power control groups. Therefore, after the gating function, no information is lost, but no redundant information is sent. The resulting signal at the output of the Data Burst Randomizer has a 50% duty cycle for the 7200 bps rate.

Selecting which power control group to transmit out of every pair is determined by a precise complicated procedure intended to randomize the gating pattern from one subscriber to the next. As shown in Figure 2–12, the last 14 bits of the long-code sequence (which is discussed in the next section), used to spread the second-to-last power control group in the frame immediately prior to the current one, is used as an input to formulate the current gating sequence. For the 7200 bps rate, these bits, labeled $\{b_0, b_1, b_2, \ldots b_{13}\}$, are used to control which power control groups are sent in the current frame. For the 7200 bps code rate, IS-95 specifies that the following power control groups are to be transmitted:

$$b_0, 2+b_1, 4+b_2, 6+b_3, 8+b_4, 10+b_5, 12+b_6, 14+b_7$$

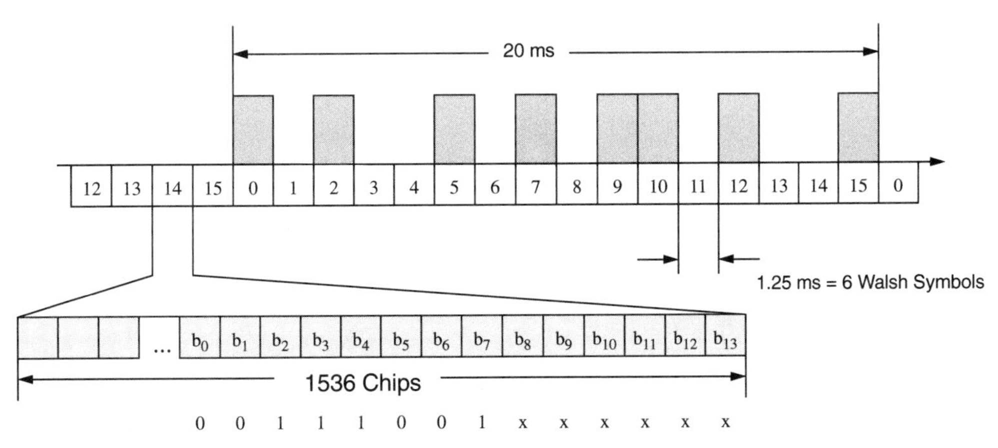

$$\{b_0, 2+b_1, 4+b_2, 6+b_3, 8+b_4, 10+b_5, 12+b_6, 14+b_7\} = \{0, 2, 5, 7, 9, 10, 12, 15\}$$

Figure 2–12 Illustration of power control gating for the 7200 bps data rate. Note that only certain 1.25 ms segments of the 20 ms frame are transmitted.

As a result of the Data Burst Randomizer, users are not transmitting at the same time, and co-channel reverse link interference is thus randomized over time. Furthermore, each user has smaller average transmitted power. Lower average transmitted power reduces the interference to nearby base stations. Also, there is reduced power consumption at the mobile since the phone is not always transmitting.

2.4.6 Long Code Spreading

After the Data Burst Randomizer, the signal is spread by the *long code*, as shown in Figures 2–4 and 2–5. The long code is used for two purposes:

- *Channelization* - The base station separates reverse channel traffic by using the long code. As discussed in the next section, every subscriber communicating with the base station uses the same quadrature spreading code and offset (except for propagation delays), so the long code is used to identify both access and traffic channels.
- *Privacy* - The long code for traffic channels can be derived in two ways. One technique uses the *Electronic Serial Number* (ESN) of the subscriber to generate the long code and is therefore publicly known if the ESN is known. Another technique generates the long code using keys that are known only to the base station and subscriber unit, providing a level of privacy and preventing simple despreading.

The long code is a particular type of Pseudo Noise sequence called a *Maximal Length Sequence*, also called an *M-sequence*. An M-sequence can be generated by a Linear Feedback Shift Register, as shown in Figure 2–13.

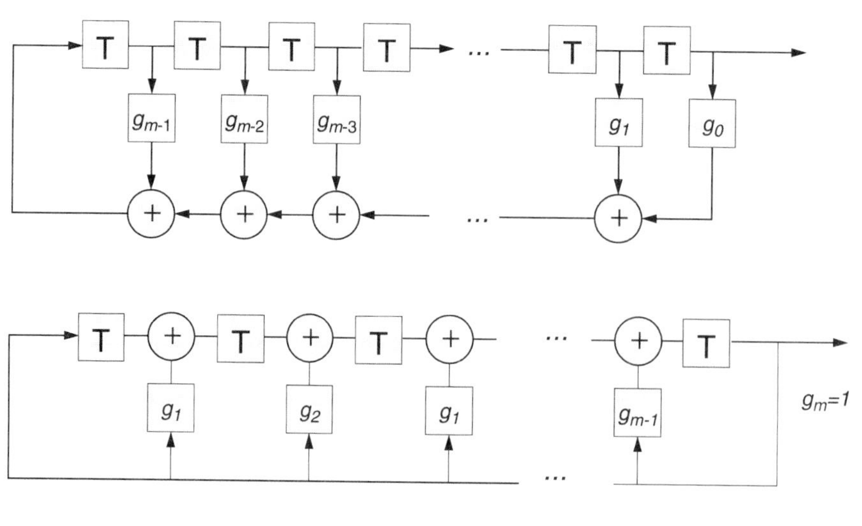

$$G(x) = g_m X^m + g_{m-1} X^{m-1} + \ldots + g_1 X + g_0$$

Figure 2–13 Two M-sequence generators using Linear Feedback Shift Registers (LFSRs). When $G(x)$ is the generator polynomial for an M-sequence, the two generators shown here will produce the same sequence. However, for the same initial conditions, the two sequences will be offset from each other. Note that g_i represents a "1" or "0" multiplier gain.

The summation blocks in Figure 2–13 are not arithmetic sums, but are instead *eXclusive-OR* (XOR) gates, which are summarized in Table 2–11. The XOR function serves a similar function to arithmetic summation in binary arithmetic. In general, when adding two binary digits, polynomials, or sequences, we will take the addition operator to represent the XOR function unless otherwise indicated.

Table 2–11 The exclusive-or (XOR) function.

a	b	a+b
0	0	0
0	1	1
1	0	1
1	1	0

M-sequences have a number of properties which make them very useful for Direct Sequence spreading.

- First of all, M-sequences have a very well-defined *autocorrelation function*. As discussed in Chapter 1, DS-CDMA receivers must use a locally generated version of the same PN-sequence that is time-synchronized with the received PN-sequence from the transmitter. In order to allow the receiver to lock to the incoming PN-sequence, it is necessary for the inner product between two identical PN-sequences to have a large magnitude when the sequences are aligned and a small magnitude when the sequences are not aligned. The inner product between two PN-sequences $p_i(t)$ and $p_j(t)$ is

$$R_{ij}(\tau) = \int_{t-T_w}^{t} p_i(t)p_j(t-\tau)dt \tag{2.7}$$

  M-sequences have the unique property that the autocorrelation function, $R_{ij}(\tau)$, takes on a well-defined, deterministic function. If T_c is the chip period, then let the integration window, $T_w = MT_c$. Here M is the length of the PN sequence, which is always equal to $2^m - 1$ for M-sequences. Then the autocorrelation function, $R_{ij}(\tau)$, between the two sequences is shown in Figure 2–14.
- Each period of the M-sequence has exactly one occurrence of a sequence of $m - 1$ consecutive zeros. We define the start of the M-sequence as the first "one" following $m - 1$ consecutive zeros.
- From the LFSR implementation shown in Figure 2–13, the shift registers may be initially loaded with any value, except for a series of m zeros. The "all-zeros" initial state will cause the LFSR to produce an infinite series of zeros, called the *trivial sequence*. In any other series, which will be an M-sequence, no series of m zeros occurs.

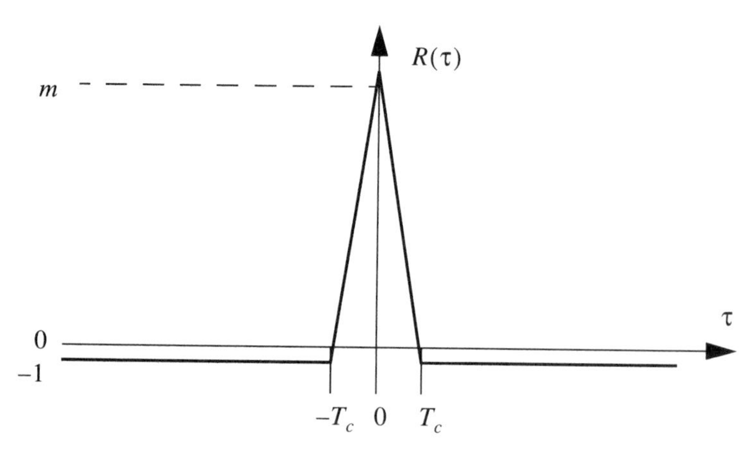

Figure 2–14 Autocorrelation function for a Maximal Length sequence.

In IS-95 the long code is derived from an M-sequence with $m = 42$. The generator polynomial is

$$G(x) = x^{42} + x^{35} + x^{33} + x^{31} + x^{27} + x^{26} + x^{25} + x^{22} + x^{21} + x^{19} + x^{18} + x^{17} + x^{16} + x^{10} + x^{7} + x^{6} + x^{5} + x^{3} + x^{2} + x^{1} + 1 \tag{2.8}$$

The sequence produced by this generator polynomial has a length of $2^{42} - 1$ chips, or a cycle time of approximately 41.4 days (hence the name, "long code"). However, the output of the LFSR is not used directly for spreading. Instead, the long code comprises an altered version of this sequence which is produced using a *Long Code Mask*. A combined structure which includes an LFSR and the Long Code Mask is illustrated in Figure 2–15.

Each gate of the Long Code Mask in Figure 2–15 controls whether one of 42 taps in the LFSR is added to produce the long code. Each tap represents a different offset in the M-sequence produced by (2.8). The sum of two or more versions of the same M-sequence, offset by different amounts, yields yet another M-sequence.

The Long Code Mask is derived using a number of different means. Access channels, which respond to pages and initiate calls, use the Long Code Mask derived from the paging and access channel numbers, along with the Base Station ID. For traffic channels, there are two different types of Long Code Mask.

The traffic channel mask can be *public* or *private*. The public mask is a specific permutation of the mobile's *Electronic Serial Number* (ESN). The private mask uses encryption techniques, thus preventing eavesdropping. The PN chip generated is a function of the mask bits and the outputs of the long code register.

2.4.7 Quadrature (Short Code) Spreading

The final stage in generating the baseband reverse link signal is the *Quadrature Spreading* operation. As shown in Figure 2–16, the output of the Data Burst Randomizer, spread by the

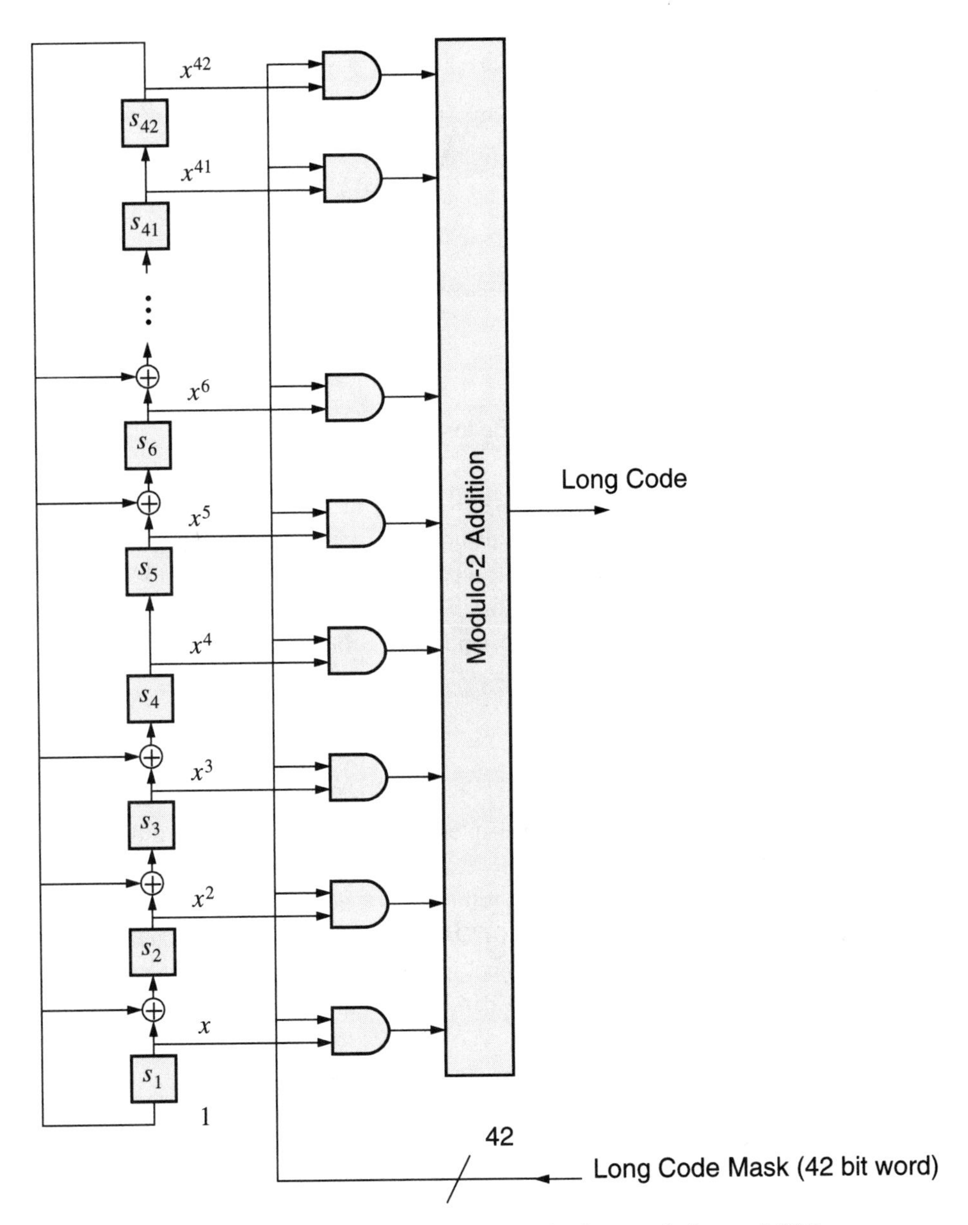

Figure 2–15 The Long Code Mask is used to derive the long code from a LFSR.

long code, is split into *In-phase* (I), and *Quadrature* (Q) components, each of which is spread separately.

The M-sequences generated by these polynomials have length $2^{15} - 1$. An extra zero is inserted at the end of the short code sequence to make the total length $2^{15} = 32768$ chips. Each

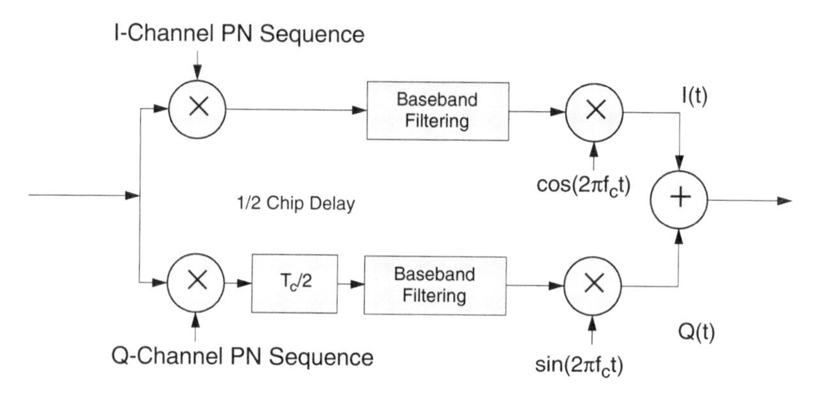

Figure 2–16 The Quadrature Spreading stage for the reverse channel.

of these sequences is $32768/(1.2288 \text{ Mcps}) = 26.667$ ms long. Since these sequences are much shorter than the 41.4 day long code sequence, they are called the *short codes*. There are exactly 75 repetitions of the short code every 2 seconds.

The two PN-sequences used for Quadrature Spreading, like the long code, are based on M-sequences. These sequences are based on a 15-tap LFSR, using the generator polynomials:

$$G_i(x) = x^{15} + x^{13} + x^9 + x^8 + x^7 + x^5 + 1 \tag{2.9}$$

and

$$G_q(x) = x^{15} + x^{12} + x^{11} + x^{10} + x^6 + x^5 + x^4 + x^3 + 1 \tag{2.10}$$

As mentioned at the beginning of this chapter, every base station uses a different Pilot PN Sequence Offset Index. An index of 0 corresponds to a PN sequence that has 15 zeros (14 from the LFSR plus the inserted zero) immediately followed by a one at 00:00:00 on January 5, 1980. Every base station derives its time estimate of *CDMA system time* from the satellite-based *Global Positioning System* (GPS). The pilot channel indices refer to sets of 64 chips, so that there are 512 offset indices. Every subscriber unit derives its timing information from the received pilot signal.

In quadrature spreading, the use of *Offset Quadrature Phase Shift Keying* (OQPSK) allows for power-efficient non-linear amplifiers. Only one of the I and Q signal components is changed at a given time, as illustrated in Figure 2–18. Since the magnitude of the signal does not pass through zero, more efficient transmit amplifiers with relatively low dynamic range can be used, while maintaining signal fidelity.

2.4.8 Reverse Access Channels

An access channel allows signaling, not associated with voice traffic, to be transmitted from the mobile station to the base station. Examples of access channel messages include registration and call origination. The reverse access channel transmitter is illustrated in Figure 2–19.

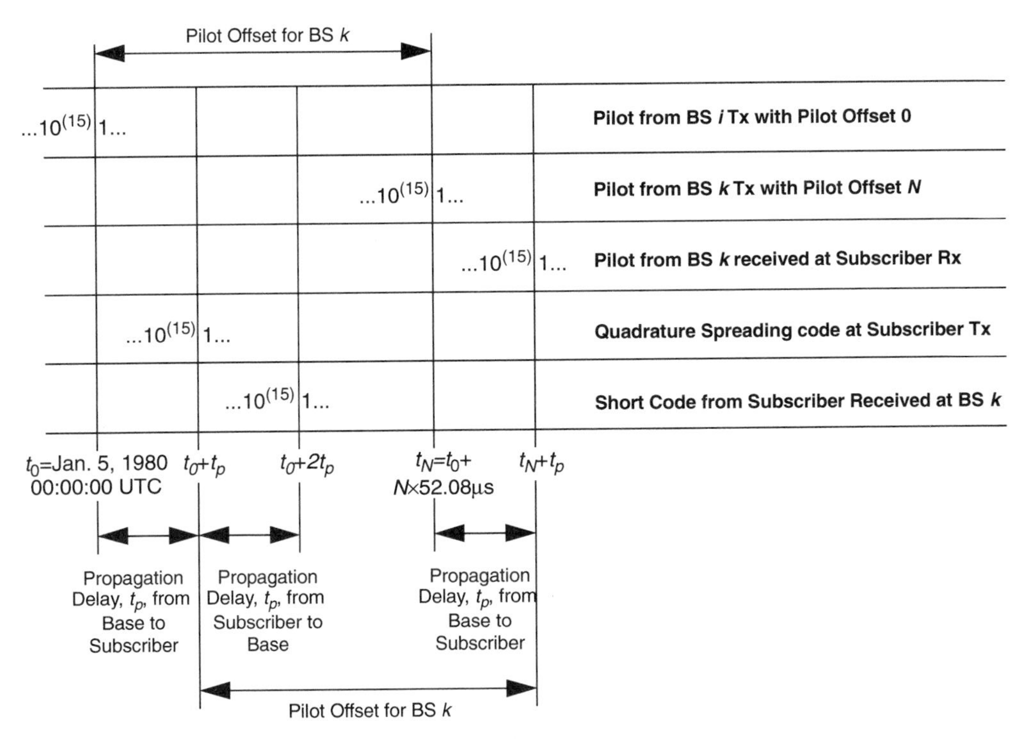

Figure 2–17 Relative offset of the short code at different points in a CDMA system. Every base station uses a different PN offset from the zero-offset code. The zero-offset PN sequence has its first 1 after fifteen zeros (denoted by $10^{(15)}1$) at Jan. 5, 1980 00:00:00 UTC. A base station with pilot offset N transmits its first 1 after fifteen zeros exactly $N \times 52.08$ μs after Jan. 5, 1980 00:00:00 UTC. The short code transmitted by the subscriber unit is synchronized to the zero-offset PN sequence; however it is delayed by the one-way propagation delay to the nearest base station.

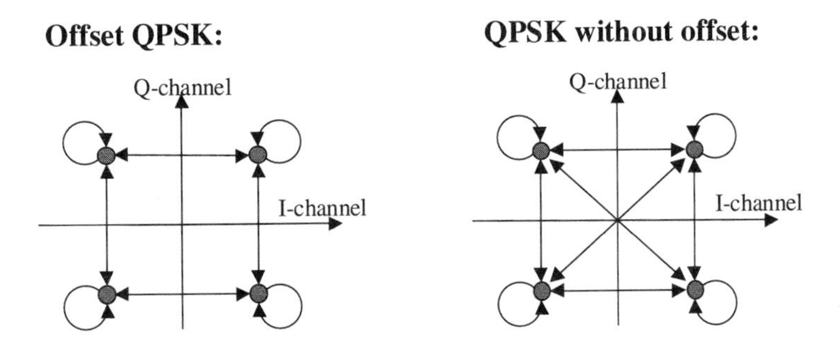

Figure 2–18 Constellation diagrams of the signal after quadrature spreading on (a) the uplink, where the I and Q sequences are offset from one another by half a chip, and (b) the downlink, where the I and Q sequences are transmitted with no offset [Rap96a, Ch. 5].

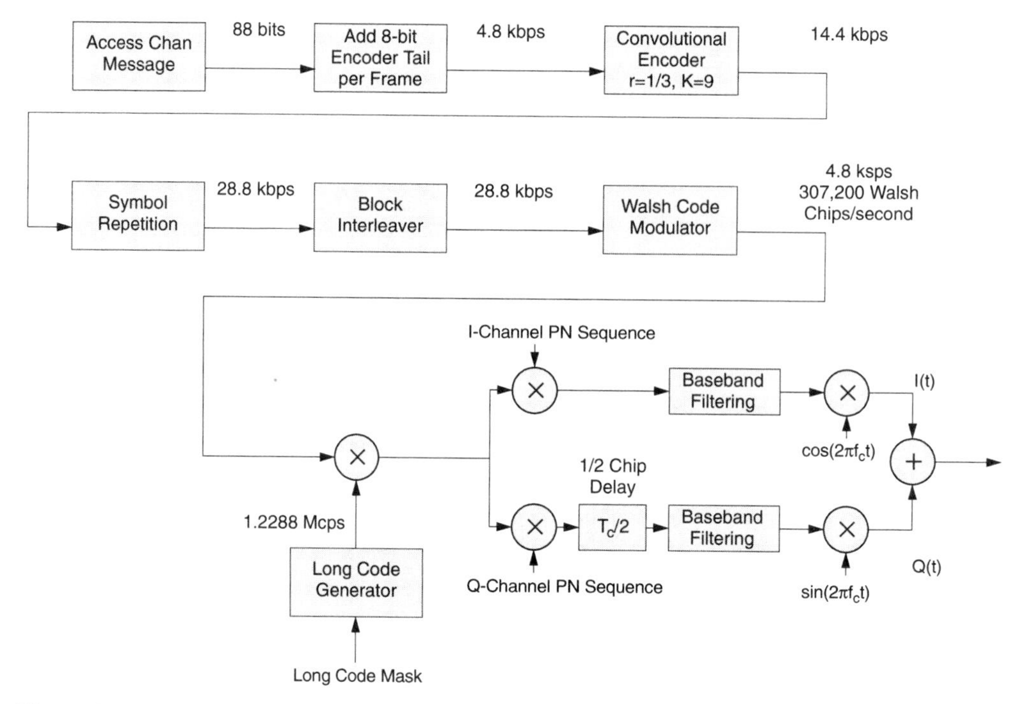

Figure 2–19 The reverse access channel. Like the traffic channels, each access channel frame is 20 ms long.

The reverse access channel is very similar to the Rate Set 1 transmitter; however, a key exception is that even though transmission is at 4.4 kbps, so that repeated Walsh symbols are generated, there is no Data Burst Randomizer, so that the redundant Walsh symbols are transmitted.

Each mobile unit uses a *Hashing function* to generate random parameters that help to prevent subscribers from sending access messages on the same access channel simultaneously. These functions include:

- A mobile randomly chooses one of the active access channels associated with a particular paging channel and uses a long code matched to the selected access channel.
- Each mobile randomly delays its long and short code spreading on access attempts by 0 to 511 chips relative to nominal system time.

2.4.9 Reverse Access - Interaction Between Signals at the Base Station

As we have seen throughout this chapter, complex procedures involving careful timing and selection of PN-codes help to separate signals at the base station. A fundamental issue with CDMA systems is the so-called *Near-Far* effect. Regardless of how well we design the sequences, if one signal arrives at the base station with a power level which is much stronger than another, it can be difficult to extract the lower level signal, as indicated by (1.38). Smart antennas can have a significant impact in this area, as we will see later; however, it is very important to use *power control* to adjust the level of signals received at the base station to the same level to maximize the ability of the base station to extract every available signal simultaneously.

As discussed in Section 1.3.3, power control takes two forms. *Open Loop Power Control* is used by the subscriber to lower its transmitter power level in inverse proportion to the strength of the downlink signal received from a base station. In *Closed Loop Power Control*, the base station specifically directs a subscriber to increase or decrease its transmitter power. Therefore, we nominally expect the signals from most subscribers to arrive near a specified power level.

Reverse link signals from different users arrive at the base station with different signal characteristics:

- Each subscriber has multiple signal components associated with it. Each signal component has its own power level, path delay, phase, and *Doppler shift*.[1]
- Uplink signals are asynchronous. Each uplink signal arrives with different long codes, and short codes that differ in offset by up to twice the maximum propagation delay between the base station and subscriber.
- The E_b/N_0 for each subscriber is estimated at the base station every 1.25 ms, or once during each power control group. The base station immediately informs the subscriber to increase or decrease its power level. After the base station receives a 1.25 ms power control group, it responds within 2 power control groups thereafter. The subscriber unit must then react to a power control command within 500 μs, so that there is a maximum delay of 2 ms between transmission and receipt of base station power control feedback. This allows the Closed Loop Power Control system to remove the impact of small-scale fading on the received signal from a subscriber, for vehicle speeds below 35 km/hr on the reverse link. On the forward link, power control is only effective at tracking small-scale fading for vehicle speeds below 3 km/hr.
- *Soft Handoff* - A key feature of IS-95 is Soft Handoff, which means that a subscriber can communicate with more than one base station simultaneously. In this mode, the subscriber unit receives multiple versions of the downlink signal from different base stations, and it uses its Rake receiver to lock onto and combine the different components. On the reverse link, each base station in the *active set* receives the signal from the subscriber, and demodulated frames are combined at the Base Station Controller (BSC) to take advantage of this *macro-diversity*. When in Soft Handoff, if any base station in the active set orders the subscriber to reduce its power, it must do so, but it may increase power only when directed simultaneously by all base stations. This means that

1. Doppler shift is a change in frequency induced by the motion of the receiver. Depending on the direction of motion of the subscriber relative to the base station, and depending on whether the component is a direct path or a multipath component, Doppler shift can lead to a change in frequency of up to $\pm vf_c/c$, where v is the speed of the vehicle in m/s,v f_c is the carrier frequency in Hz, and c is the speed of light, 3×10^8 m/s.

a subscriber in Soft Handoff generally transmits at a lower power level than it would if it were communicating with only one base station.

We have now described the general characteristics of the CDMA reverse link. The CDMA reverse link is a carefully balanced system, designed to manage subscriber power very efficiently to maximize battery life and capacity, while maintaining range and link quality. Now we turn our attention to the CDMA forward link.

2.5 IS-95 Forward Channel Signals

The forward channel transmitted by each base station shares many components with the reverse link. Both directions use Walsh Functions, Long Code Spreading, and Short Code Quadrature Spreading. However, these functional blocks are used much differently in the forward and reverse directions. These differences are shown in Table 2–12. An essential difference is that while long codes are used for channelization on the uplink, the Walsh Functions are used for downlink channelization. Because Walsh functions are orthogonal with each other when phase-aligned, symbol-aligned, and chip-aligned, and since a different Walsh function is assigned to each downlink channel, downlink signals transmitted from the same source contribute very little interference to each other at the mobile receiver, provided that channel distortion is small.

On the asynchronous reverse link, since it is not possible to align Walsh Functions transmitted by different subscribers, they are not used to separate reverse link signals from different sources; instead, the different Walsh Functions transmitted by a user provide *Channel coding*, giving enhanced link performance by transforming six information bits into one of 64 symbols, where each symbol contains sixty four chips.

Channelization on the reverse link is implemented using the long codes. Since each reverse link signal is spread with a long code with a different offset (using the Long Code Mask) from other reverse link signals, the base station can identify reverse link signals intended for different access and traffic channels by the long code offset. Since channelization on the forward link is accomplished using Walsh Functions, the long code is used to encrypt the forward link without spreading.

2.5.1 Forward Channel Transmitter Structures

The structure of the forward pilot, sync, and paging channels are shown in Figure 2–20. Each forward channel is spread using a Walsh Code at 1.2288 Mcps, and short code spreading is applied as shown in Figure 2–21.

2.5.2 The Pilot Channel

As shown in Figure 2–1, 64 channels can be formed on the same 1.25 MHz CDMA downlink carrier assignment using Walsh Code Modulation. The pilot signal, shown in Figure 2–20, is a stream of zero bits spread only by the short PN code, described in Section 2.4.7. It is code

Table 2–12 Functions of the short code, long code, and Walsh Functions on the forward and reverse links.

Function	Reverse Link	Forward Link
Quadrature Short Code Spreading	• Used for quadrature spreading. • I and Q channels offset from each other by half a chip.	• Each BS transmits a portion of its power on an unmodulated carrier spread with the short code. Offset of the short code identifies the BS. • I and Q channel spreading codes are chip-aligned.
Long Code	• Long Code Spreading, at 1.2288 Mcps is used to select a reverse traffic channel or access channel. (*Channelization*) • Used for encryption if private Long Code Masks are used. • Used to randomize power gating.	• Used to encrypt data. • Identifies placement of power control bits in each power control group. • Every 64th chip of the long code sequence is used.
Walsh Functions	Channel coding uses an alphabet of 64 orthogonal sequences.	Each Walsh Function is assigned to a Pilot, sync, paging, or traffic channel. (*Channelization*)

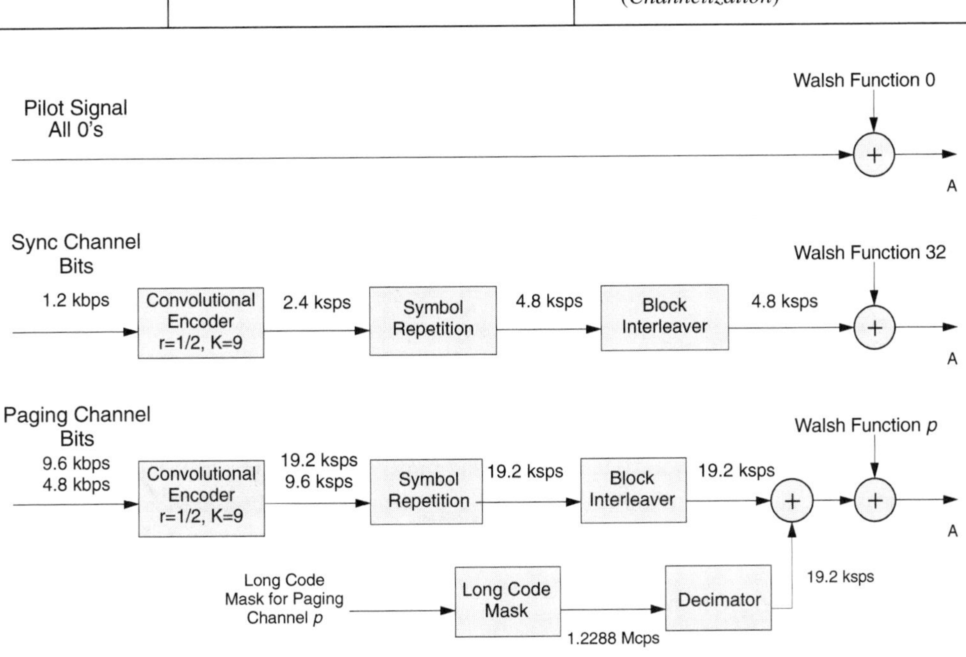

Figure 2–20 Structure of the pilot, sync, and paging channels on the forward link.

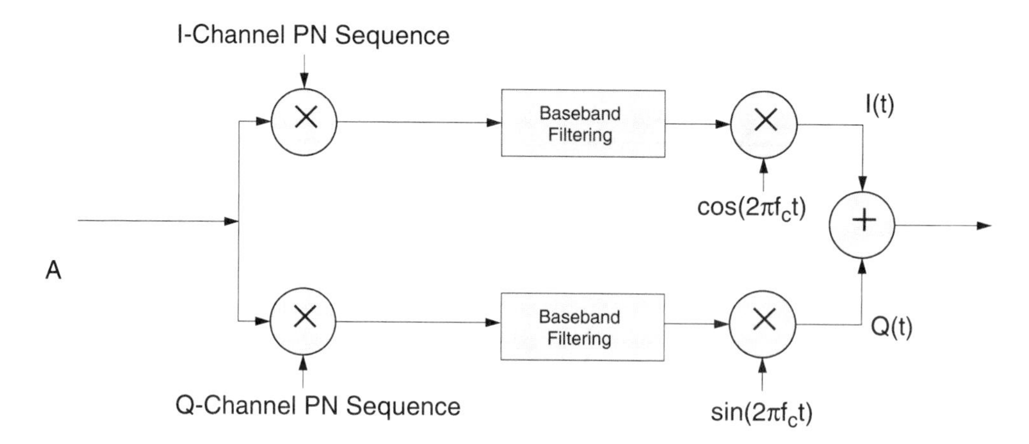

Figure 2–21 All forward link channels, including the pilot, sync, paging, and traffic channels, are spread using the short code before transmission. At A, a logical "0" is mapped into an analog valve of +1, while a logical "1" is mapped to an analog valve of –1.

channel 0, but the Walsh Function zero is a sequence of all zeros. As described in Section 2.2, each base station transmits its pilot with a different offset relative to the pilot with PN-Offset 0, shown in Figure 2–17.

The subscriber unit is able to lock onto the pilot more easily than the other forward link channels because it is unmodulated and transmitted at a higher power level than other channels. The subscriber uses a Rake Finger to determine the PN-offset, phase, and frequency offset of the pilot signal. The pilot channel is used to easily demodulate information contained on the sync, paging, and traffic channels, since it provides a coherent reference which the mobile uses for coherent demodulation.

Because the pilot is so critical to downlink performance, the CDMA base station typically transmits 15-25% of its power in the pilot signal, even though this channel carries no information [Qua93].

2.5.3 The Sync Channel

After determining the strongest pilot signal, the subscriber unit demodulates the sync channel. The sync channel constantly broadcasts information at 1200 bps. This information includes system ID and network ID, detailed timing information, pilot offset index, and paging channel data rate for the base station. From the sync channel, any subscriber unit can determine the exact PN time offset which is part of the modulated 1200 bps message.

2.5.4 Paging Channels

Paging channels convey information from the base station to the mobile station, including the *systems parameters message*, *access parameters message*, *CDMA channel list message*, and *channel assignment message.* Typically, one paging channel is used, although up to seven are permitted. A single 9600 bps paging channel can support 180 pages per second [Kur97]. This channel is used to notify the subscriber unit of an incoming call, as well as important channel lists such as the active set.

2.5.5 Forward Traffic Channels

The forward traffic channels carry downlink data, messages, and voice for each subscriber. Like the reverse channel, there are different structures on the forward link to support Rate Set 1 and Rate Set 2, as shown in Figures 2–22 and 2–23.

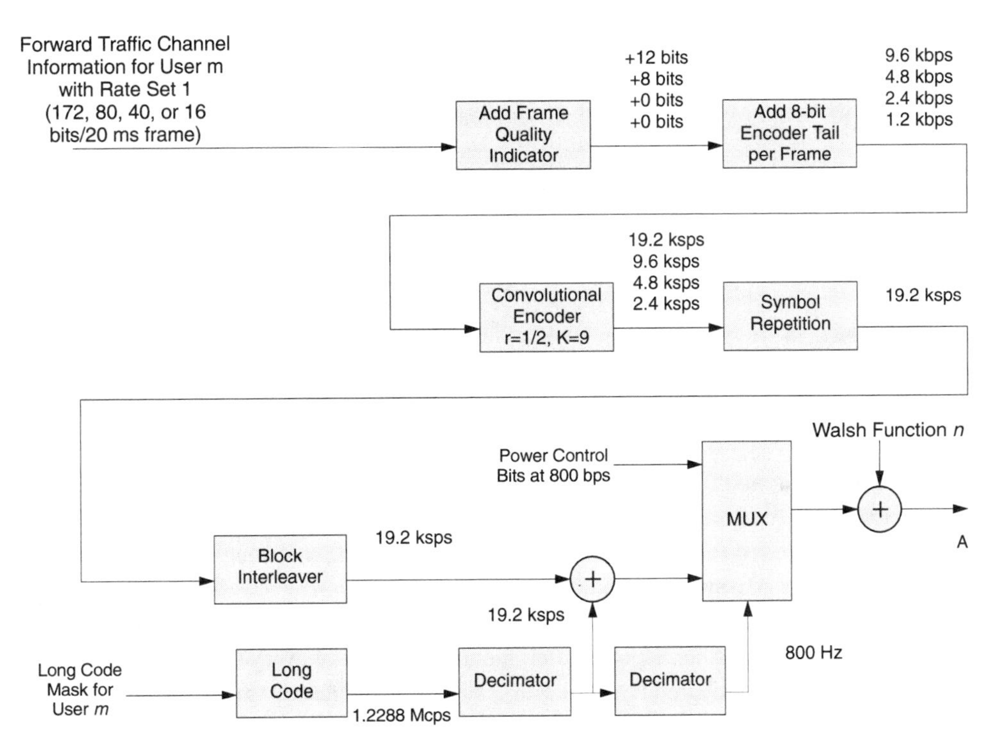

Figure 2–22 Structure of the Forward Traffic Channel for Rate Set 1.

The Frame Quality Indicator, Encoder Tail, Convolutional Encoder and Symbol Repetition are essentially identical to their reverse link counterparts. The 1/2 rate convolutional encoder used on the downlink is the same as the convolutional encoder used on the Rate Set 2 reverse channel; however, here it is used for both Rate Set 1 and Rate Set 2.

Symbol repetition on the forward link results in a rate of 19.2 ksps for Rate Set 1 and 28.8 ksps for Rate Set 2. The block interleaver for both rate sets requires an input rate of 19.2 ksps, so

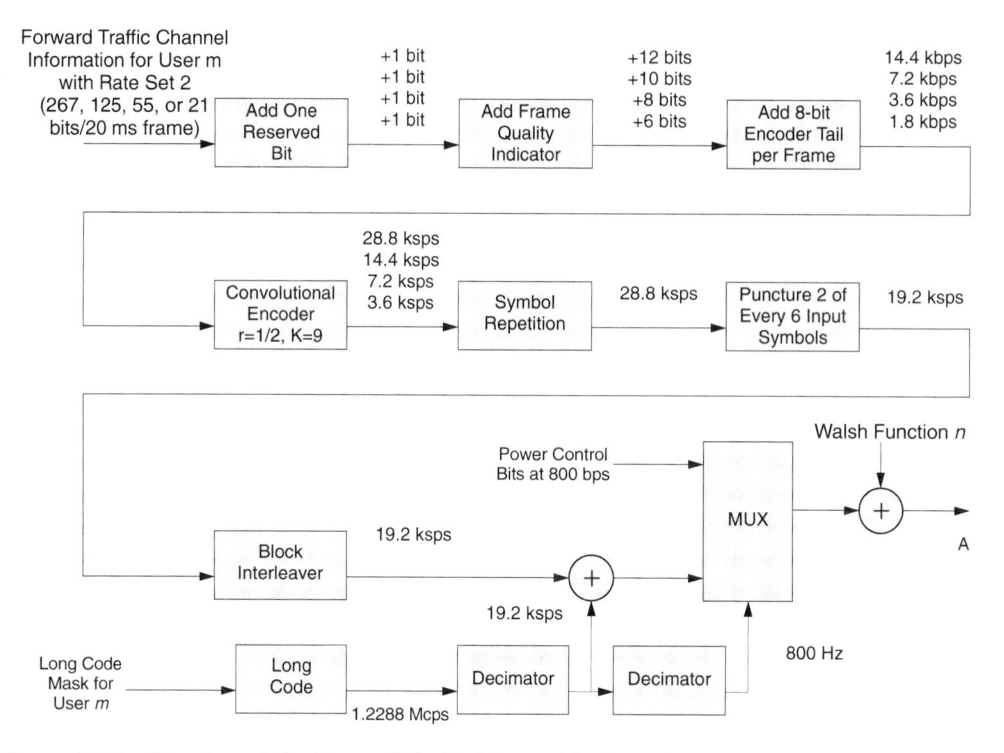

Figure 2–23 Structure of the Forward Traffic Channel for Rate Set 2.

additional manipulation of the Rate Set 2 structure is required to reduce the symbol rate. This is done by puncturing the 28.8 ksps data out of the symbol repetition block in Rate Set 2.

In puncturing used for Rate Set 2, the third and fifth bits out of each six bit symbol repetition block are deleted. Examining the convolutional encoder structure shown in Figure 2–8, and noting that symbols out of the convolutional encoder are transmitted with c_0 first, followed by c_1, we see that puncturing corresponds to eliminating two out of every three symbols produced by the g_0 polynomial, while retaining all of the symbols produced by the g_1 polynomial. Considering the convolutional encoder and puncturing together, for the 13.35 kbps vocoder rate in Rate Set 2, for every six information bits into the convolutional encoder, eight symbols remain after puncturing, resulting in a rate 3/4 code. The rate 3/4 code has less error correcting capability than the rate 1/2 code used for Rate Set 1 on the forward link. Key parameters for the Rate Set 1 and Rate Set 2 forward link transmitters are provided in Tables 2–13 and 2–14.

Another major difference between the forward and reverse links lies in how the long code is used. On the forward link, the long code is used for scrambling to make eavesdropping difficult. On the forward link, the long code is generated in the same manner as the reverse link, described in Section 2.4.6. However, instead of using the entire long code for spreading, on the forward link, only every 64th chip is used to scramble the 19.2 kbps data stream out of the block

Table 2–13 Forward channel parameters for Rate Set 1.

Parameter	Data Rate (bps)				Units
	9600	4800	2400	1200	
Bits per 20 ms Frame out of Vocoder	172	80	40	16	bits/20 ms Frame
Frame Quality Indicator Bits	12	8	0	0	bits/20 ms Frame
Total bits per Frame with 8-bit Encoder Tail	192	96	48	24	bits/20 ms Frame
Bit Rate Into 1/3 Rate Coder in each 20 ms Frame	9600	4800	2400	1200	bps
Rate out of 1/2 Rate Coder	19.2	9.6	4.8	2.4	ksps
Repeated Code Symbol Rate	19.2	19.2	19.2	19.2	ksps
Effective Code Rate	1/2	1/2	1/2	1/3	Info. bits/channel symbol
Walsh Chip Rate	1.2288	1.2288	1.2288	1.2288	Mcps
Walsh Chips/Repeated Code Symbol	64	64	64	64	Walsh Chips/Repeated Sym.
PN Chips/Repeated Code Symbol	64	64	64	64	PN Chips/Repeated Sym.

interleaver, as illustrated in Figures 2–22, 2–23, and 2–24. This scrambling is easily reversed if the receiver knows the Long Code Mask; however, it provides a level of encryption, complicating downlink voice interception when the Long Code Mask is unknown.

2.5.6 The Power Control Subchannel

Every downlink traffic channel has a *power control subchannel* associated with it. As described in Section 2.4.9, power control is essential in CDMA systems to combat the *Near-Far* effect. Closed Loop Power Control messages are carried on the power control subchannel at a rate of 800 bits per second for each subscriber. Closed Loop Power Control is comprised of the *inner closed loop* and the *outer closed loop*.

The inner closed loop is based on the E_b/N_0 estimate at the base station. The base station orders the mobile to step power ± 1 dB. Eight hundred power control messages are sent per second. In the outer closed loop, the base station controls the *frame error rate* (FER) of all users.

The outer closed loop extends control. The trade-off between FER and capacity is controlled by outer closed loop power control. In the inner closed loop, the SNR of the mobile is

Table 2–14 Forward channel parameters for Rate Set 2.

Parameter	Data Rate (bps)				Units
	14400	7200	3600	1800	
Bits per 20 ms Frame out of Vocoder	267	125	55	21	Bits/20 ms Frame
Frame Quality Indicator Bits	12	10	8	6	Bits/20 ms Frame
Total Bits per Frame with 8-bit Encoder Tail and Erasure Bit	288	144	72	36	Bits/20 ms Frame
Bit Rate into 1/2 Rate Coder in each 20 ms Frame	14400	7200	3600	1800	bps
Rate out of 1/2 Rate Coder	28.8	14.4	7.2	3.6	ksps
Repeated Code Symbol Rate	28.8	28.8	28.8	28.8	ksps
Symbol Rate after Puncturing	19.2	19.2	19.2	19.2	ksps
Effective Code Rate	3/4	3/4	3/4	3/4	Info. Bits/Channel Symbol
Walsh Chip Rate	1.2288	1.2288	1.2288	1.2288	Mcps
Walsh Chips/Repeated Code Symbol	64	64	64	64	Walsh Chips/Repeated Sym.
PN Chips/Repeated Code Symbol	64	64	64	64	PN Chips/Repeated Sym.

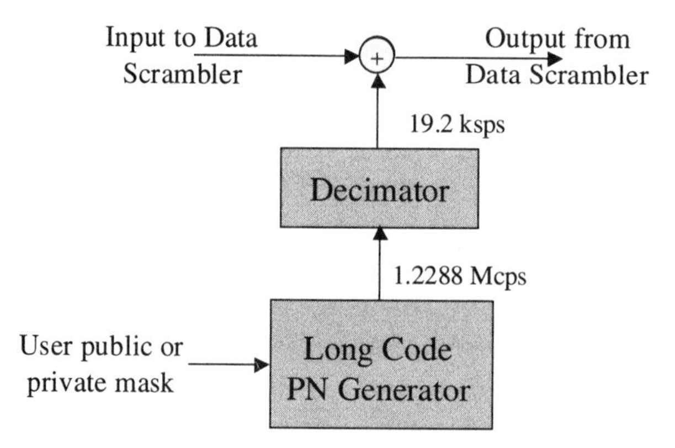

Figure 2–24 The forward channel is scrambled using every 64th chip of the long code.

compared to an adjustable threshold. Increasing the threshold reduces the FER. The FER is estimated using the frame selector at the MSC. The system can optimize itself to equalize the FER rather than the SNR of its subscribers, allowing the system to further reduce the transmitter power for subscribers in soft handoff.

The power control subchannel is inserted onto the forward channel by actually replacing traffic channel symbols. One power control bit is sent every 1.25 ms, or once per power control group. Recall from Figures 2–22 and 2–23 that after data scrambling, symbols at a rate of 19,200 sps are input to the MUltipleXer (MUX), which combines the traffic channel with the power control subchannel.

For Rate Set 1, two symbols, lasting a total of 104.167 μs, in each 1.25 ms power control group are replaced by the power control bit. There are a total of 24 modulation symbols per power control group, but the power control bit is always inserted in place of one of the first 16 symbols. The exact location within these 16 symbols is derived from the last four bits of the 24-bit-long decimated long code used to scramble each frame. Since the subscriber unit knows the Long Code Mask, it knows the location of the power control bits that are inserted into the downlink data stream.

To understand how the base station adjusts the power level used to transmit power control commands, we must first define a few terms.

The energy per bit, or E_b, is the energy transmitted per symbol, E_s, divided by the number of bits per symbol. For example, using the 4800 bps rate with Rate Set 1, there are 4 "information bits" for every 19.2 ksps symbol transmitted over the channel. For this discussion and in the CDMA standard, FQI bits and encoder tail bits are counted as information bits, along with the output of the vocoder (see Tables 2–13 and 2–14). We will let η represent the number of channel symbols used to transmit each information bit.

For Rate Set 1, each power control bit replaces two traffic channel symbols, lasting a total of 104.167 μs (see Figure 2–22). The power control bit is transmitted at a power level of at least $P_s/(2\eta)$, where P_s is the power level used to transmit traffic channel symbols. Therefore, at a vocoder rate of 4800 bps, each power control bit will be transmitted at a power level that is at least 1/8th of the power level of the traffic stream. Then the energy per power control bit is equal to the energy per bit of the traffic channel.

In Rate Set 2, multiplexing of the power control subchannel is slightly different (see Figure 2–23). Rather than replacing two traffic channel symbols, starting at one of 16 offsets from the beginning of each power control group, only a single symbol is replaced. Therefore, on Rate Set 2, each power control bit inserted into the downlink traffic stream is only 52.083 μs long and is transmitted at a power level of $3P_s/(4\eta)$. For the 7200 bps data rate in Rate Set 2, $\eta = 8/3$, so that each power control bit is transmitted at power level of at least $(9/32)P_s$, with an energy of $3E_b/4$.

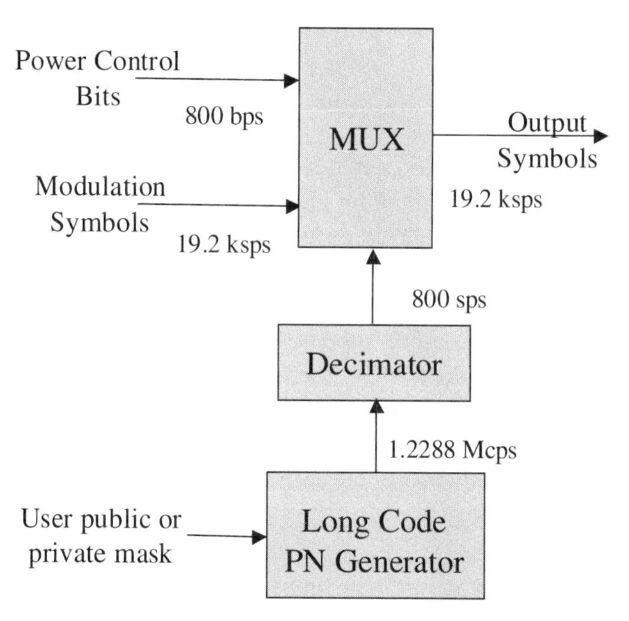

Figure 2–25 The every 64th chip of the long code, or 24 chips per 1.25 ms power control group, is used to scramble data at the traffic input to the MUX. The last 4 chips of these 24 chips are used to derive a starting location, numbered 0-15, for power control bit insertion. The inserted power control bits replace any downlink traffic information.

2.5.7 Downlink Power Control

The power allocated to each channel on the downlink can be adjusted by the base station to provide the best overall wireless coverage to all active subscribers. It is necessary to measure the downlink FER for outer forward link Closed Loop Power Control. The mobile station keeps track of the FER and then sends it to the base station. One counter keeps track of the total number of frames and another keeps track of the total number of bad frames. The mobile station reports these two numbers to the base station in a *Power Measurement Report Message* (PMRM). In Rate Set 2, this function is enhanced using the Frame Erasure bit which allows the mobile to provide an indication of the downlink Frame Error Rate once every 20 ms.

2.6 IS-95 Evolution and cdma2000

The IS-95 standard defines a well-designed air interface for voice signals. However, as noted in Section 1.2.5, Third Generation systems will support many other types of services, including broadband data and video, which cannot be squeezed into the 14.4 kbps maximum data rate currently supported in IS-95.

To address these shortcomings in the near term, an updated version of IS-95 called IS-95B has been proposed. IS-95B will support user data rates up to 76.8 kbps in Rate Set 1 and 115.2 kbps in Rate Set 2. The IS-95B standard will support a number of features that are not in the cur-

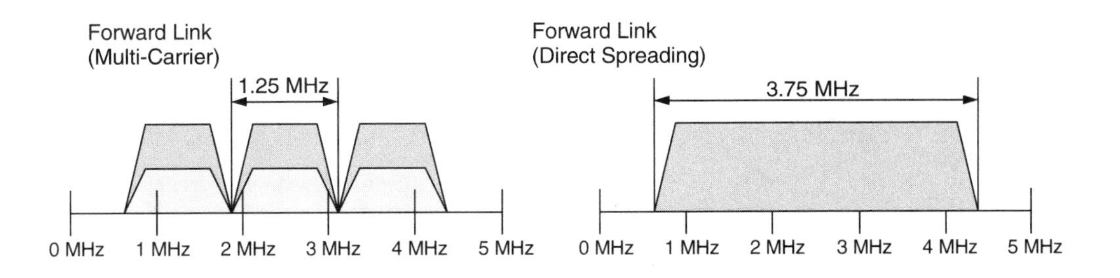

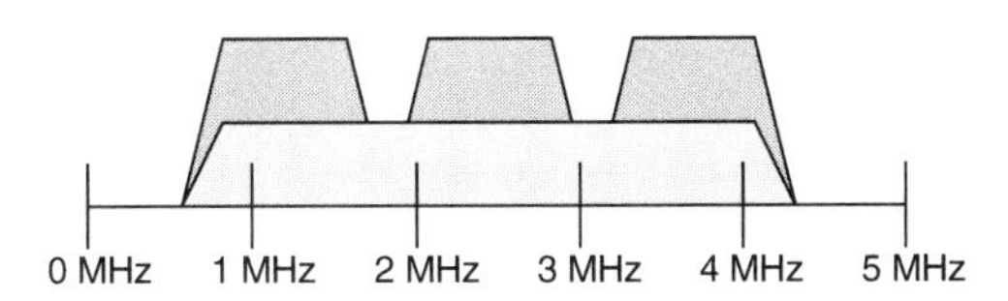

Figure 2–26 The cdma2000 forward link can consist of either multi-carrier or direct spread signals. The multi-carrier approach can be more fully integrated with an existing IS-95 system. On the uplink, a mix of direct spread and multi-carrier approaches can be used. A deployment in a 5 MHz band (in each direction) is shown here. For deployment in larger spectrum allocations, larger numbers of carriers or higher spreading rates can be used [TIA98].

rent standard such as dynamic soft handoff thresholds, improvement in search window optimization, and handoff during access [Tie98].

In the long term the cdma2000 standard, introduced in Section 1.2.5, will provide much greater flexibility, while retaining some level of compatibility with IS-95 systems [TIA98]. The cdma2000 standard will support both multi-carrier and direct spread approaches with chip rates of $N \times 1.2288$ Mcps where N is 1, 3, 6, 9, or 12. The forward link is spread using one of the two approaches, i.e., using N carriers where each carrier is spread at 1.2288 Mcps, or using a single carrier with a spreading rate of $N \times 1.2288$ Mcps . The reverse link can support a mix of the two spreading approaches. The cdma2000 standard will also support other new features, including a pilot for the reverse link, a number of forward link transmit diversity schemes, and support of auxiliary carriers to support downlink beamforming, which is discussed further in Section 4.5.

2.7 Summary

In this chapter, we provided a detailed overview of the IS-95 CDMA standard. We discussed the air interface in considerable detail, with emphasis on the structure of signals produced at the subscriber unit and base station. Key issues that impact the performance of CDMA networks were also presented.

In Chapter 3, a framework will be developed to study smart antenna systems, and in Chapters 4 and 5, these topics will be combined to illustrate how CDMA networks can be enhanced by smart antenna technology.

CHAPTER 3

Introduction to Smart Antennas: Spatial Processing for Wireless Systems

Smart antennas offer a broad range of ways to improve wireless system performance. In general, smart antennas, as shown in Figure 3–1, have the potential to provide enhanced range and reduced infrastructure costs in early deployments, enhanced link performance as the system is built-out, and increased long-term system capacity. In this chapter, general smart antenna concepts are covered which can be applied to a broad range of wireless technologies, including FDMA, TDMA, and CDMA systems. In Chapter 4, we discuss smart antenna techniques which are specific to CDMA. First, we consider a brief summary of the key benefits of smart antenna systems, followed by a discussion of the fundamentals of smart antenna technology.

3.1 Key Benefits of Smart Antenna Technology

Smart antennas provide enhanced coverage through range extension, hole filling, and better building penetration. Given the same transmitter power output at the base station and subscriber unit, smart antennas can increase range by increasing the gain of the base station antenna. From Chapter 1, the uplink power received from a mobile unit at a base station is given by

$$P_r = P_t + G_s + G_b - PL \tag{3.1}$$

where P_r is the power received at the base station, P_t is the power transmitted by the subscriber, G_s is the gain of the subscriber unit antenna, and G_b is the gain of the base station antenna. On the uplink, if a certain received power, $P_{r,min}$, is required at the base station, by increasing the gain of the base station, G_b, the link can tolerate greater path loss, PL. Using (1.59), we write

$$PL(d) = \overline{PL}(d_0) + 10n\log\left(\frac{d}{d_0}\right) + X_\sigma \tag{3.2}$$

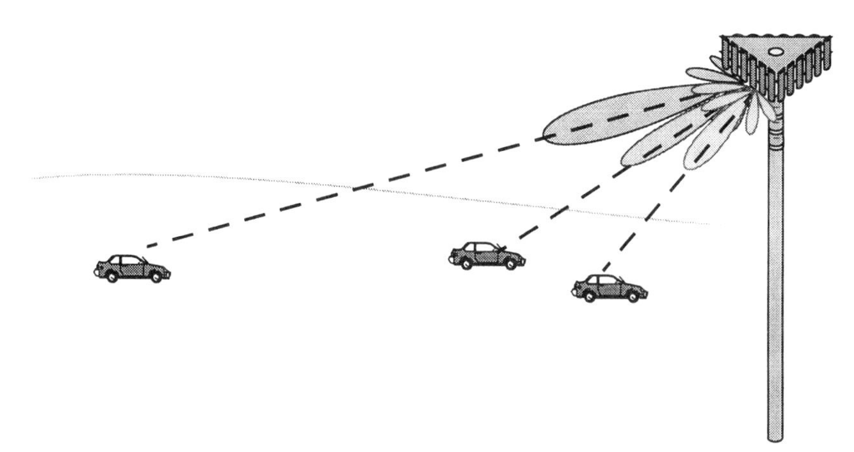

Figure 3–1 Smart antenna systems can form a different beam for each subscriber on the uplink and downlink, minimizing the impact of noise and interference for each subscriber and base station.

Therefore, by increasing the tolerable path loss, we can increase the reception range, d, of the base station. Since smart antennas can allow higher gain compared to conventional antennas, smart antennas systems can provide *range extension*, which is discussed in greater detail in Chapter 4. In order to improve the range on the downlink, we can use smart antennas at the subscriber receiver or at the base station transmitter. Since smart antennas are not usually feasible at mobile and portable subscriber terminals, we may consider downlink beamforming at the base station to increase range in balanced systems. Smart antennas may play a role in subscriber equipment for fixed wireless applications, as shown in Chapter 5.

Through range extension, initial deployment costs to install a wireless system can be reduced. When initially deploying cellular wireless networks, systems are often designed to meet coverage requirements. Even with only a few customers in a system, sufficient number of base stations must be deployed to provide coverage to critical areas. As more customers are added to a cellular network, system capacity can be increased by decreasing the coverage range of base stations and adding additional cell sites. In this later phase, revenue from a large base of subscribers can offset the costs of installing additional base stations; however, in early deployment, to meet initial coverage requirements, a number of base stations must be installed without the customer revenue to support these base stations. Smart antennas can ease this problem by allowing larger early cell sizes. However, the additional cost of using smart antenna systems over conventional technologies must be taken into account when calculating the economic benefit of smart antenna systems.

Smart antennas provide robustness to system perturbations and reduced sensitivity to non-ideal behavior. As described in Chapter 2, CDMA systems require power control to ensure that all of the signals arriving at a base station are at approximately the same power level. Smart antennas help to isolate the uplink signals from different users, reducing the power control requirements or mitigating the impact of imperfect power control. CDMA wireless systems are also particularly sensitive to the geographical distribution of subscribers. Smart antennas refocus coverage patterns to deal with *hot spots*, areas with temporarily high subscriber densities.

Link quality can be improved through multipath management. Multipath in radio channels can result in fading or time dispersion. Smart antennas help to mitigate the impact of multipath or even exploit the diversity inherent in multipath.

Smart antennas can improve system capacity. Smart antennas can be used to allow the subscriber and base station to operate at the same range as a conventional system, but at lower power. This may allow FDMA and TDMA systems to be rechannelized to reuse frequency channels more often than systems using conventional fixed antennas, since the carrier-to-interference ratio is much greater when smart antennas are used. In CDMA systems, if smart antennas are used to allow subscribers to transmit less power for each link, then the Multiple Access Interference is reduced, which increases the number of simultaneous subscribers that can be supported in each cell.

Smart antennas can also be used to spatially separate signals, allowing different subscribers to share the same spectral resources, provided that they are *spatially-separable* at the base station. This *Space Division Multiple Access* (SDMA) allows multiple users to operate in the same cell, on the same frequency/time slot provided, using the smart antenna to separate the signals. Since this approach allows more users to be supported within a limited spectrum allocation, compared with conventional antennas, SDMA can lead to improved capacity.

3.2 Introduction to Smart Antenna Technology

Smart antennas use an array of low gain antenna elements which are connected by a combining network. An arbitrary array of elements is shown in Figure 3–2. Here, ϕ is the azimuthal angle and θ is the elevation angle of a plane wave incident on the array. The horizon is represented by $\theta = \pi/2$.

To simplify the analysis of antenna arrays, we make the following assumptions:

- The spacing between array elements is small enough that there is no amplitude variation between the signals received at different elements.
- There is no mutual coupling between elements.
- All incident fields can be decomposed into a discrete number of plane waves. That is, there are a finite number of signals.
- The bandwidth of the signal incident on the array is small compared with the carrier frequency

For a plane wave incident on the array from direction (θ, ϕ), the difference in phase between the signal component incident on array element m and a reference element at the origin is

$$\Delta\psi_m = \beta\Delta d_m = \beta(x_m \cos\phi \sin\theta + y_m \sin\phi \sin\theta + z_m \cos\theta) \tag{3.3}$$

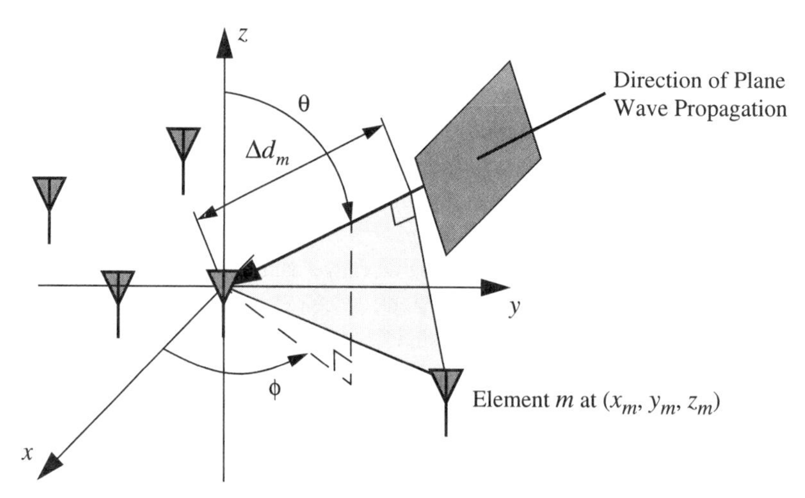

Figure 3–2 Geometry used to determine the Direction-Of-Arrival of a plane wave incident on an arbitrary array of antenna elements.

where $\beta = 2\pi/\lambda$ is the *phase propagation factor*. The term λ denotes the wavelength, given by c/f, where c is the speed of light, 3×10^{8} m/s, and f is the carrier frequency in Hz.

In general, the array may consist of a number of antenna elements distributed in any desired pattern; however, the array is frequently implemented as a *linear equally spaced* (LES), uniform circular, or uniformly spaced planar array of similar, co-polarized, low gain elements which are oriented in the same direction. A simple M-element LES antenna array, oriented along the x-axis, with an element spacing of Δx, is illustrated in Figure 3–3.

The baseband complex envelope representation of the array is shown in Figure 3–3. Each branch of the array has a weighting element, w_m. The *weighting element*, w_m, has both a magnitude and a phase associated with it.

Consider a plane wave incident on the array from an angle (θ, ϕ) relative to the axis of the array. We represent the modulation of the plane wave using the baseband complex envelope, $s(t)$. Assume for the moment that all of the array elements are noiseless isotropic antennas which have uniform gain in all directions. Using (3.3), with $x_m = m\Delta x$, the signal received at antenna element m on the LES is

$$u_m(t) = As(t)e^{-j\beta m\Delta d} = As(t)e^{-j\beta m\Delta x\cos\phi\sin\theta} \tag{3.4}$$

where A is the arbitrary gain constant. The signal $z(t)$ at the array output is

$$z(t) = \sum_{m=0}^{M-1} w_m u_m(t) = As(t)\sum_{m=0}^{M-1} w_m e^{-j\beta m\Delta x\cos\phi\sin\theta} = As(t)f(\theta, \phi) \tag{3.5}$$

The term $f(\theta, \phi)$ is called the *array factor*. The array factor determines the ratio of the received signal available at the array output, $z(t)$, to the signal, $As(t)$, measured at the reference

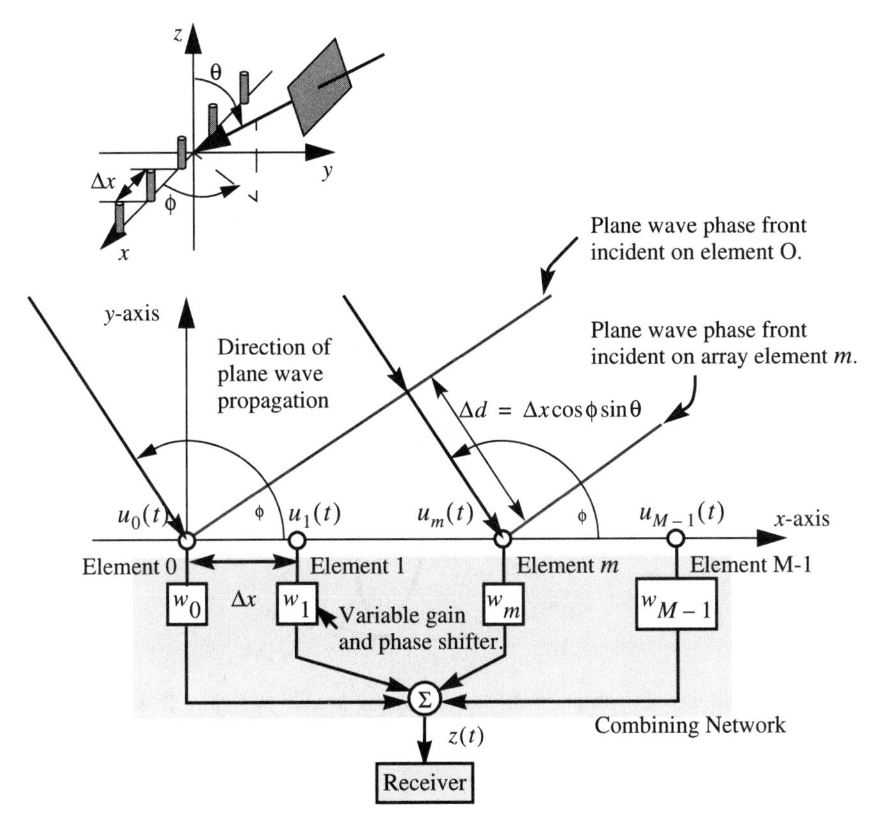

Figure 3–3 A baseband complex envelope model of a linear equally spaced array oriented along the *x*-axis, receiving a plane wave from direction (θ, ϕ).

element, as a function of Direction-Of-Arrival, (θ, ϕ). By adjusting the set of weights, $\{w_m\}$, it is possible to direct the maximum of the main beam of the array factor in any desired direction, (θ_0, ϕ_0).

The received power at the array output is

$$P_r = \frac{1}{2}|z(t)|^2 = \frac{1}{2}|As(t)|^2|f(\theta, \phi)|^2 \tag{3.6}$$

To show how the weights, $\{w_m\}$, can be used to change the antenna pattern of the array, let the mth weight be given by

$$w_m = e^{j\beta m \Delta x \cos\phi_0} \tag{3.7}$$

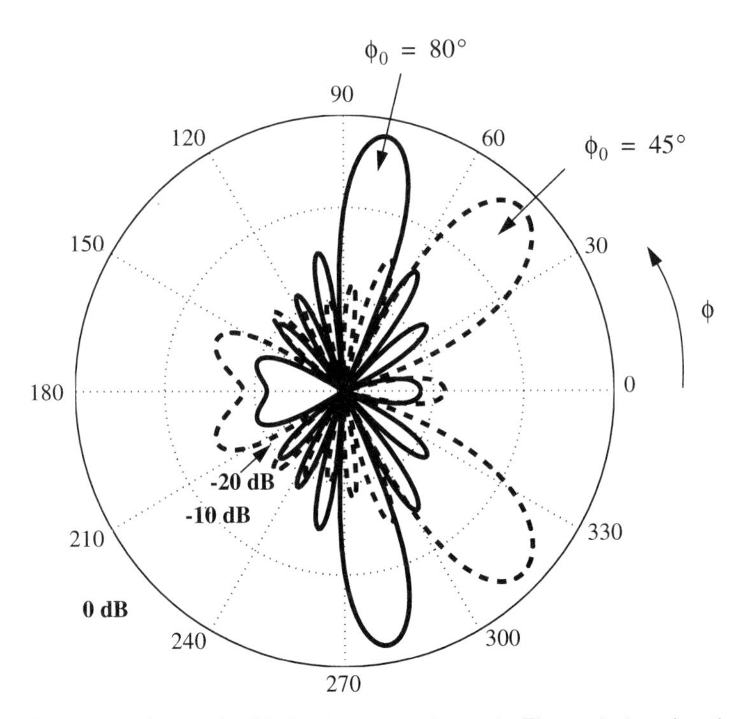

Figure 3–4 Plot of the array factor, in dB, for the array shown in Figure 3–3, using the weights described by (3.7) for ϕ_0 equal to 45° and 80°.

Then the array factor is

$$f(\theta, \phi) = \sum_{m=0}^{M-1} e^{-j\beta m\Delta x(\cos\phi\sin\theta - \cos\phi_0)} \tag{3.8}$$

$$= \frac{\sin\left(\frac{\beta M\Delta x}{2}(\cos\phi\sin\theta - \cos\phi_0)\right)}{\sin\left(\frac{\beta\Delta x}{2}(\cos\phi\sin\theta - \cos\phi_0)\right)} \cdot e^{-j\frac{\beta\Delta x}{2}(\cos\phi\sin\theta - \cos\phi_0)}$$

Consider the case in which a wave is incident on the array shown in Figure 3–3 in the x-y (horizontal) plane so that $\theta \approx \pi/2$. This is a reasonable approximation for many cellular and PCS smart antenna applications. The array factor is shown in Figure 3–4 for ϕ_0 equal to 45° and 80°. By varying a single parameter, ϕ_0, the beam can be steered to any desired direction.

It is useful to make the substitution

$$\cos\psi = \cos\phi\sin\theta \tag{3.9}$$

where ψ is the angle of incidence of the plane wave measured from the x-axis, in the same way that θ is measured from the z-axis in Figure 3–3. This reflects the fact that a linear array of isotropic elements has an antenna pattern which is circularly symmetric about the axis of the array.

In the general case, the array factor pattern is a function of both θ and ϕ. If the field pattern of each array element is $g_a(\theta, \phi)$ and all elements are identical and oriented in the same direction, then the total field pattern of the array is given by

$$F(\theta, \phi) = f(\theta, \phi)g_a(\theta, \phi) \tag{3.10}$$

This is called the principle of pattern multiplication [Stu81]. Each element also has a polarization state associated with it, as does any incident plane wave. It is assumed that the plane waves incident on each element are co-polarized so that there is no polarization loss in the interaction between the plane wave and each element.

In working with array antennas, it is very convenient to make use of vector notation. Define the *weight vector* as

$$\boldsymbol{w} = \begin{bmatrix} w_0 & \dots & w_{M-1} \end{bmatrix}^H \tag{3.11}$$

where the superscript H represents the *Hermitian transpose*, which is a *transposition* combined with complex conjugation.

The signals from each antenna element are grouped in a *data vector*

$$\boldsymbol{u} = \begin{bmatrix} u_0(t) & \dots & u_{M-1}(t) \end{bmatrix}^T \tag{3.12}$$

Then the array output, $z(t)$, from (3.5) can be expressed as the inner product of the array weight vector, $\boldsymbol{w}$, and the data vector, $\boldsymbol{u}(t)$,

$$z(t) = \boldsymbol{w}^H\boldsymbol{u}(t) \tag{3.13}$$

The array factor in a direction (θ, ϕ) is

$$f(\theta, \phi) = \boldsymbol{w}^H\boldsymbol{a}(\theta, \phi) \tag{3.14}$$

The vector $\boldsymbol{a}(\theta, \phi)$ is called the *steering vector* in direction (θ, ϕ). Given a plane wave incident from a direction (θ, ϕ) as illustrated in Figure 3–2, the steering vector, $\boldsymbol{a}(\theta, \phi)$ describes the phase of the signal available at each antenna element relative to the phase of the signal at the reference element (element 0). Using (3.3), the steering vector is

$$\boldsymbol{a}(\theta, \phi) = \begin{bmatrix} 1 & a_1(\theta, \phi) & \dots & a_{M-1}(\theta, \phi) \end{bmatrix}^T \tag{3.15}$$

where

$$a_m(\theta, \phi) = e^{-j\beta(x_m\cos(\phi)\sin(\theta) + y_m\sin(\phi)\sin(\theta) + z_m\cos(\theta))} \tag{3.16}$$

A set of steering vectors, either measured or calculated, over all values of θ and ϕ, is called the *array manifold*. In addition to being useful in the analysis of arrays, knowledge of the array manifold is critical in direction finding, downlink beamforming, and other aspects of array operation. The angle pair (θ, ϕ) is called the *Direction-Of-Arrival* (DOA) of the received plane wave. For simplicity, unless otherwise noted, we assume that multipath components arrive at the base station in the horizontal plane, $\theta = \pi/2$, so that the azimuthal direction, ϕ, completely specifies the DOA.

In general, the utility of an array is determined by a number of factors. The size, or *aperture*, of the array determines the maximum gain that the array can achieve as discussed in Section 1.5. On the other hand, the number of elements determines the number of *degrees of freedom* that one has in designing array patterns. For LES arrays, these two quantities are related. If the element spacing exceeds $\lambda/2$ in an LES array, then grating lobes can appear, giving the array undesired beams, which may amplify noise or interference [Stu82].

Often, however, it is desirable to attempt to obtain a longer aperture using a given number of elements than is possible by using half-wavelength spaced arrays. This is frequently driven by the fact that RF hardware associated with each antenna element can be expensive and bulky. Through non-uniform element spacing, it is possible to obtain linear and planar array geometries that can yield much larger apertures than half-wavelength spaced arrays for the same number of elements. Due to the large aperture, these long baseline or *sparse arrays* can achieve significantly smaller beamwidths than a half-wavelength spaced array of similar complexity at the expense of size. Through careful numerical design, it is possible to obtain sparse array geometries that have good peak-to-sidelobe ratios over a range of steering angles.

3.3 The Vector Channel Impulse Response and the Spatial Signature

In Chapter 1, it was noted that multipath in the radio channel can lead to fading and time dispersion in the radio channel. In Chapter 6, a detailed discussion of models for the spatial radio channel, along with methods for measuring and characterizing these channels, are presented.

A discrete model is used to characterize the channel, where each multipath component is considered to be a plane wave, arriving from a discrete direction at a discrete time delay. For a particular subscriber, the channel between the portable transmitter and the base station receiver is modeled using the *Vector Channel Impulse Response* (VCIR),

$$\boldsymbol{h}(\tau, t) = \sum_{i=0}^{L-1} \boldsymbol{a}(\theta_i, \phi_i)\alpha_i(t)\delta(\tau - \tau_i) \tag{3.17}$$

where α_i, τ_i, and (θ_i, ϕ_i) are the complex amplitude, path delay, and Direction-Of-Arrival of the ith multipath component. There are a total of L multipath components. The complex amplitude of the ith component is written as a function of time and may be expressed as

$$\alpha_i(t) = \rho_i e^{j(2\pi f_i t + \psi_i)} \tag{3.18}$$

where ρ_i represents the path gain for the ith component, f_i is the Doppler shift due to the motion of either a mobile unit or scatterers in the environment, and ψ_i is a fixed phase offset. In general, all of the variables of the VCIR can vary with time, the position of the user, and the velocity of the user. As the user moves over a small local area, a distance of several wavelengths, we consider the number of components, L, to be fixed, and for each component, the DOA (θ_i, ϕ_i), the path gain (ρ_i), the Doppler shift (f_i), the phase offset (ψ_i), and the delay (τ_i), are assumed to be approximately constant.

The vector channel impulse response relates the transmitted signal to the signal received at each antenna element of the array. Given a transmitted signal, $s(t)$, the data vector is

$$\boldsymbol{u}(t) = \left[u_0(t) \ \ldots \ u_{M-1}(t)\right]^T \tag{3.19}$$

$$= s(t) * \boldsymbol{h}(\tau, t) + \boldsymbol{n}(t) = \int_{-\infty}^{t} s(\lambda)\boldsymbol{h}(t-\lambda, t)d\lambda + \boldsymbol{n}(t)$$

$$= \sum_{i=0}^{L-1} \boldsymbol{a}(\phi_i)\alpha_i(t)s(t-\tau_i) + \boldsymbol{n}(t)$$

where $\boldsymbol{n}(t)$ is the vector representing the noise introduced at each antenna element.

In channels where the differences between the path delays of multipath components are very small relative to the symbol period of $s(t)$, we make the approximation that $\tau_i \approx \tau_0$. Then the vector channel impulse response may be expressed as

$$\boldsymbol{u}(t) = s(t-\tau_0)\sum_{i=0}^{L-1} \boldsymbol{a}(\phi_i)\alpha_i(t) + \boldsymbol{n}(t) = s(t-\tau_0)\boldsymbol{b}(t) + \boldsymbol{n}(t) \tag{3.20}$$

where the term $\boldsymbol{b}(t)$ is called the *spatial signature* of the narrowband (flat fading) channel,

$$\boldsymbol{b}(t) = \sum_{i=0}^{L-1} \boldsymbol{a}(\phi_i)\alpha_i(t) = \sum_{i=0}^{L-1} \boldsymbol{a}(\phi_i)\rho_i e^{j(2\pi f_i t + \psi_i)} \tag{3.21}$$

Note that $\boldsymbol{b}(t)$ represents a multiplicative channel rather than a convolutional channel. If the channel is frequency selective, such that the multipath delays are on the order of or exceed the chip (symbol) duration [Rap96a], the VCIR of (3.17) must be used.

The concepts outlined in this section form the fundamental basis for both spatial processing receivers and transmission beamforming systems. When arrays are used at the base station, spatial processing receivers are used to enhance and separate reverse link signals. In some applications, spatial processing receivers can also be used at the subscriber end of the link.

Spatial processing, when used for transmission at the base station, is called downlink beamforming or forward link beamforming.

3.4 Spatial Processing Receivers

Spatial processing receivers can be implemented in a number of different ways, using both analog and digital components, as shown in Figure 3–5. For base stations supporting multiple simultaneous uplink signals, the RF/IF spatial processing structure in Figure 3–5(a) is less

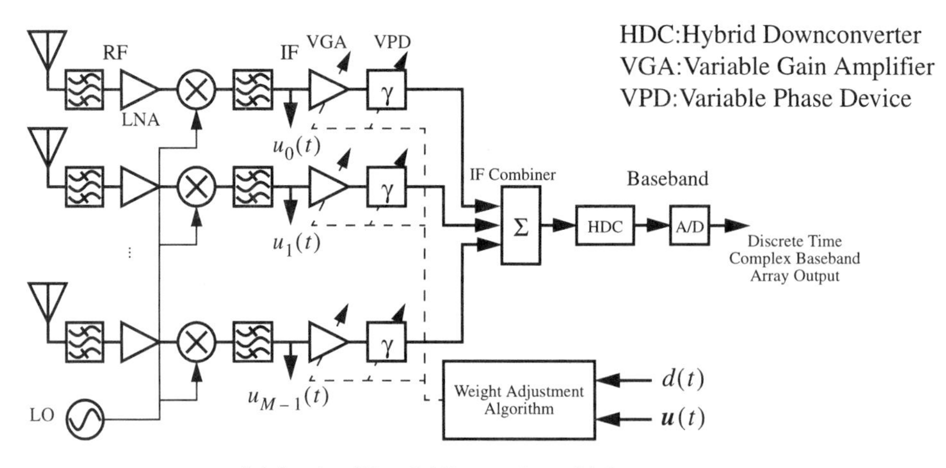

(a) Analog IF weighting and combining

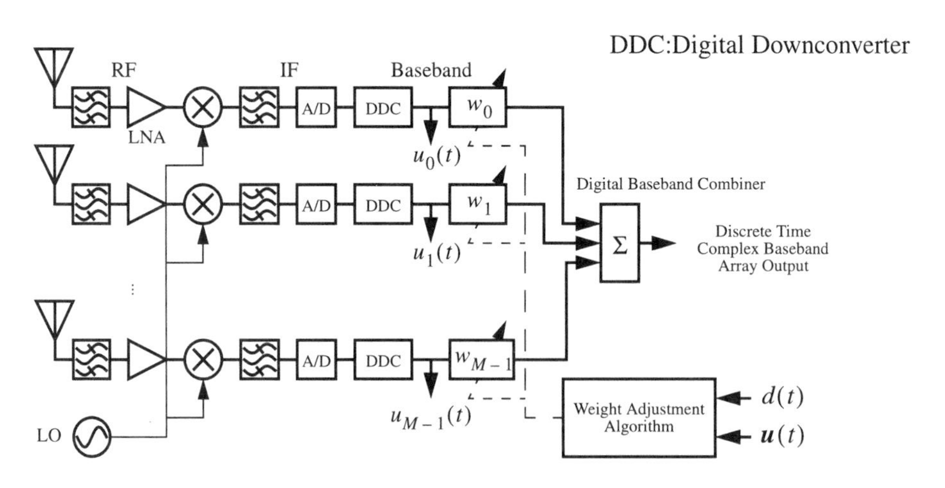

(b) Digital complex baseband weighting and combining

Figure 3–5 Array-based spatial processing can be performed at an RF or IF frequency, using analog components, or it can be performed digitally at baseband. Note that $d(t)$ is an estimate or replica of the desired signal at the array output.

attractive than the digital system in Figure 3–5(b), whereas a separate RF beamforming network is required for each independent beam when analog components are used. The digital system can form multiple simultaneous beams, one for each *Signal Of Interest* (SOI), whereas a separate RF beamformer network is required for each independent beam when analog components are used.

An alternative structure for smart antenna systems combines RF/IF beamforming with digital spatial processing by using two sections, a fixed *BeamForming Network* (BFN), and a time-variable adaptive array processor. This structure is illustrated in Figure 3–6.

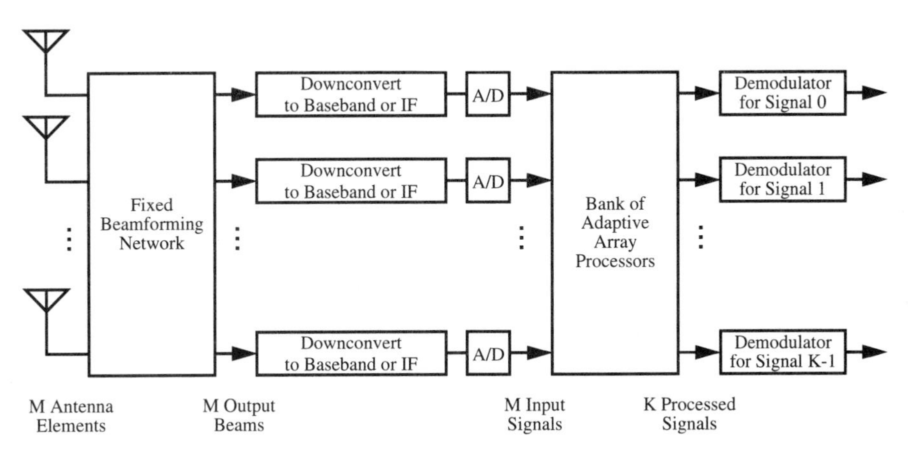

Figure 3–6 In general, a smart antenna system may make use of a fixed beamforming network, a bank of adaptive array processors, or both. The fixed beamforming network forms M beams from M low gain elements. A bank of adaptive array processors, extracts K signals from the M input signals. If spatial processing is the only means of signal separation, then $K \leq M$. For CDMA systems, other signal separation techniques can be applied.

3.5 Fixed Beamforming Networks

The BFN is characterized by an M-row matrix T, where the vector of signals at the output of the matrix, $\boldsymbol{y}(t)$, is related to the vector of signals at the array input, $\boldsymbol{u}(t)$, by

$$\boldsymbol{y}(t) = \boldsymbol{T}^H \boldsymbol{u}(t) \tag{3.22}$$

The nth output of the BFN corresponds to an array weight vector contained in the nth column of $\boldsymbol{T}$. Often, a beamforming network is used to produce M beams from M elements. The $M \times M$ *BeamForming Network Matrix* (BFNM) is given by

$$\boldsymbol{T} = \left[\boldsymbol{w}_0 \ \boldsymbol{w}_1 \ \dots \ \boldsymbol{w}_{M-1}\right] \tag{3.23}$$

The beams are orthogonal if the weight vector corresponding to each beam is orthogonal to the weight vector corresponding to every other beam. If the matrix $\boldsymbol{T}$ is $M \times M$ with orthogonal columns, then the BFN has some special properties which are useful both in switched beam systems and when the fixed BFN is followed by a bank of adaptive array processors. Figure 3–8 illustrates the beam patterns produced by an orthogonal beamforming network. Relatively simple techniques are available to implement beamforming networks using cascaded 90° hybrid couplers as in the *Butler matrix* [But61][Col85]. These networks, which are available as off-the-shelf components, are typically implemented at an RF stage. Figure 3–7 shows a simple 4×4 beamforming matrix as described by Butler [But61][Ana95]. The fixed beamforming matrix is bi-directional, which means that each port corresponding to a particular receive beam pattern can also be used to transmit, using the same beam pattern.

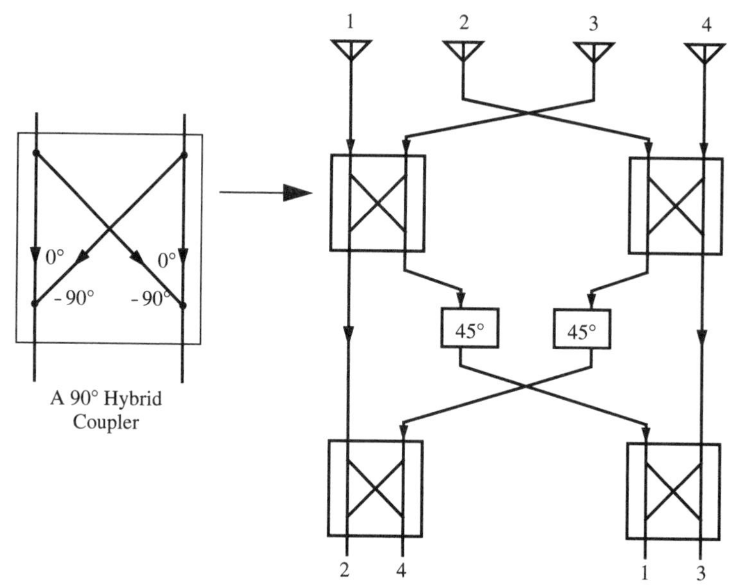

Figure 3–7 A Butler matrix for fixed beamforming [But61][Col85]. (Reprinted with permission from *Electronic Design*, April 12, 1961. Copyright 1961, Penton Media Inc., 611 Route 46 W, Hasbrouck Heights, NJ 07604 (U.S.A.) and R. E. Collin, *Antennas and Radiowave Propagation*, McGraw-Hill, NY, 1985. Reproduced with permission of The McGraw-Hill Companies.)

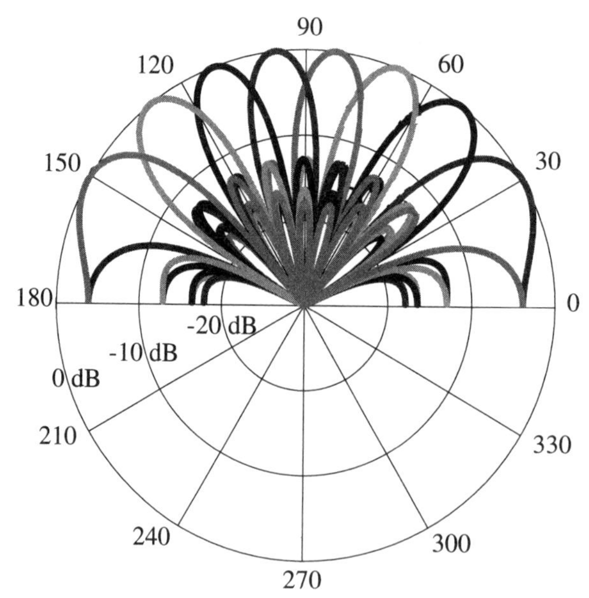

Figure 3–8 A set of 8 beams is produced using a fixed beamforming network with orthogonal beams. Note that the maximum of each beam pattern corresponds to a null in the direction of each of the other beam maxima.

Table 3–1 Element phasing, beam direction, and inter-element phasing for the Butler matrix of Figure 3–7.

	Element 1	Element 2	Element 3	Element 4	Beam Direction	Inter-Element Phasing
Port 1	-45°	-180°	45°	-90°	138.6°	-135°
Port 2	0°	-45°	-90°	-135°	104.5°	-45°
Port 3	-135°	-90°	-45°	0°	75.5°	45°
Port 4	-90°	-45°	-180°	-45°	41.4°	135°

It can easily be shown that the weight vectors corresponding to each port in Table 3–1 are mutually orthogonal. Beam patterns for a larger 8×8 beamforming network matrix are shown in Figure 3–8.

Instead of relying on a fixed BFN, smart antenna systems may use a bank of *adaptive array processors*. Each adaptive array processor is a time-variable network which uses a weight vector $\mathbf{w}_{k,i}$ to extract signal k at time index i. Usually, there is feedback from the demodulator stage to adapt the weight vector. When an adaptive array processor is used, the fixed BFN is optional. If the BFN matrix is $M \times M$ with orthogonal columns, then it can be shown that any adaptive array solution that could be obtained without the BFN can be obtained with the BFN. In other words, in the static case, the BFN does not degrade the static performance of the adaptive array processor. In fact, the BFN can improve the performance of the adaptive array processor by providing a certain amount of spatial *pre-selection*, in which interference arriving from directions away from the desired signal is reduced before applying the adaptive array processor. This eases dynamic range requirements on the downconversion and sampling systems and helps the adaptive array initially acquire signals. In particular, since the dynamic range is a major limitation in designing very wideband analog-to-digital converters, the use of spatial pre-selection may significantly reduce the cost and difficulty of implementing adaptive array systems.

3.6 Switched Beam Systems

In smart antenna systems which use only the fixed BFN, a switch is used to select the best beam to receive a particular signal. The switched beam system illustrated in Figure 3–9 is relatively simple to implement, requiring only a beamforming network, an RF switch, and control logic to select a particular beam. By selecting an output, one of the M predetermined weight vectors is used as shown by (3.23). A separate *beam selection* must be made for each receiver. The mechanism for performing beam selection is highly dependent on whether we are considering an FDMA, TDMA, or CDMA system; however, it is possible to consider switched beam approaches to each of these multiple access methods.

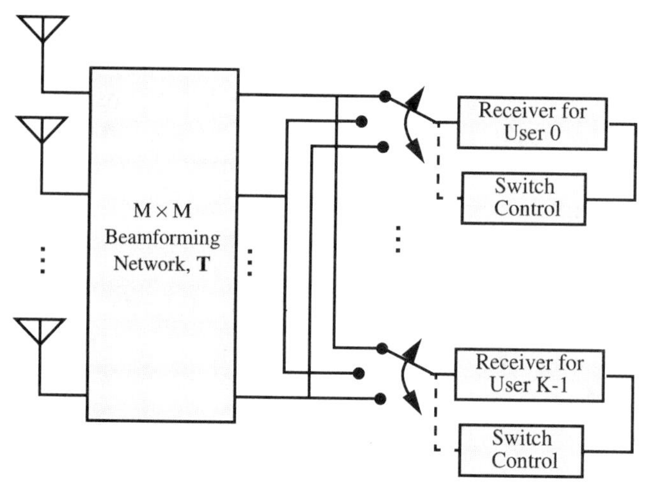

Figure 3–9 A switched beam network uses a beamforming network to form M beams from M array elements.

Switched beam systems offer many of the advantages of more elaborate smart antenna systems at a fraction of the complexity and expense. There are, however, several limitations to switched beam arrays. First of all, the system is not able to provide any protection from multipath components which arrive with *Directions-Of-Arrival* (DOAs) near that of the desired component. As shown in [Lib95][Lib96b], systems based on fixed BFNs only are much more sensitive to the angular distribution of multipath components than systems based on adaptive array processors. A second disadvantage of switched beam systems is that they are typically unable to take advantage of path diversity by combining coherent multipath components. Finally, the received power level from the user fluctuates as a subscriber travels in an arc about the base station due to *scalloping*. Scalloping is the roll-off of the antenna pattern as a function of angle as the DOA varies from the boresight of each beam produced by the BFN. Typically, BFNs provide beams which cross at 4 dB points. Thus a subscriber's signal strength varies as the user moves from the center of the beam to the edge of the coverage region of a particular beam.

Despite these disadvantages, switched beam systems are popular for several reasons. They provide some of the range extension benefits obtained from more elaborate systems. Depending on the propagation environment, switched beam systems offer some reduction in delay spread, as shown in [Lib95][Lib96b], which provides the capability to deploy low tier PCS systems in high-antenna height, high subscriber speed environments. Switched beam systems require only moderate interaction with the base station receiver, compared with adaptive antenna systems, which appeals to manufacturers wishing to supply "bolt-on" or "applique" solutions. Finally, since this is a relatively low technology approach, the engineering costs associated with implementing these systems may be much lower than those associated with more complicated systems.

3.7 Adaptive Antenna Systems

By increasing the complexity of the array signal processing, it is possible to achieve greater performance improvements than are attainable using switched beam systems. In an adaptive array, as shown in Figure 3–10, the weight vector $\boldsymbol{w}_{k,i}$ is adjusted, or adapted, to maximize the quality of the signal that is available to the demodulator for signal k at time index i.

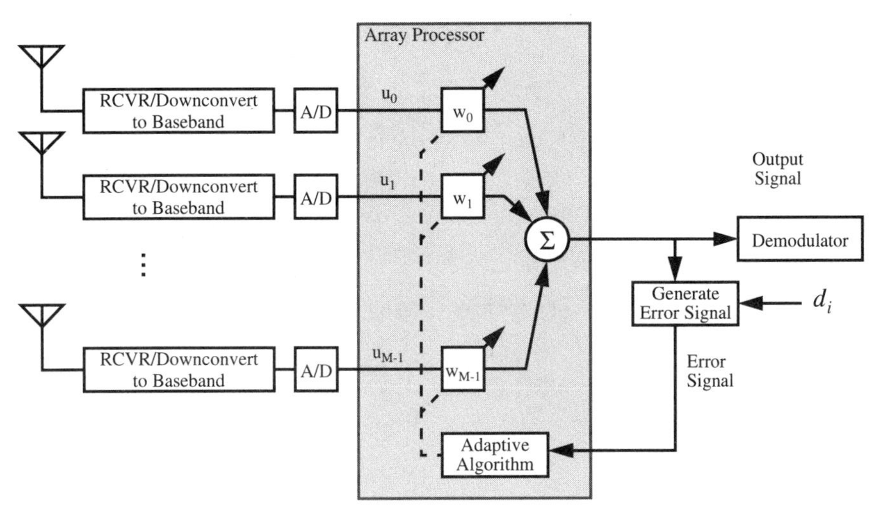

Figure 3–10 An adaptive array structure. Note that d_i represents an estimate or replica of the desired signal for the ith user at the array output.

In optimal beamforming techniques, a weight vector is determined which minimizes a cost function. Typically, this cost function is inversely associated with the quality of the signal at the array output, so that when the cost function is minimized, the quality of the signal is maximized at the array output. Two of the most popular techniques which have been applied extensively in communication systems are the *Minimum Mean Square Error* (MMSE) and *Least Squares* (LS) criteria. In both of these techniques, the square of the difference between the array output, $z(t) = \boldsymbol{w}^H{}_k\boldsymbol{u}_i(t)$, and $d_k(t)$, a locally generated estimate of the desired signal for the kth subscriber, is minimized by finding an appropriate weight vector, $\boldsymbol{w}_k$. MMSE solutions are posed in terms of ensemble averages and produce a single weight vector, $\boldsymbol{w}_k$, which is optimal over the ensemble of possible realizations of the stationary environment. This is the approach used in classical Wiener filter theory.

In the MMSE approach, the cost function to be minimized is

$$J(\boldsymbol{w}_k) = \mathrm{E}[|\boldsymbol{w}_k^H\boldsymbol{u}_i - d_{k,i}|^2] \tag{3.24}$$

where $d_{k,i} = d_k(iT_s)$ and $\boldsymbol{u}_i = \boldsymbol{u}(iT_s)$ where T_s is the sampling period [Hay91].

The cost function is the expected value (taken over the ensemble of realizations of $\boldsymbol{u}_i$) of the square error between the array output for the kth signal and the desired version of that signal at time index i. We can rewrite (3.24) as

$$J(\boldsymbol{w}_k) = \boldsymbol{w}_k^H E[\boldsymbol{u}_i \boldsymbol{u}_i^H]\boldsymbol{w}_k - E[d_{k,i}\boldsymbol{u}_i^H]\boldsymbol{w}_k - \boldsymbol{w}_k^H E[\boldsymbol{u}_i d_{k,i}^*] + E[d_{k,i} d_{k,i}^*] \tag{3.25}$$

In general, we minimize a vector function by determining a location where the *gradient* of the function goes to zero. The gradient of a function of a complex vector, is defined as

$$\nabla f(\boldsymbol{w}) = \begin{bmatrix} \frac{\partial}{\partial a_0} f(\boldsymbol{w}) \\ \cdots \\ \frac{\partial}{\partial a_{M-1}} f(\boldsymbol{w}) \end{bmatrix} + j \begin{bmatrix} \frac{\partial}{\partial b_0} f(\boldsymbol{w}) \\ \cdots \\ \frac{\partial}{\partial b_{M-1}} f(\boldsymbol{w}) \end{bmatrix} \tag{3.26}$$

where $w_i = a_i + jb_i$. Using this definition we find that

$$\nabla(\boldsymbol{w}^H \boldsymbol{A} \boldsymbol{w}) = 2\boldsymbol{A}\boldsymbol{w} \tag{3.27}$$

$$\nabla(\boldsymbol{w}^H \boldsymbol{c}) = 2\boldsymbol{c} \tag{3.28}$$

$$\nabla(\boldsymbol{c}^H \boldsymbol{w}) = \boldsymbol{0} \tag{3.29}$$

Therefore, we can write

$$\nabla J(\boldsymbol{w}_k) = 2E[\boldsymbol{u}_i \boldsymbol{u}_i^H]\boldsymbol{w}_k - 2E[\boldsymbol{u}_i d_{k,i}^*] = 2\boldsymbol{R}\boldsymbol{w}_k - 2\boldsymbol{p} \tag{3.30}$$

where $\boldsymbol{R}$ is the correlation matrix of the data vector,

$$\boldsymbol{R} = \mathrm{E}[\boldsymbol{u}_i \boldsymbol{u}_i^H] \tag{3.31}$$

and $\boldsymbol{p}$ is the cross-correlation between the data vector and the desired signal,

$$\boldsymbol{p} = \mathrm{E}[\boldsymbol{u}_i d_{k,i}^*] \tag{3.32}$$

Setting the gradient of the cost function equal to zero, we find that the solution for $\boldsymbol{w}_k$ which minimizes $J(\boldsymbol{w}_k)$ is

$$\boldsymbol{w}_k = \boldsymbol{R}^{-1}\boldsymbol{p} \tag{3.33}$$

The solution to (3.32) is the optimal antenna array weight vector in the MMSE sense [Hay91].

Rather than solving (3.33) directly, *adaptive* techniques are often used with an iterative approach which provides an updated weight vector, $\boldsymbol{w}_{k,i}$, after each computation. Typically, these algorithms have a per-step complexity which is much lower than the direct solution of (3.33), and can track non-stationary channels.

An adaptive solution which minimizes the cost function is

$$\boldsymbol{w}_{k,i+1} = \boldsymbol{w}_{k,i} - \frac{1}{2}\mu \nabla J(\boldsymbol{w}_{k,i}) \tag{3.34}$$

where μ is the *convergence factor* which controls the rate of adaptation. Classes of algorithms such as (3.34) are called *Stochastic Gradient* techniques because they use the gradient of the mean square error function to update the weight vector. Substituting (3.30) into (3.34), we have

$$\begin{aligned} w_{k,i+1} &= w_{k,i} - \mu(Rw_{k,i} - p) \\ &= w_{k,i} - \mu(\mathrm{E}[u_i u_i^H] w_{k,i} - \mathrm{E}[u_i d_{k,i}^*]) \end{aligned} \tag{3.35}$$

If the expectation operator is removed from (3.35), we have

$$\begin{aligned} w_{k,i+1} &= w_{k,i} - \mu u_i (u_i^H w_{k,i} - d_{k,i}^*) \\ &= w_{k,i} - \mu u_i e_{i,k}^* \end{aligned} \tag{3.36}$$

where $e_{i,k} = w_{k,i}^H u_i - d_{k,i}$ is the instantaneous error between the array output and the desired response. Equation (3.36) is called the Least Mean Square error algorithm and is a widely used technique for iteratively updating adaptive arrays.

An alternative approach to adaptive filtering is to minimize the error between the output of the array and a desired response over a finite number of time samples, rather than the ensemble average. This leads to Least Squares solutions. In Least Squares, a set of P snap shots of the data vector, u_i are collected. We define the Least Squares cost function as

$$J(w_k) = \left| \sum_{m=0}^{P-1} w_k^H u_m - d_{k,m} \right|^2 \tag{3.37}$$

The gradient operator, as defined in (3.26), is used to obtain

$$\nabla J(w_k) = 2 \sum_{m=0}^{P-1} \sum_{n=0}^{P-1} u_m u_n^H w_k - 2 \sum_{m=0}^{P-1} \sum_{n=0}^{P-1} u_m d_{k,n}^* \tag{3.38}$$

We define the data matrix as

$$A = \begin{bmatrix} u_0 \; u_1 \; \dots \; u_{P-1} \end{bmatrix} \tag{3.39}$$

and the desired signal vector as

$$d_k = \begin{bmatrix} d_{k,0} \; d_{k,1} \; \dots \; d_{k,P-1} \end{bmatrix}^T \tag{3.40}$$

Then the solution which forces the LS gradient function to zero is

$$w_k = (A^H A)^{-1} A^H d_k \tag{3.41}$$

A central problem with both LS and MMSE approaches is that they require knowledge of, or estimation of, the desired spatial filter output. For LS and MMSE approaches, this is accomplished by sending a known *training sequence* periodically, which is known to both the transmitter or receiver. Training sequences use valuable spectral resources for a purpose that conveys no information content from the transmitter to receiver and can be sent only periodically.

An alternative to the training sequences required in MMSE and LS approaches is to use decision-directed learning. In *decision-directed adaptation*, an estimate of the desired signal samples, $d_{k,i}$, is reconstructed based on the output of the array and the signal demodulator. Since the desired signal is generated locally, without prior knowledge of the transmitted data, training sequences are not required. However, when errors occur at the demodulator, the reconstructed estimate of the desired signal is poor, and adaptive algorithms that use this estimate can lead toward an incorrect weight solution, further degrading the quality of the demodulated signal. A combination of training sequences and decision direction can be used together to improve this situation. Training sequences can be used for initial adaptation, allowing a weight vector to be formed that reduces the impact of noise and interference. After initial adaptation, decision direction can be used to track time-variation in the radio channel.

There are a broad class of *blind adaptive algorithms* which allow operation without training sequences. These techniques use underlying characteristics of the signal structure to update spatial processing weight vectors. Blind adaptive algorithms, optimal spatial filtering and other adaptive spatial processing techniques are discussed in greater detail in Chapter 7.

Before proceeding, it is useful to note that we can use the structure of Figure 3–10 in parallel to form K weight vectors (i.e., K demodulated signals), as shown in Figure 3–11. The number of weight vectors, K, can be as large as needed; however, if there are M antenna elements, only M orthogonal weight vectors can be formed. Each of the weight vectors can be formed independently or jointly. One approach for joint adaptation of the weight vectors was proposed by Agee in [Age89]. In this approach, each weight vector is adapted independently, using a blind adaptive algorithm, subject to the constraint that each weight vector is orthogonal to other weight vectors for signals incident on the array. Using this approach, or some variant, the adaptive array can serve as a spatial separator for signals.

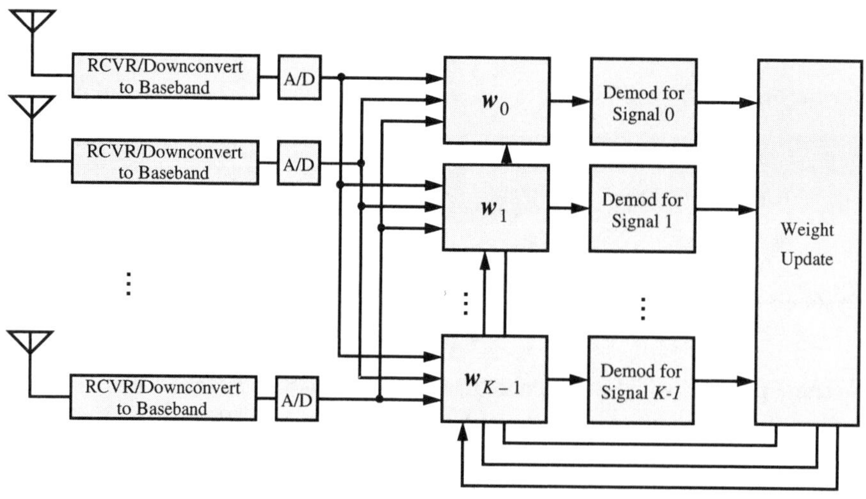

Figure 3–11 A multi-target adaptive array. Weight vectors for each subscriber signal can be updated jointly or independently.

3.8 Wideband Smart Antennas

The bandwidth of signals incident on the array has a significant impact on the ability of the array to reject interference. A narrowband array constructed as shown in Figure 3–3 is able to form nulls only at a single frequency [Com88]. This is because the steering vectors associated with the array are a function of both the frequency of the incident signal and the spacing of elements.

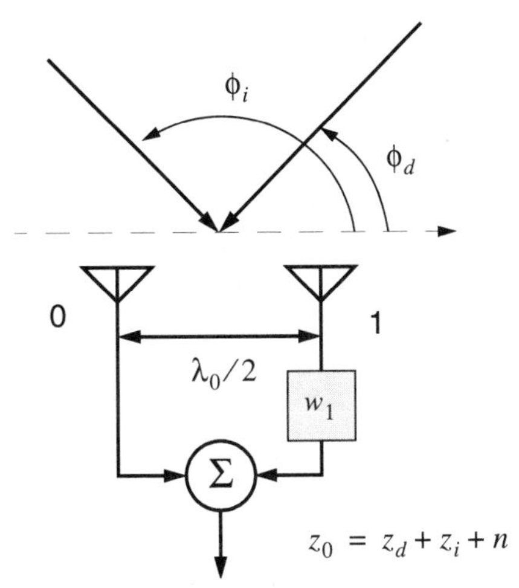

Figure 3–12 A narrowband array to null an interferer incident from angle ϕ_i [Com88].

Consider a two-element array, with elements spaced at $\lambda_0/2 = c/(2f_0)$, as shown in Figure 3–12. Assume that we adjust the complex weight, w_1, to null a signal incident from angle ϕ_i at a frequency f_0. The signal at the output of the array, z_0, is the sum of the contribution of the desired signal z_d, the interferer z_i, and noise n. If the *Power Spectral Density* (PSD) of the interfering signal, measured at element 0, is flat, with a level S_{i0} watts/Hz, then the PSD of z_i at the output of the array is

$$\begin{aligned} S_i(f) &= S_{i0}\left|\boldsymbol{w}^H\boldsymbol{a}(\phi_i)\right|^2 \qquad (3.42) \\ &= S_{i0}\left|1 + w_0 e^{-j\beta\Delta x\cos\phi_i}\right|^2 \\ &= S_{i0}\left|1 + w_0 e^{-j\frac{2\pi}{\lambda}\frac{\lambda_0}{2}\cos\phi_i}\right|^2 \\ &= S_{i0}\left|1 + w_0 e^{-j\pi\frac{f}{f_0}\cos\phi_i}\right|^2 \end{aligned}$$

The weight, w_0, is adjusted to null the interfering signal at its center frequency of f_0, therefore $w_0 = -e^{j\pi\cos\phi_i}$. Then

$$S_i(f) = S_{i0}\left|1 - e^{-j\pi\left(\frac{f}{f_0} - 1\right)\cos\phi_i}\right|^2 \tag{3.43}$$

$$= S_{i0}\left|e^{j\frac{\pi}{2}\left(\frac{f}{f_0} - 1\right)\cos\phi_i} - e^{-j\frac{\pi}{2}\left(\frac{f}{f_0} - 1\right)\cos\phi_i}\right|^2$$

$$= S_{i0}\left|e^{j\frac{\pi}{2}\left(\frac{f}{f_0} - 1\right)\cos\phi_i} - e^{-j\frac{\pi}{2}\left(\frac{f}{f_0} - 1\right)\cos\phi_i}\right|^2$$

$$= 4S_{i0}\sin^2\left(\frac{\pi}{2}\left(\frac{f}{f_0} - 1\right)\cos\phi_i\right) \tag{3.44}$$

Now, consider the case where the desired signal is 50 dB above the noise floor, and the interfering signal is 30 dB stronger than the desired signal. We assume that $\phi_d = 90°$ and $\phi_i = 0°$. A carrier frequency of 1900 MHz is used for this example. By integrating (3.44) from $f_o - B/2$ to $f_o - B/2$, we can evaluate the total interference available at the output of the array. The Carrier-to-Interference-and-Noise-Ratio (CINR) at the output of the array degrades as the bandwidths of the incident signals increase. In Table 3–2, we see that if the bandwidths of the signals are each 10 MHz, the CINR is degraded by 27 dB, relative to the case of 100 kHz bandwidths.

Table 3–2 Reduction in CINR as a function of the bandwidth of signals incident on the narrowband array in Figure 3–12.

Bandwidth of Desired and Interfering Signals	CINR
10 kHz	50.0 dB
100 kHz	49.8 dB
1 MHz	41.8 dB
10 MHz	22.5 dB

A *wideband* array is an adaptive array system which combines spatial filtering with temporal filtering. In this type of system, illustrated in Figure 3–13, a tapped-delay-line is used on each branch of the array [Com88].

The tapped-delay-line allows each element to have a phase response that varies with frequency, compensating for the fact that lower frequency signal components have less phase shift for a given propagation distance, whereas higher frequency signal components have greater phase shift as they travel the same length. This structure can be considered to be an *equalizer*,

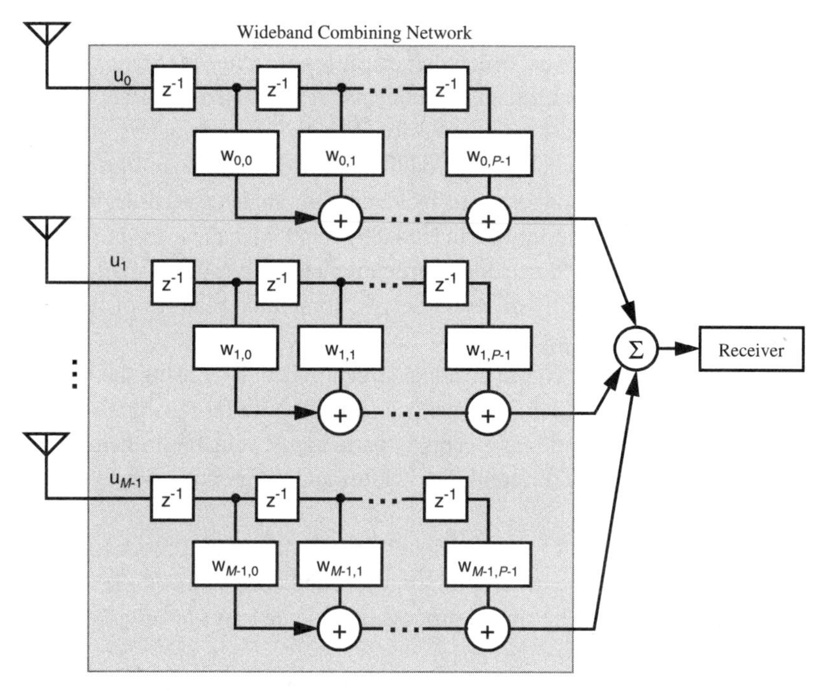

Figure 3–13 A wideband adaptive array combines spatial and temporal adaptive filtering.

which makes the response of the array the same across different frequencies. The frequency response of the tapped delay line for element m is given by

$$W_m(e^{j\omega}) = \sum_{p=0}^{P-1} w_{m,p} e^{-j\omega p} \tag{3.45}$$

For a two-dimensional array, with elements in the x-y (horizontal) plane, the frequency dependent antenna pattern is given by

$$f_u(\omega, \phi) = \sum_{m=0}^{M-1} W_m(e^{j\omega}) e^{-j\frac{\omega}{c}x_m \cos(\phi) - j\frac{\omega}{c}y_m \sin(\phi)} \tag{3.46}$$

While it can be easily shown that it is not possible to achieve an arbitrary antenna pattern across all frequencies, it is possible to use tapped delay lines to "flatten" the spatial response as a function of frequency. Even if the bandwidth of signals incident on the array is very small relative to the center frequency, so that bandwidth degradation is not a factor, the wideband array, also called a *space-time*, *spatio-temporal*, or *two-dimensional* processor, can be extremely valuable.

As described in Chapter 1, multipath in radio channels can result in time-delayed replicas of the incident signal at the receiver, which can cause *InterSymbol Interference* (ISI), reducing received signal quality. If these multipath components are separated in angle, but with approximately the same path delay, then a one-dimensional spatial array structure can be effective at removing or combining these components. Conditions for doing this are discussed at length in Chapter 8. If the multipath components are incident from angles which are very close to one another, a temporal equalizer, as described in [Proa89][Rap96a], can be used to suppress delayed multipath components. The two dimensional structure is able to capture energy from multipath components arriving at significantly different delays, combining features of both a spatial processor and a temporal equalizer.

Another advantage of the two-dimensional structure is that by using the tapped-delay-line as an adaptive interference rejecting filter, we can suppress interfering signals in the frequency domain as well as in the spatial domain. This is particularly valuable in heterogeneous signal environments where signals with different bandwidths and different carrier frequencies may be incident on the array.

An important special case of the wideband array is the spatial processing CDMA Rake receiver, which offers some of the benefits of the wideband array at a complexity level which is closer to the narrowband array. The spatial processing Rake receiver is described in Chapter 4.

3.9 Spatial Diversity, Diversity Combining, and Sectoring

Multipath in the radio channel can lead to fading of the received signal, particularly when the bandwidth of the signal is small compared to the coherence bandwidth of the channel. This is a problem which has traditionally been dealt with using spatial diversity at the base station receiver. Spatial selection diversity provides an effective and economical means of reducing narrowband fading.

As shown in Chapter 8, adaptive antenna arrays, under certain circumstances, have the ability to combine coherent multipath components, resulting in a reduction of narrowband fading. Unfortunately, if there are many interferers, or if the number of significant multipath components is large, or if the angular separation between components is small, or when the noise level is high, the ability of the adaptive antenna system to reduce flat fading may be limited.

For switched beam antenna systems, the ability to reduce narrowband fading may be limited, even under the best of circumstances. Therefore, in addition to the general structure shown in Figure 3–6, it may be desirable to implement spatial diversity in a smart antenna system. One scenario is illustrated in Figure 3–14. In this system, beamforming is performed independently by the two arrays, and the selection diversity system chooses the array output with the highest signal quality.

In addition to spatial diversity, most existing cellular and PCS systems make use of sectoring, in which each cell is divided into several angular regions. Typically, each cell is divided into 3 sectors which are 120° wide or 6 sectors which are 60° wide. For a given cell size, it is possible to reuse spectral resources more frequently by sectoring, thus allowing a larger traffic density to

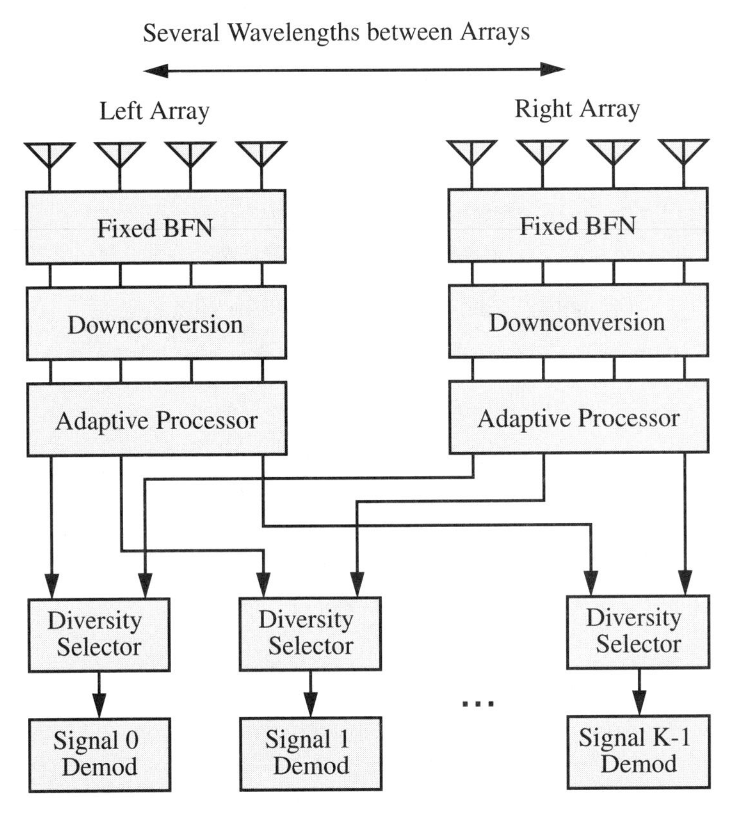

Figure 3–14 A spatial selection diversity system can use two arrays which are separated by several wavelengths to provide additional protection against narrowband fading.

be supported in each cell. In FDMA and TDMA systems, sectoring results in a loss in trunking efficiency due to the small number of channels in each sector, but there is a net increase in system capacity because of improved frequency reuse [Rap96a]. In CDMA systems, sectoring does not result in a loss in trunking efficiency, but it does require increased equipment at the base station. Sectoring allows each base station to handle much more traffic at a single base station than could be supported by using omni-directional coverage. Moreover, by dividing the cell into sectors, directional antennas, which have higher gain than omni-direction antennas, can be used at the base station. This increases range, which in turn increases cell size.

It is useful to consider how smart antenna systems can be deployed in sectorized antenna schemes. For several reasons, it is often beneficial to use directional antenna elements in arrays. For example, using a linear array of omni-directional elements, the array forms "mirror-image" beams on each side of the axis of the array. In addition, a linear array has decreased angular separation capabilities for angles near the end-fire direction of the array. Therefore, a convenient way to combine sectoring with smart antenna systems is to use a mast head configuration as shown in Figure 3–15. In this system, each sector makes use of a left and right array for diversity. Individual directional elements can be used to achieve a suitable front-to-back ratio and inter-sector beam pattern isolation. In this configuration, it is not necessary to use the arrays at directions very close to end-fire, so problems that occur in these regions are avoided.

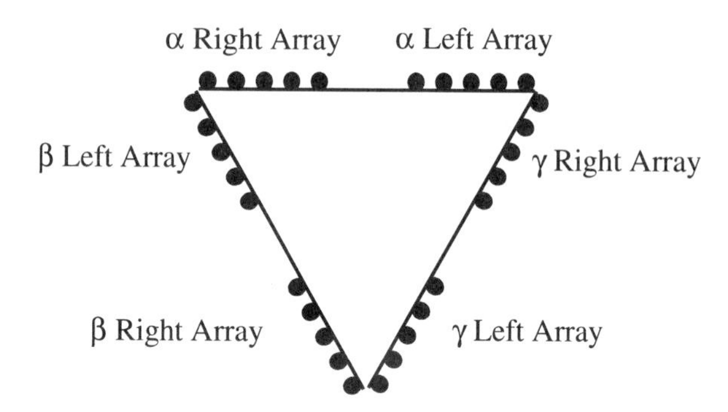

Figure 3–15 A top view of a three sector antenna array arrangement in which each sector contains two array systems for spatial diversity. Each element in the array is designed to provide coverage to isolate each 120° sector from adjacent sectors.

In addition to switched beam systems and adaptive array systems, two related array-based technologies also exist. The first technology is the *phased array* system, in which the magnitude of each weight, $|w_m|$, is fixed and only the phase of each weight, $\angle w_m$, can be adjusted. This allows a beam pattern to be steered in any desired direction as shown in Section 3.2. These techniques can be implemented using commonly available RF hardware and have been used in radar systems for some time. Phased array systems offer the ability to continuously scan in angle, as in adaptive array systems, but at a level of complexity that is similar to switched beam systems. Phased array systems do not have the ability to place nulls independent of the direction of the main beam and there is no mechanism to coherently combine multipath as can be done with the adaptive array. Also, it is more difficult to implement a continuously variable phased array system in a multi-user environment compared to a switched beam system.

Another related technology is the *Maximal Ratio Combiner* (MRC), which adjusts both the magnitude and phase of weights in the combining network to maximize the signal-to-noise-ratio (SNR) at the output of the combiner. In an interference-free environment, the MRC array may in fact be implemented as an adaptive array system; however, there are differences. For example, the MRC array may be implemented using array elements which are widely separated. The adaptive array typically uses element separations of a fraction of a wavelength to avoid grating lobes. Usually this is not a concern in the MRC combiner, which may be designed as a spatial diversity system rather than a beamforming system. Another difference between an MRC array and a fully adaptive array is that it is possible to implement an MRC array which does not use feedback from the array output to adjust the amplitude weighting of each branch, as illustrated in Figure 6.1(b) on page 192 of [Par89]. In such an MRC implementation, the weighting applied to each diversity branch is adjusted independently from the other branches according to the estimated signal voltage in that branch divided by the noise power. Co-phasing in an MRC may be performed independently from the weighting process. While an MRC array can achieve optimal performance in the presence of noise, it does not provide the ability to reject interference or multipath. Since the amplitude weighting of the MRC array depends only on the power of the signal and noise in each branch, it may be simpler to implement the weighting as an applique solution as compared to adaptive array approaches. Traditionally, MRC arrays have been analyzed in interference-free environments with multipath accounted for only by modeling independent narrowband fading at each array element [Lib95]. In noise-limited environments without uncorrelated multipath, the MRC array provides performance comparable to the adaptive array. When interference and multipath are considered, an adaptive array may have significant advantages over an MRC system which is implemented using independent branch weighting and co-phasing.

3.10 Digital Radio Receiver Techniques and Software Radios for Smart Antennas

In this section, we consider Digital Radio Receiver Techniques and Software Radio concepts as they apply to smart antennas. Smart antenna systems, particularly those with adaptive spatial processing, typically incorporate extensive digital processing. Therefore, it is natural to consider digital processing techniques for other receiver functions. Like the term *Smart Antenna*, the term *Software Radio* encompasses a broad range of technologies with differing capabilities. Generally, a software radio implements key receiver functions digitally in software, including

- Downconversion,
- Channelization,
- In-phase and Quadrature signal component extraction, and
- Demodulation

Software radios offer a number of advantages over traditional radio receiver designs that use either analog components, Application Specific Integrated Circuits (ASICs), or other fixed digital hardware. Software radios can be dynamically reprogrammed to handle different air interfaces or to support upgrades. This makes software radios ideal for environments where different signal formats may be of interest, such as in dual-mode base stations that must support both AMPS and CDMA.

A key function in adaptive spatial processing is the digitization of received signals. To minimize the impact of variation in analog components, it is desirable to digitize the signal at the highest feasible *Intermediate Frequency* (IF). Modern A/D converters are available with front-end bandwidths of several hundred MHz [Raz98]. However, the processing power required to support multiple channels of data sampled at several hundred MHz can be prohibitively expensive. *Bandpass Sampling* offers a solution to these problems. In bandpass sampling, aliasing is

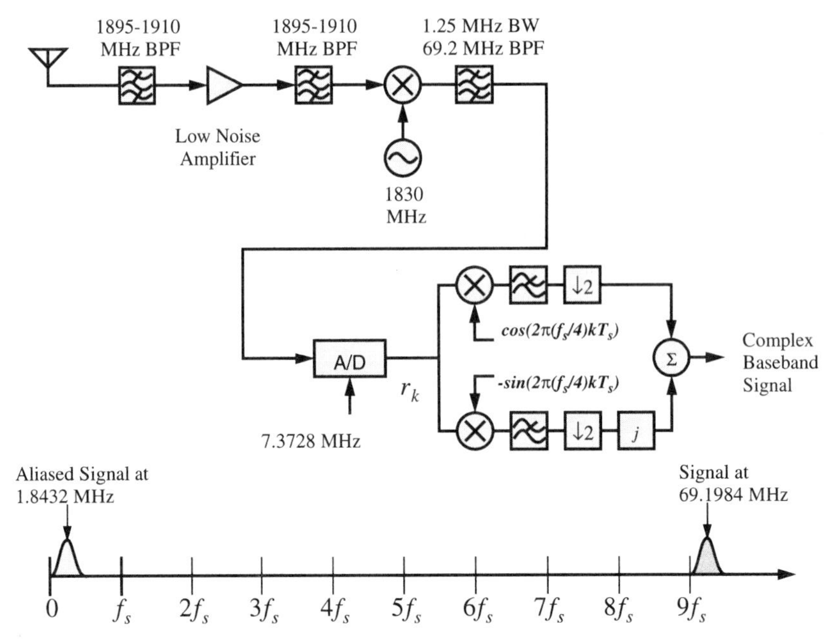

Figure 3–16 A bandpass sampling receiver. In this example, the 69.1984 MHz IF signal is sampled at a sample rate of 7.3728 Msamples/second. In the digital representation of the signal, aliasing provides a copy of the signal at a center frequency of 1.8432 MHz.

exploited by carefully filtering the IF signal, then undersampling it at an appropriate rate. This is illustrated in Figure 3–16.

The next step, after digitizing the received signal, is to recover the In-phase (I) and Quadrature (Q) signal envelopes from the IF signal. An efficient technique for this is *Digital DownConversion* (DDC) [Sau90]. We can represent the sampled IF signal from the A/D converter as

$$r_k = AR_k \cos(2\pi f_{if} kT_s + \nu_k) \tag{3.47}$$

where k is the sample number, A is an arbitrary gain, R_k is the amplitude of the sampled IF signal, f_{if} is the receiver frequency, T_s is the sampling interval, and ν_k is the phase of the sampled IF signal.

The I channel is formed by multiplying by $l_{I,k} = \cos(2\pi(f_s/4)kT_s)$ and the Q channel is formed by multiplying by $l_{Q,k} = -\sin(2\pi(f_s/4)kT_s)$. Note that since $f_s = 1/T_s$, these become

$$l_{I,k} = \cos\left(\frac{\pi k}{2}\right) = 1, 0, -1, 0, \ldots$$

$$l_{Q,k} = -\sin\left(\frac{\pi k}{2}\right) = 0, -1, 0, 1, \ldots \tag{3.48}$$

The downconverted signal is given by

$$p_k = p_{I,k} + jp_{Q,k} \tag{3.49}$$

$$= r_k \cdot l_{I,k} + jr_k \cdot l_{Q,k}$$

$$= \frac{AR_k}{2}\left(e^{j\left(2\pi f_{if}kT_s + v_k - \frac{\pi k}{2}\right)} + e^{j\left(-2\pi f_{if}kT_s - v_k - \frac{\pi k}{2}\right)}\right)$$

which has a frequency component centered at $f_{if} - f_s/4$ and another at $-f_{if} - f_s/4$ as illustrated in Figure 3–17. Another image of the original signal also appears at $-f_{if} + 3f_s/4$. If $f_{if} = f_s/4$, the frequency components appear at $-f_s/2$, 0, and $f_s/2$. A low pass filter, as shown in Figure 3–18, must be used to eliminate the undesired components.

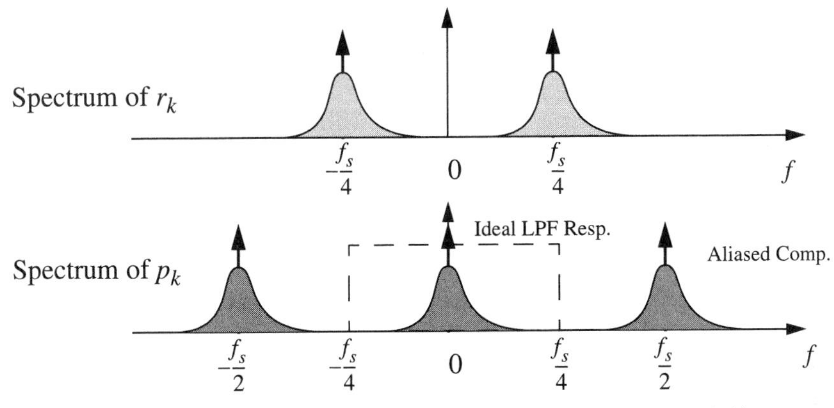

Figure 3–17 Spectrum of the signal before and after down-conversion by $f_s/4$.

To illustrate how this process can be implemented, consider a DDC system designed to recover the I- and Q-components of the signal sampled in Figure 3–16. In order to eliminate the undesired spectral component, the I and Q channels are each filtered by a 64th order, linear-phase FIR filter (a filter with 64 weights) having the frequency response illustrated in Figure 3–18. This filter provides 80 dB of stopband attenuation. The filter is designed to reject frequency components greater than $f_s/4$, eliminating the undesired images at $\pm f_s/2$. This filter has a linear phase and a 1 dB cut-off frequency of approximately 650 kHz, allowing it to pass an RF signal with a bandwidth of 1.3 MHz without distortion. This method is practical for IS-95 demodulation.

Note that if the low pass filter sufficiently eliminates all signal content for frequencies between $f_s/4$ and $3f_s/4$, then it is possible to resample the resulting signal at a sample rate

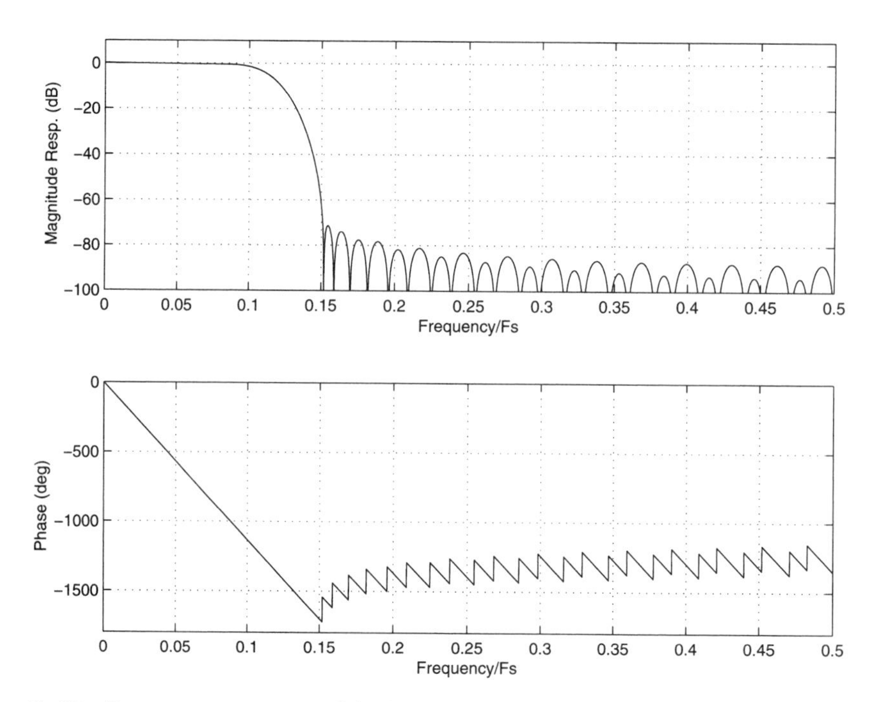

Figure 3–18 Frequency response of the low pass filter used to eliminate the undesired component of the downconverted signal. The sampling frequency is $F_s = 7.3728$ MHz.

which is half of the original sampling rate. This decimation process helps to conserve storage and processing resources.

An efficient implementation of this filter may be obtained by taking into account the downconversion prior to the filter and the decimation at the output of the filter. To illustrate this, consider a 3rd order filter having the transfer function

$$H(z) = b_0 + b_1 z^{-1} + b_2 z^{-2} + b_3 z^{-3} \tag{3.50}$$

The coefficients b_i are real valued for a linear phase filter, so that the real and imaginary portions of the complex-valued sequence p_k may be filtered separately using entirely real arithmetic. The real portion of p_k, which corresponds to the I channel, is designated as $p_{I,k}$,

$$p_{I,k} = r_k l_{I,k} = \begin{cases} r_k & k = 0, 4, 8, \ldots \\ -r_k & k = 2, 6, 10, \ldots \\ 0 & k = 1, 3, 5, \ldots \end{cases} \tag{3.51}$$

The filter given in (3.50) is shown in Figure 3–19.

Writing (3.50), in terms of r_k rather than $p_{I,k}$ for several successive samples, provides insight into a technique for efficiently combining the process of downconversion with filtering.

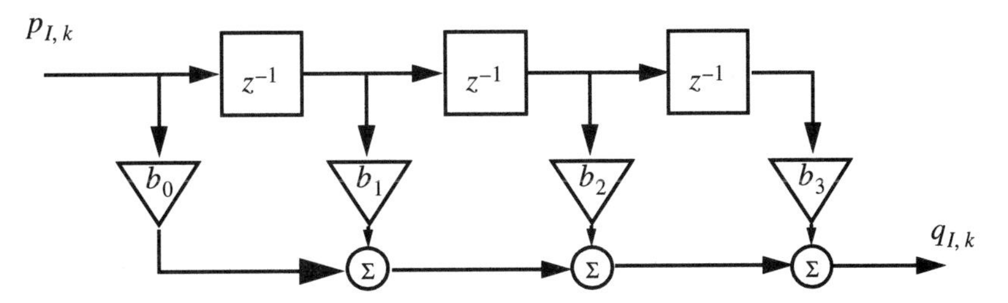

Figure 3–19 The FIR filter given by (3.49).

$$
\begin{array}{lll}
(a) & q_{I,k} = b_0(r_k) + b_1(0) + b_2(-r_{k-2}) + b_3(0) & k = 0, 4, 8, \ldots \\
(b) & q_{I,k} = b_0(0) + b_1(r_{k-1}) + b_2(0) + b_3(-r_{k-3}) & k = 1, 5, 9, \ldots \\
(c) & q_{I,k} = b_0(-r_k) + b_1(0) + b_2(r_{k-2}) + b_3(0) & k = 2, 6, 10, \ldots \\
(d) & q_{I,k} = b_0(0) + b_1(-r_{k-1}) + b_2(0) + b_3(r_{k-3}) & k = 3, 7, 11, \ldots
\end{array}
\quad (3.52)
$$

Examining (3.52), we see that we may combine the downconversion process with the filtering process by using two different filters and switching between the two for successive samples at the output as shown in Figure 3–20.

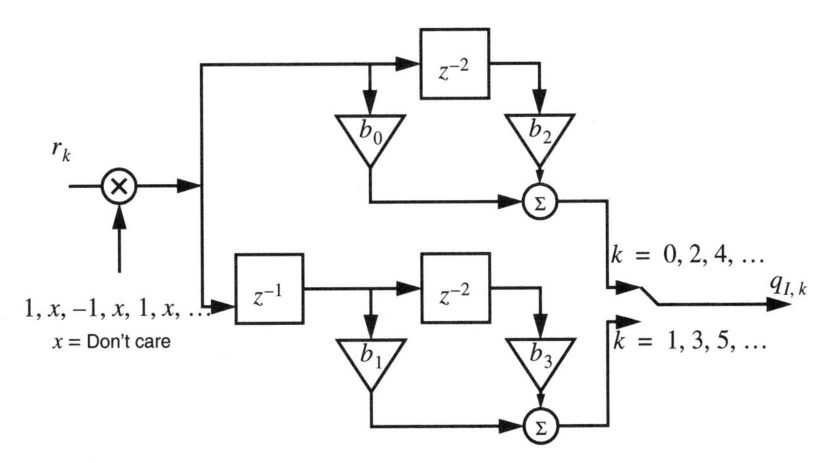

Figure 3–20 Combining the steps of downconversion with filtering.

We may also combine the decimation step by realizing that decimation by a factor of two may be achieved by simply ignoring the lower filter block in Figure 3–20 and taking only the even samples. Furthermore, we can move the decimation process before the filter, reducing the second order delay of the filter to a first order delay. Thus, a filter structure which implements the downconversion, filtering, and decimation may be efficiently implemented as shown in Figure 3–21.

Similarly, the Q channel may be efficiently recovered using the structure shown in Figure 3–22.

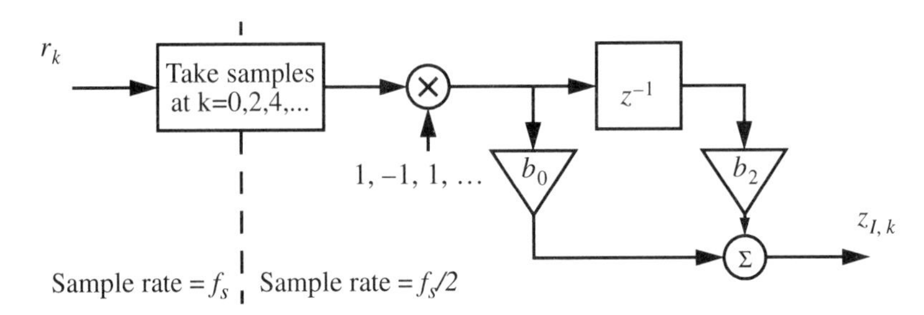

Figure 3–21 A filter structure to implement downconversion, 3rd order filtering, and decimation for the I channel.

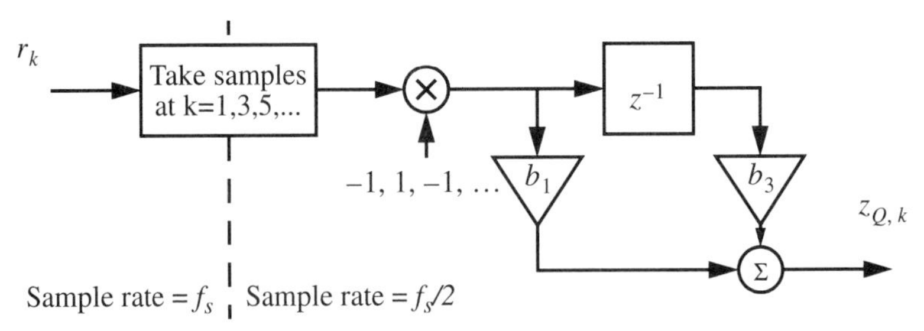

Figure 3–22 A filter structure to implement downconversion, 3rd order filtering, and decimation for the Q channel.

Using these techniques, the 64th order digital filter, with frequency response shown in Figure 3–18, is implemented using the complexity of a 32nd order filter.

In this example, we have considered a decimation factor of 2, the minimum needed to recover I and Q from an IF signal. Higher decimation factors are also often used, in which case the filter $H(z)$ must be sufficiently sharp to reject signal components above $f_s/(2D)$, where D is the decimation factor.

Instead of pre-multiplying the signals by the alternating {1,-1,1,-1,...} sequence as shown in Figures 3–21 and 3–22, we can alternatively use different taps for each iteration of the filter operation. For example, in the I-channel filter of Figure 3–21, we can use the taps $\{b_0, -b_1\}$ for $k = 0, 4, 8, \ldots$, and $\{-b_0, b_1\}$ for $k = 2, 6, 10, \ldots$, in place of pre-multiplication. This results in a *polyphase* filter. Polyphase filters use a time-varying filter response that is able to shift frequency components in addition to filtering them. This is precisely the behavior that we are exploiting in Digital DownConversion (DDC) [Tie91][Sau90].

More complex polyphase filters can also be used for *channelization*. For base stations that receive signals on multiple carrier frequencies, software radio systems can sample a block of spectrum, including multiple carriers, and can use polyphase filters to simultaneously separate and downconvert signals in a manner similar to the DDC process described here [Vai92].

3.11 Transmission Beamforming

When the uplink, or reverse link, is the limiting link, smart antenna systems can be used at the base station receiver to improve overall system performance. In *unbalanced* wireless systems, where reverse link signals are significantly weaker than the forward link signals, smart antenna base station receivers can be extremely valuable at improving reverse link performance. For balanced systems, improvements in range and capacity also depend on improvements in the downlink.

Unlike FDMA and TDMA systems, there are a wealth of link-dependent parameters in CDMA systems, such as the amount of power in the downlink signal devoted to the pilot signal and the percentage of coverage area where soft handoff is supported, that allow CDMA system designers to carefully balance uplink and downlink performance, with and without smart antennas. It is conceivable to use smart antennas only on the uplink in a CDMA system, to spatially separate uplink signals and increase reverse link range, thus allowing fewer subscribers to be in soft handoff. Since fewer downlink channels are required to handle soft handoff, overall system capacity can be increased. However, to realize the true benefit of smart antennas, balanced improvements in both the uplink and downlink should be considered.

One approach to improving downlink performance is to use spatial processing at the subscriber end of the link, but because of limited space, processing power, and the dynamic nature of the local environment near portable units, today only limited spatial processing can be performed at the subscriber. An important exception is found in Wireless Local Loop systems, where the availability of power and space make smart antenna subscriber terminals feasible.

For portable wireless systems, signal quality available at the subscriber can be improved by changing the way in which it is transmitted from the base station. Intuitively, it is wasteful for the base station to transmit one-to-one radio signals in directions other than those which contribute to the signal received by the desired subscriber. Such stray radio power contributes only interference to other co-channel subscribers. In downlink beamforming, the objective is to form a beam pattern, using an antenna array, which provides adequate signal quality to the desired subscriber while minimizing the interference transmitted in other directions. A multi-signal downlink beamforming system is illustrated in Figure 3–23, where each transmitted signal has its own weight vector, $\boldsymbol{w}_k$.

A number of issues complicate downlink beamforming. Perhaps most importantly, multipath has a major impact on downlink beamforming schemes. In the best case, when the propagation channel between the subscriber and the base station is perfectly known, it is desirable to take advantage of path diversity, transmitting along multiple paths to the subscriber while suppressing power transmitted in directions which do not reach the user. However, perfect knowl-

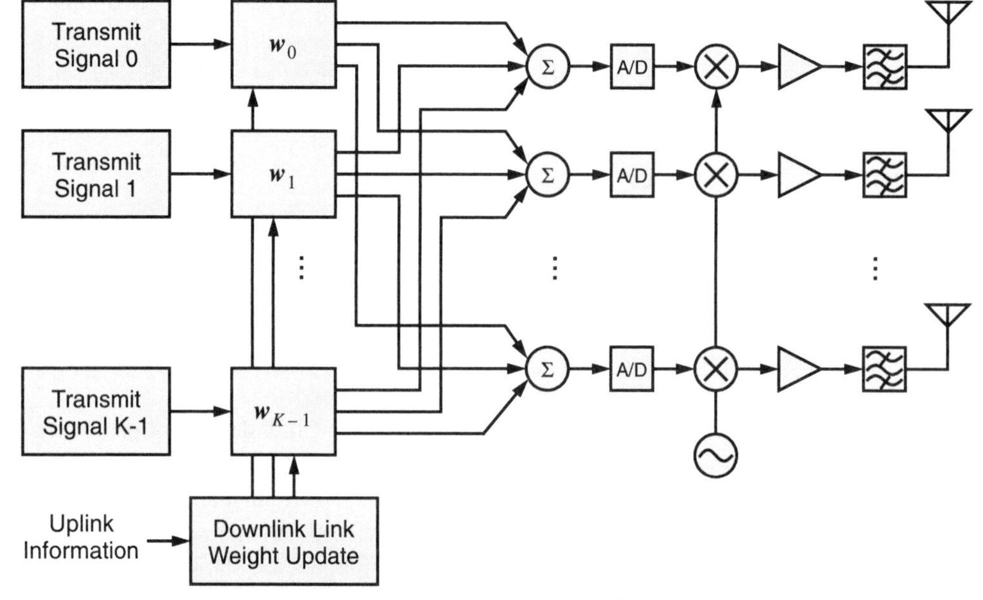

Figure 3–23 A downlink beamforming system, capable of forming K simultaneous beams.

edge of the channel is never available, and the time-varying nature of the channel complicates this approach. If the goal is to select a single direction for transmission, then it is critical to select the best path. If the subscriber moves and loss along a particular path increases rapidly, then the base station must detect this and quickly redirect power to a new path if available.

When downlink beamforming is used for capacity enhancement rather than range extension, it is necessary to limit the Effective Isotropic Radiated Power (EIRP) so that, as the gain of the downlink antenna array increases in a particular direction, the total transmitted power is reduced accordingly. If this is not done, then as the array increases its gain in one direction, it may actually worsen the downlink CIR for a co-channel user in a nearby cell.

The simplest approach to downlink beamforming is to transmit with the same pattern which was used by the uplink smart antenna receiver. In narrowband systems, for example, we extract the spatial signature $\boldsymbol{b}(t)$, given in (3.21). In the absence of interference, as we show in Chapter 5, the optimal weight vector needed to extract a signal with spatial signature $\boldsymbol{b}(t)$ is $\boldsymbol{w} = a\boldsymbol{b}(t)$, so it is possible to use this same weight vector for both receive spatial processing and transmit beamforming. This approach has the most potential in the case of Time Division Duplexed (TDD) systems where the uplink and downlink channels share the same frequency channel. In TDD systems, weights adapted on the uplink may be used to transmit a signal back to a particular user, provided that the channel remains stationary between uplink and downlink transmissions.

One issue complicating this simple downlink beamforming approach is that the spatial response from a subscriber may change very rapidly as the subscriber moves over a small distance, as a result of constructive and destructive interactions among multipath components. If the spatial processor is not able to track these changes rapidly, then downlink beamforming could actually result in a degradation of the signal delivered to the subscriber. By noting that the DOA of a single dominant path, if present, may vary more slowly than the spatial signature, $\boldsymbol{b}(t)$, a more robust downlink beamforming approach can be developed.

In Frequency Division Duplexed (FDD) systems, the uplink and downlink channels use different frequency bands. In FDD systems, open-loop downlink beamforming is based on measurements of the uplink channel. Licensed US high tier PCS systems operating in the 2 GHz band (1850-1910 MHz/1930-1990 MHz) use FDD where the uplink and downlink signals are separated by 80 MHz. As a first issue, we note that the array factor given in (3.8) is a function of frequency. Thus, if the same weights used to receive are used to transmit, a different array pattern results. It may be necessary to adjust the array weights to account for this change.

More importantly, there are differences in the channel at the uplink and downlink frequencies. Depending on the characteristics of the channel, the path gain, ρ_i, and the DOA, ϕ_i, may be similar or different at two different carrier frequencies. If multipath components are due to specular reflections, then there may be very little difference in ρ_i and ϕ_i for each path at the uplink and downlink frequencies. On the other hand, the carrier phase change along each path will be significantly different at the two frequencies, meaning that the spatial signature measured on the uplink may not be the ideal, or even a good, choice to use for weighting the transmitted signal. As shown in Chapter 8, if the difference between the delays of two multipath components is very small, in the absence of strong noise and interference, the receiving array forms a beam that captures power from both components and co-phases the signals to coherently combine them. If the uplink pattern is applied on the downlink, then, due to different electrical path lengths resulting from the different carrier frequencies, it is possible that, at the subscriber unit, rather than being co-phased, components may combine destructively, resulting in a fade at the subscriber unit.

Another method for downlink beamforming, which can be applied to both TDD and FDD systems, is to use Direction-Finding (DF) techniques to determine the Directions-Of-Arrival of components at the receiver [Hay91]. DF methods are discussed in detail in Chapter 9. If a strong direct path is present and can be identified, then it is reasonable to form a downlink beam for a particular user in a single direction, toward the subscriber.

If a wideband array or Rake receiver is used, then it is possible to use Time-Of-Arrival information to assist in determining the direct path. If no LOS path can be identified, the DOA information may be used to direct a broad beam in the general direction of the subscriber. As discussed in Chapter 9, typically array-based DF techniques are based on the MUltiple SIgnal Classification (MUSIC) algorithm and its many derivative approaches [Muh96a][Hay91]. These *subspace-based* techniques are generally applicable if the number of strong multipath components is small compared with the number of array elements. The degree of correlation among components is also very important in determining the ability of the array to determine the Directions-Of-Arrival of individual paths. In heavily overloaded environments, accurate DF using subspace techniques becomes difficult.

Recently, new DF approaches have been proposed which take advantage of adaptive beamforming to determine Directions-Of-Arrival [Tal94][Muh96a]. In these approaches, a blind multi-target adaptive algorithm is used to extract up to M orthogonal spatial signatures from M uncorrelated signals. From each spatial signature, up to $2M/3$ individual multipath components are extracted by applying spatial smoothing to the data correlation matrix, $\boldsymbol{R}$, of equation (3.31). These DF methods, which may be used to estimate the angle-of-arrival, are described in Chapter 9. DF approaches are also useful in geolocation systems which provide a estimate of the position of wireless subscribers. When used in combination with ranging methods that estimate the propagation delay between the subscriber transmitter and multiple base stations, accurate position location can be achieved. In Chapter 10, we provide a fundamental treatment of position location techniques based on ranging and Angle-Of-Arrival estimation. Such geolocation systems will be required to comply with E-911 emergency positioning requirements [Ken96][Miz96b][Ree98][Rap96b][Aat97a][Rap98]. When DOA and positioning information is being estimated, this information can be used to assist in the uplink beamforming process as well, illustrating the close link between beamforming and positioning.

Even using DF techniques, it may not be possible to adequately localize the subscriber in complicated multipath environments. Therefore, iterative techniques which use feedback from the subscriber may be necessary for accurate downlink beamforming [Nag94b][Lia95][Ger93]. Such methods may hold promise both for downlink beamforming and position estimation.

Thus, downlink beamforming strategies can be categorized into three types. In the first type, a downlink beam pattern is formed which matches the uplink beam pattern. Because of the problems discussed in this section, this technique is potentially ineffective and risky, particularly in FDD systems. In the second category are techniques which identify one or more specific paths and form one beam or multiple beams to enhance the signal received at the portable unit. If multiple beams are used to provide path diversity, then great care must be exercised to ensure that interference seen by co-channel users is not actually increased by the array. Finally, there are techniques which rely on feedback from the mobile unit to assist in downlink beamforming. The feedback-based techniques can potentially suffer due to round trip propagation and processing delays and may also require modification of existing air-interface standards.

3.12 Array Calibration

In downlink beamforming and Direction-Finding systems, array calibration is critical. Array calibration is necessary in order to ensure that the signal processing algorithms know the array response for any Angle-Of-Arrival. In transmission beamforming, the array response must be known so that proper phase and amplitude weights may be applied to achieve a beam in the desired direction. Array calibration involves accurate measurement of the phase and amplitude of the RF channel associated with each element of the array. For Direction-Finding systems, phase inaccuracies of a few degrees or less can result in decreased accuracy and reduced ability to detect and resolve low level multipath components, particularly when using the super-resolution approaches described in Chapter 9.

A basic form of array calibration is the collection of array manifold information, $\{\boldsymbol{a}(\phi_i)\}$, which maps a steering vector to each Direction-Of-Arrival. This is done by measuring the response of the array on a calibrated antenna range with controlled or limited multipath.

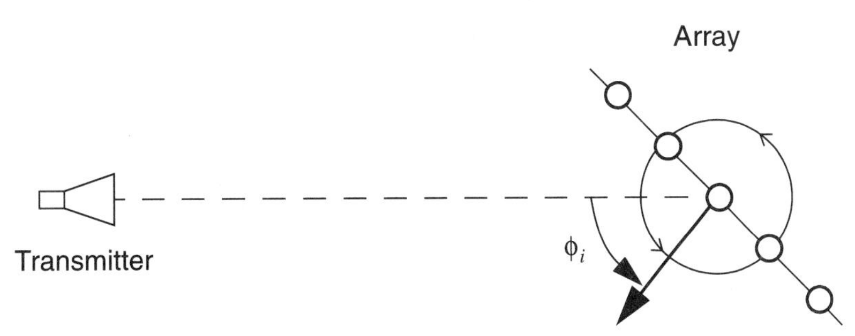

Figure 3–24 When measuring the array manifold, the array is rotated, relative to a fixed source, under controlled multipath conditions.

Array manifold measurements may be performed by rotating the array relative to a fixed source at closely spaced angles, ϕ_i, as shown in Figure 3–24. The array response at each angle, $\boldsymbol{u}(\phi_i)$, is recorded and normalized to estimate the steering vectors, $\boldsymbol{a}(\phi_i)$. Alternatively, for very large arrays, it may be more convenient to move the source as the array remains fixed.

For field-deployed arrays, it is often not convenient to collect array manifold information at closely spaced angles. Once a rigid array is deployed the relative geometry of the elements is not likely to change; however, the phase and gain of active elements can change as a function of temperature, and as components age. In order to estimate the unknown gain and phase of the RF channel for each array element, an alternative approach is to use a model-based approach.

In a model-based approach it is assumed that each steering vector takes the form,

$$\boldsymbol{a}(\phi) = \begin{bmatrix} 1 \\ \hat{a}_1(\phi)\gamma_1 e^{j\psi_1} \\ \vdots \\ \hat{a}_{M-1}(\phi)\gamma_{M-1} e^{j\psi_{M-1}} \end{bmatrix} \tag{3.53}$$

where $\hat{a}_i(\phi)$ is the gain-amplitude response of element i based on the known array geometry, and the term $\gamma_1 e^{j\psi}$ is the unknown narrowband channel response for the ith element. The received signal is measured for at least M Angles-Of-Arrival, $\{\phi_i\}$, in a benign propagation environment. Then the set of channel phases and amplitudes is estimated by finding the sets $\{\hat{\gamma}_i\}$ and $\{\hat{\psi}_i\}$, which provide the best match to the received data using an MMSE fit.

Wideband systems can be calibrated by repeating narrowband calibration at a number of different frequencies. Alternatively, wideband transmission sequences, often made up of known Pseudo-Random Bit Sequences, can be used for over-the-air calibration.

In order to track the gain and phase variation of each channel between over-the-air calibration measurements, it is useful to equip the array with a calibration injection system as shown in Figure 3–25. In this system a signal can be input directly into each array element. If measurements are performed frequently, the change in the response of each channel can be tracked, prolonging the validity of the over-the-air calibration measurements.

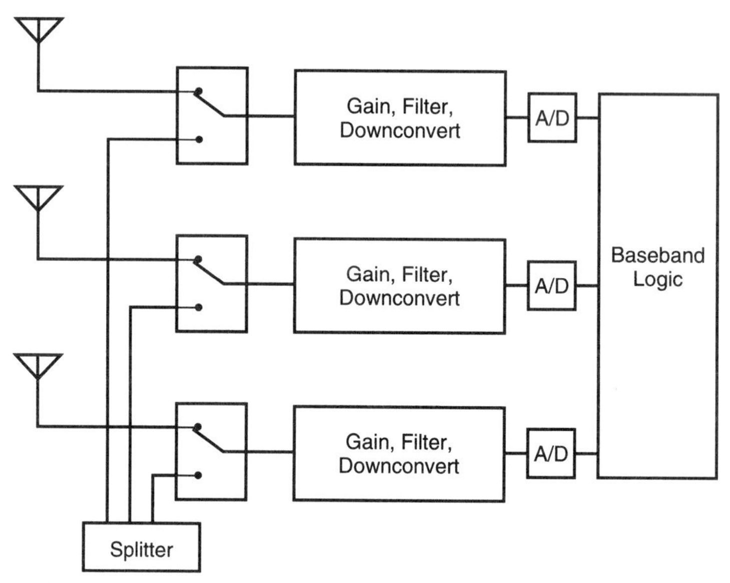

Figure 3–25 Calibration injection systems allow a known signal, possibly with a known relative phase at each injection point, to be input directly into the front-end of the receiver chain for each array element.

For downlink beamforming systems, in order to relate a weight vector with a desired transmission beam pattern, it is necessary to know the gain and phase of transmit systems, including power amplifiers and upconversion components. As with over-the-air receive calibration, transmission calibration can be performed by measuring the antenna pattern on a range as the array is rotated with respect to a receiver.

3.13 Summary

This chapter has presented key features of smart antenna technology, including switched beam systems and adaptive antenna concepts. The essential features of MMSE and LS adaptive array systems were given, and the general architectures for beamforming were illustrated. In the next chapter, specific techniques for implementing smart antennas for CDMA wireless systems are explored.

CHAPTER 4

Smart Antennas Techniques for CDMA

Chapter 3 provided an overview of smart antenna technology that can be applied to a broad range of wireless systems. In this chapter, we turn our focus to implementing smart antennas for Direct Sequence CDMA (DS-CDMA) systems and examine some of the challenges involved in applying smart antennas to CDMA. Then, in Chapter 5, we study the impact of smart antenna systems on CDMA link range and system capacity.

Spatial filtering at the CDMA Base Station Receiver can take many forms, as illustrated by the diagram in Figure 4–1. The different structures offer different combinations of complexity and performance.

Spatial processing systems for the base station may be classified as either single-user or multi-user systems. In multi-user systems, beams are formed for all users simultaneously. As described in the context of the multi-target receiver in the previous chapter, multi-user systems jointly process signals to extract K signals. Alternatively, spatial processing can be applied independently for each signal.

4.1 Non-Coherent CDMA Spatial Processors

Within the class of single-user spatial processors for CDMA, we can divide multi-antenna systems into coherent and non-coherent systems.

The reverse link in IS-95 format CDMA uses 64-ary orthogonal modulation, as discussed in Section 2.4.4. In 64-ary orthogonal modulation, each modulation symbol, or Walsh Function, is orthogonal to every other modulation symbol. A detector for orthogonal modulation is a *matched filter bank*. The matched filter bank has a matched filter for all 64 Walsh Functions. The matched filter output with the strongest signal is selected, using a comparator to recover the 6 bits corresponding to that Walsh Function. When orthogonal modulation is used, we can sum the

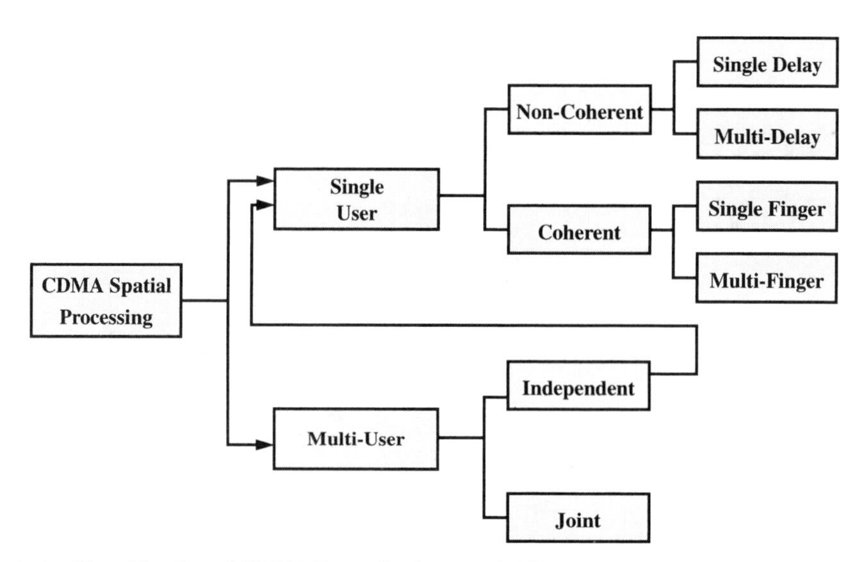

Figure 4–1 Classification of CDMA Base Station spatial filtering approaches.

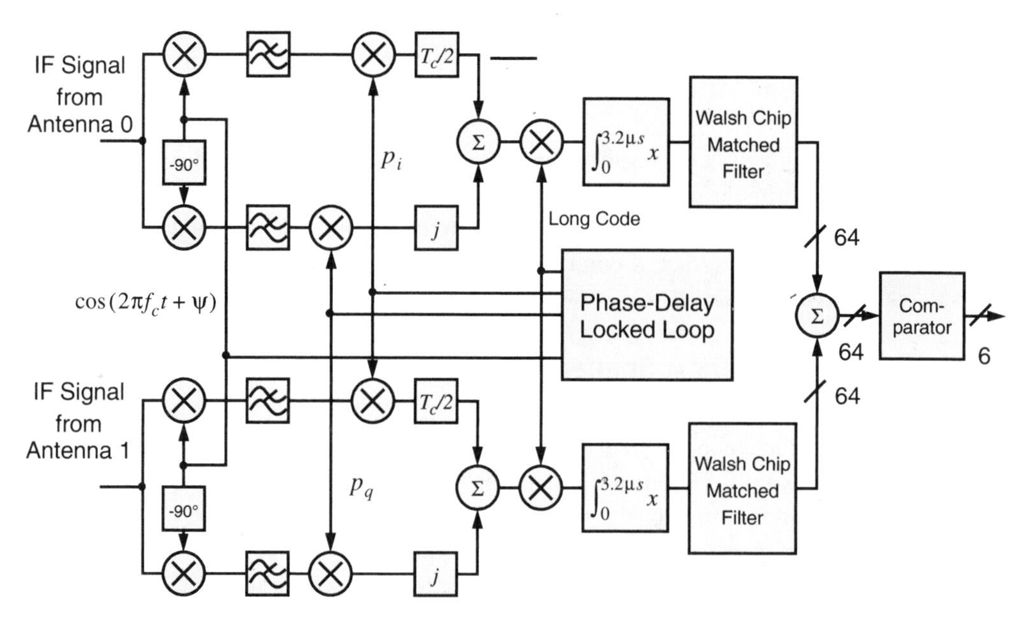

Figure 4–2 An implementation of a non-coherent multi-antenna combiner for CDMA base station receiver.

corresponding matched filter outputs from each antenna as shown in Figure 4–2. The integrator in Figure 4–2 accounts for the fact that there are 4 PN chips per Walsh chip in the IS-95 reverse link, as discussed in Section 2.4.

The structure shown here does not require adaptation and provides gain. However, the non-coherent combiner does not have the ability to null interference or manage multipath. The non-coherent combiner is essentially an equal gain diversity combiner for orthogonal signals.

By using groups of Rake fingers, where each group is locked to a different delay, and within each group, there are M branches which are combined in this non-coherent manner, we can implement the non-coherent, multi-delay, single-user receiver in Figure 4–1.

The non-coherent combiner is particularly useful for extracting short bursts from long-range subscribers. Because it requires no adaptation, the *non-coherent combiner* can help to "pull signals out of the noise", provided that PN synchronization can be achieved. PN synchronization can also be performed simultaneously on multichannel data to enhance lock-in capability.

4.2 Coherent CDMA Spatial Processors and the Spatial Processing Rake Receiver

Next, we consider the structure of a coherent-combining, single finger, single user, spatial filtering receiver, shown in Figure 4–3.

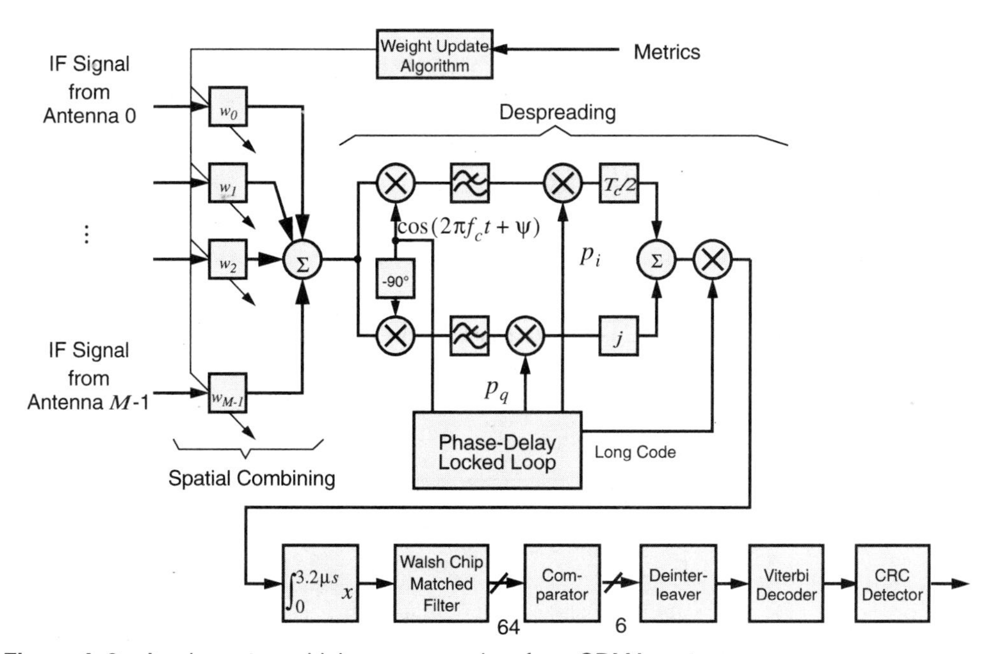

Figure 4–3 A coherent-combining array receiver for a CDMA system.

We have not shown the source of the metric used for weight update in Figure 4–3. Possibilities include examination of the constellation at the input to the *Walsh Chip Matched Filter*, examination of the Walsh Chip Matched Filter output [Lu98], or other metrics. Additional adaptive algorithms for CDMA are discussed in Chapter 8.

In CDMA base station receivers, it is also possible to reverse the order of the despreading and the spatial combiner, relative to Figure 4–3. If we despread the incoming signal first, then spatially combine signals from each branch immediately before or after the Walsh Chip Matched filter, the same spatial processing is performed as in Figure 4–3. This is due to the fact that signals from all elements of the array can be multiplied by a scalar value without distorting the spatial characteristics of the received data vector. However, the structure in Figure 4–3 has advantages for CDMA base station receivers because it requires only one despreading module for each spatial filtering receiver. If we reverse the order of spatial processing and despreading, then M despreaders are required for each spatial filtering receiver.

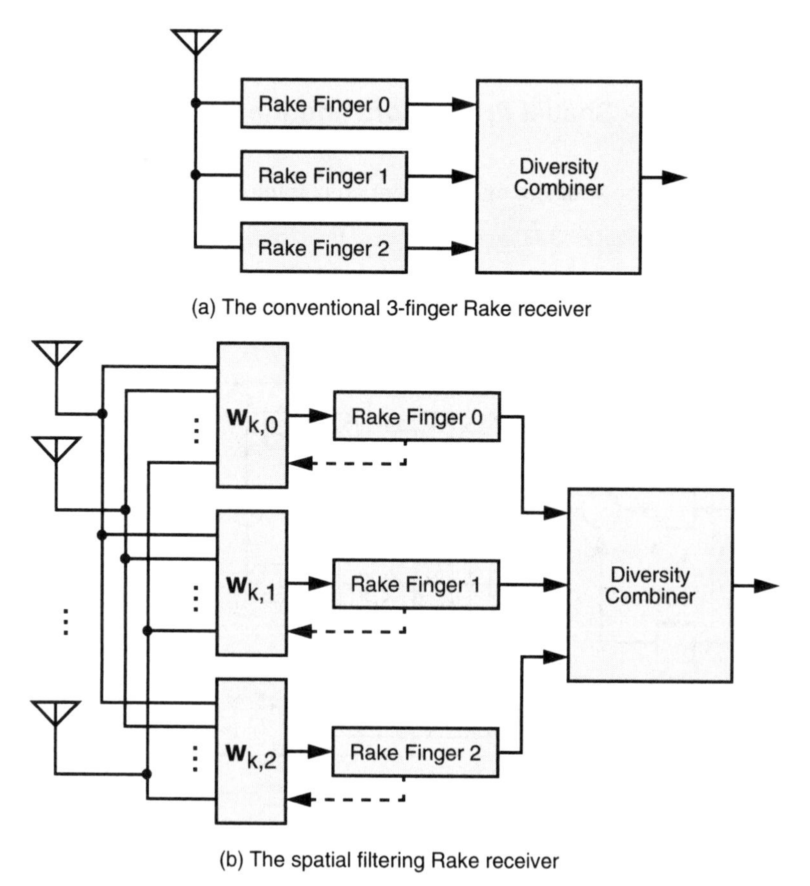

Figure 4–4 A conventional Rake receiver and a spatial filtering Rake receiver. Each Rake finger uses a different adaptive antenna pattern.

As discussed in Chapter 8, when uncorrelated multipath is present for the Signal-Of-Interest, the narrowband array attempts to place nulls in the directions of all but one of these components. Thus the narrowband array results in a reduction in interference due to uncorrelated multipath, but it is suboptimal because available path diversity is not being exploited. There are two ways in which this path diversity may be recovered, using adaptive antenna systems. This first and most general method is to use the general wideband array illustrated in Figure 3–13. This system uses P taps in a tapped delay line for each element. If the length of each tapped delay line in time, PT_s, is long enough to capture delayed multipath components, then the wideband array can capture power in components which arrive at different delays and recombine them. However, this technique can be computationally intensive.

If multipath components arrive in resolvable clusters, as shown in Figure 4–5, a second technique is to use the fact that two uncorrelated components are resolvable by a Rake receiver. In the conventional Rake receiver, each Rake finger is time-locked to a different delay to capture multipath components arriving with different path delays. A spatial filtering Rake receiver is illustrated in Figure 4–4 along with a conventional Rake receiver for comparison. The spatial filtering Rake receiver is essentially a wideband array which has only a few taps, but it is able to vary the delay between taps to match the delays of the received multipath components.

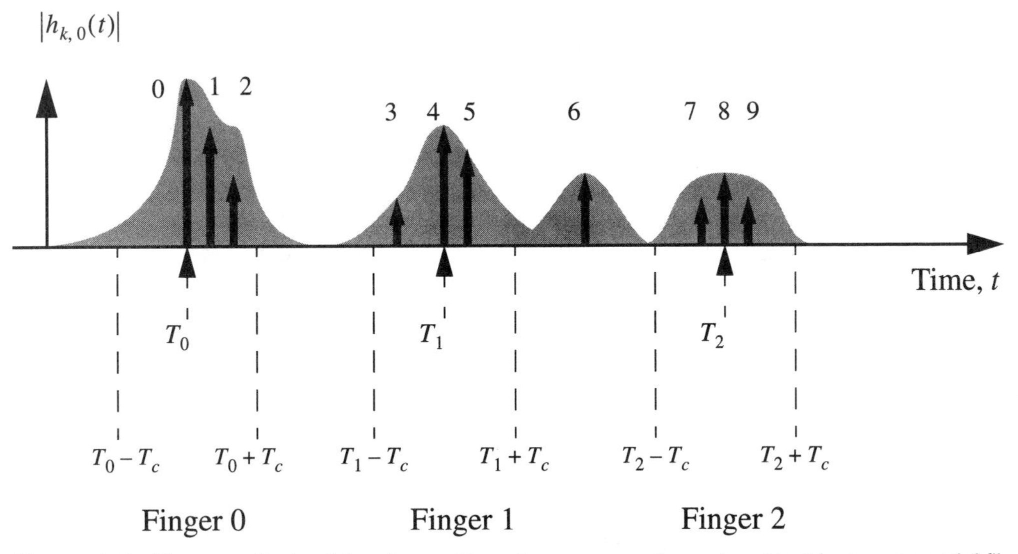

Figure 4–5 The magnitude of the channel impulse response for a signal incident on a spatial filtering Rake receiver.

In this structure, each Rake finger uses the adaptive antenna array to reject multipath components that are uncorrelated with the component to which the finger is locked. At the same time, the weight vector for each finger is adjusted to take advantage of all correlated components which arrive with path delays which are within one chip period of the path delay of the component to which the finger is locked. Diversity combining is then used to combine the output from each Rake finger.

The patterns used by each finger of the spatial filtering Rake receiver are illustrated in Figure 4–6. As shown in this illustration, the spatial response for each finger of the receiver is adjusted to maximize the *Signal-to-Interference-and-Noise-Ratio* (SINR) for that finger. In this example, the first two strong multipath components, SOI-1 and SOI-2 arrive with delays which are very close to each other, as shown in Figure 4–6 (d). Finger 0, which is synchronized with these first arriving components, forms a beam which captures and co-phases both components while forming nulls in the directions of uncorrelated (later arriving) components. Finger 1 is synchronized with the delay associated with SOI-3. Therefore, the beam pattern for Finger 1 maximizes power in the direction of SOI-3 while nulling interference from the other uncorrelated multipath components and the interference source, SNOI, as shown in Figures 4–6 a, b, and c. Finally, Finger 2 captures power from the SOI-4 multipath component while rejecting all uncorrelated signals. The outputs from all Rake fingers are then combined using diversity combining techniques.

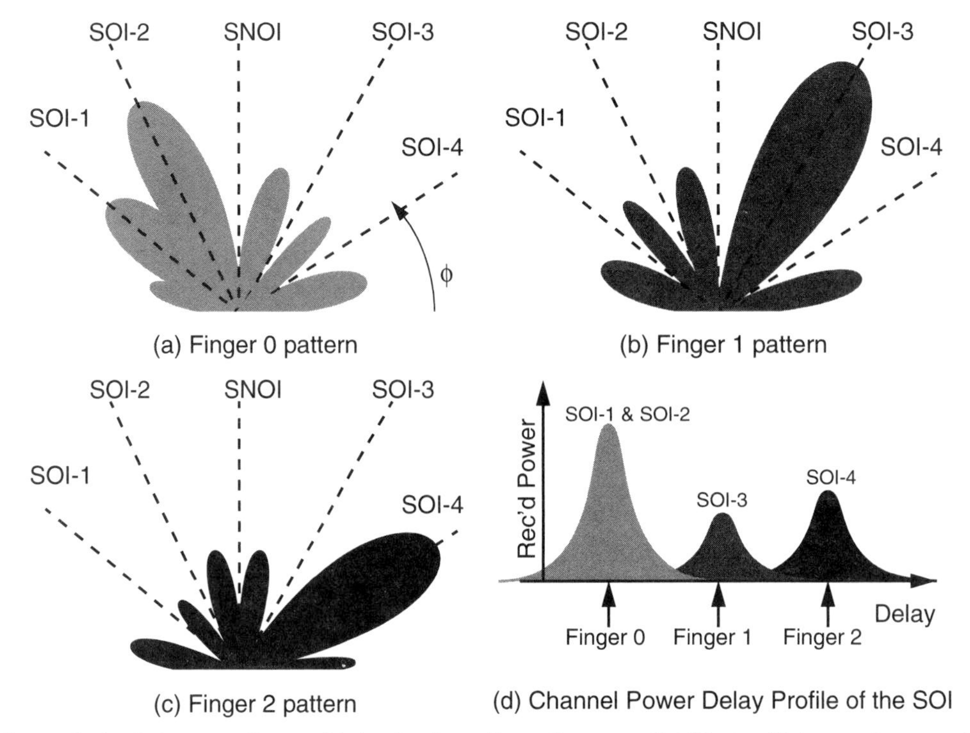

Figure 4–6 Antenna patterns obtained using a three finger spatial filtering Rake receiver receiving four components. The Signals-Of-Interest (SOIs) and Signals-Not-Of-Interest (SNOIs) are shown as a function of Direction-Of-Arrival (a – c) and time delay (d).

Figure 4–7 shows a multi-sector smart antenna receiver for the CDMA base station. Each Rake finger can select the α sector, the β sector, or the γ sector. This is the technique performed in conventional *Softer Handoff* systems for CDMA. In the system in Figure 4–7 we augment Softer Handoff by allowing each Rake finger to use the array within the selected sector to form a beam for the desired reverse link traffic channel.

4.3 Multi-User Spatial Processing

In joint multi-user detection systems, the fact that the base station is simultaneously demodulating K signals at the receiver is exploited. These K signals are the same signals that present Multiple Access Interference to every other signal at the base station receiver as described in Section 1.3.2. Therefore, if the base station can accurate identify and separate some signals, these waveforms can be subtracted or cancelled from remaining signals, reducing the MAI seen by every other receiver. This is the essential concept behind *estimation-subtraction.*

Multi-user detection can be combined with spatial processing. A structure that combines spatial processing with a *Parallel Interference Canceler* (PIC) is shown in Figure 4–8. The signal for each interfering subscriber is demodulated, and regenerated. A weighted version of the regenerated signal is subtracted from the signal for user 0 [Koh90]. In this system, the spatial processor removes the impact of spatially-separable interference, while the PIC structure minimizes remaining MAI, which may arrive from directions close to the Signal-Of-Interest. A compendium of papers [Rap98] is a convenient source of papers for modern multiuser spatial processing techniques.

In general, multi-user, joint processing systems will outperform independent spatial processors, but at a higher computational complexity. In Chapter 8, a range of optimal spatial filtering techniques is considered, including both joint and single-user approaches.

4.4 Dynamic Re-sectoring Using Smart Antennas

In CDMA systems, the ability to accommodate dynamic traffic densities is important. Areas of high traffic density are often highly time-variant. As an example, consider highway traffic near toll-collection plazas, which present high user density during the morning rush hour, but possibly light traffic later in the day. Smart antennas can help by balancing the sector-to-sector load for these "hot spots."

As described by Feuerstein in [Feu98], the fixed beamforming matrices described in Section 3.5 serve a valuable role in CDMA systems by allowing dynamic re-sectoring. Switched beam antenna systems can be used to modify nominal sector coverage. To support high traffic density (Erlangs per square kilometer), dynamic sectoring is used to reduce the area covered by a sector. In Figure 4–9, we see how dynamic re-sectoring can be used to adjust to time-varying traffic needs such as morning and evening hot spots.

Using dynamic re-sectoring, adequate signal strength and capacity can be supplied to areas as needed, without excessive coverage when it is not needed. By providing downlink coverage only where and when it is needed, the downlink power from adjacent base stations is reduced, so dynamic re-sectoring helps to reduce pilot pollution.

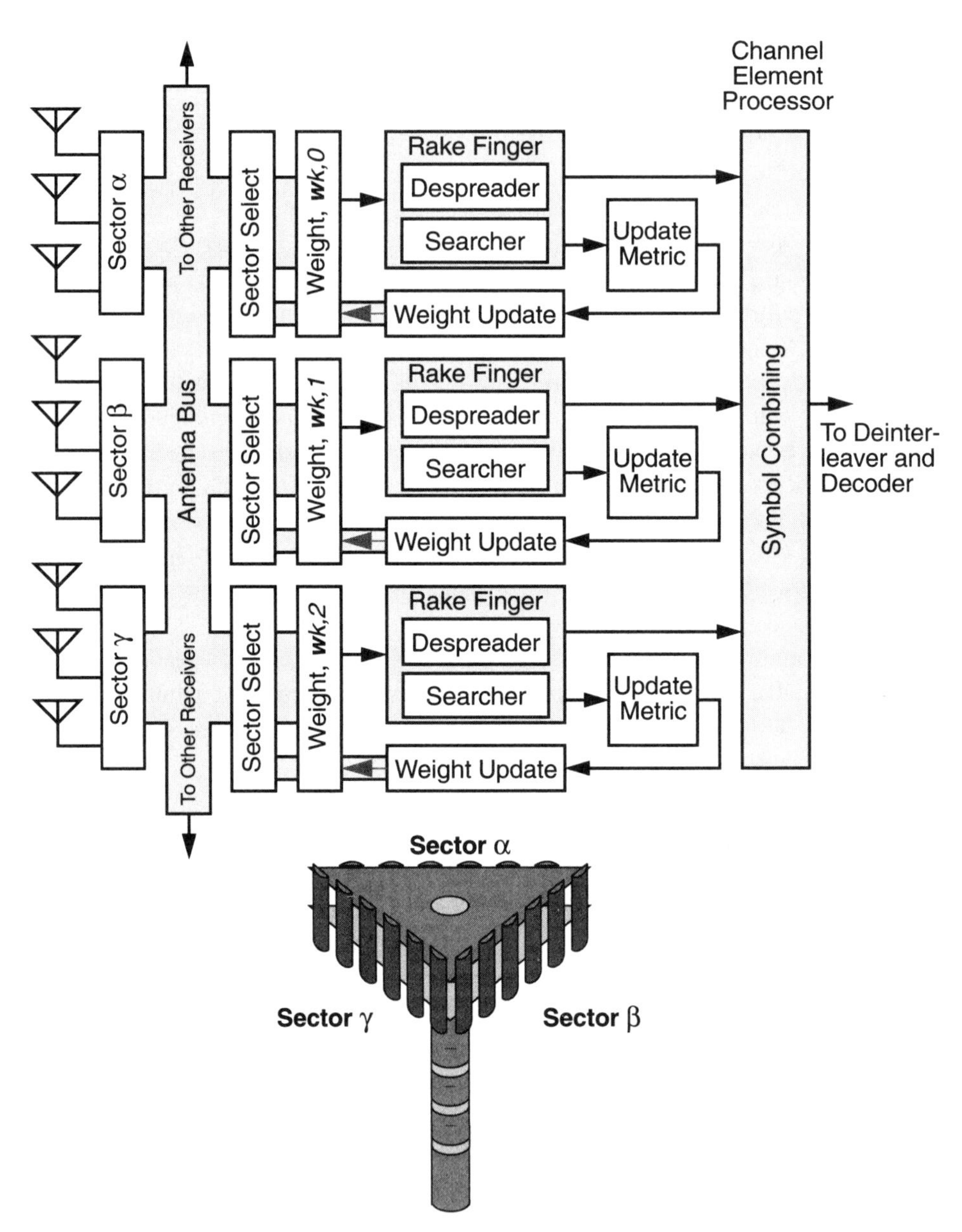

Figure 4–7 An example of a single user, coherent-combining, multi-finger receiver which takes inputs from multiple sectorized smart antennas. Each Rake finger selects one of three arrays. Then array processing is performed on the selected array to maximize the quality of the desired signal at the output of each finger. The tower configuration is illustrated, using six 120° sector panels per cell face. Each array forms a narrow beam that can be directed to any location within the 120° region covered by each cell face.

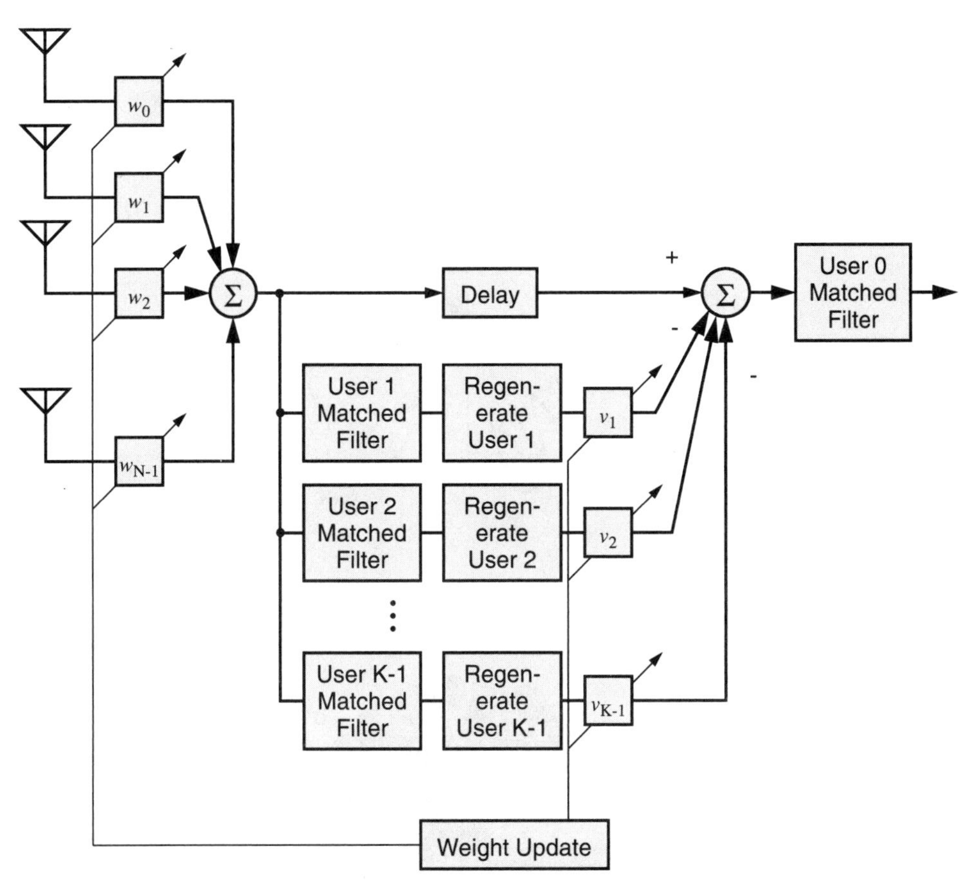

Figure 4–8 A combined spatial processor and Parallel Interference Canceler (PIC), as proposed in [Koh90]. The structure to extract user 0's signal is shown. A similar structure is constructed to extract other signals.

4.5 Downlink Beamforming for CDMA

Examining the IS-95 downlink channel structure in Figure 2–1, we note that a few of the downlink channels, the pilot, sync, and paging channel, must be supplied continuously to all areas covered by a sector. Otherwise, mobile subscribers in those areas may not detect that the sector is available to provide service. This means CDMA systems are limited in the degree to which downlink beamforming can be used to focus power only where it is needed.

This is a common issue with most major air interfaces. For example, in GSM/DCS-1900 (ANSI J-STD-007 - [Ans95a]), the *BroadCast Control Channel* (BCCH) must be supplied throughout each sector. In CDMA the amount of power in the pilot is typically between 15-25% of the total power transmitted on a carrier by a sector [Qua93]. Therefore, it is possible to improve the performance of CDMA systems significantly by spatially controlling downlink power.

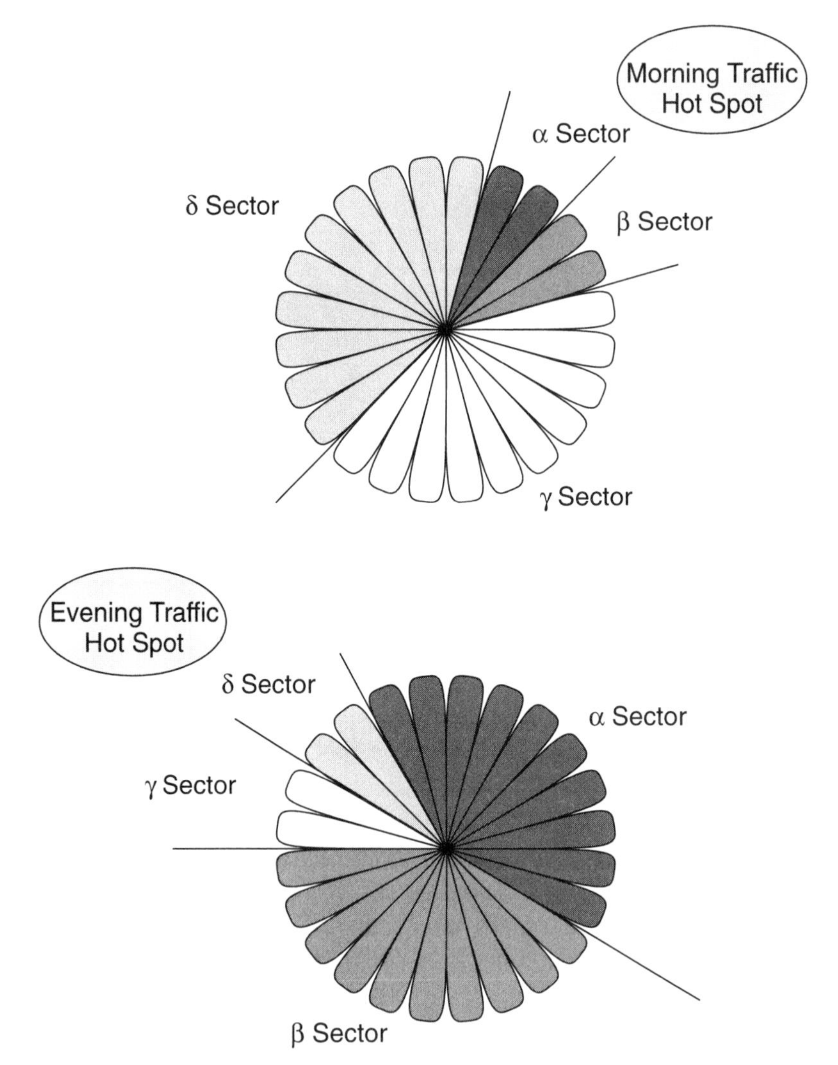

Figure 4–9 Dynamic re-sectoring using fixed beamforming networks. Hot spots represent areas of high tele-density, which frequently occur due to morning and evening commuter traffic patterns. (Reproduced with permission from *America's Network*, Vol. 102, Number 4, February 1998. Copyright by Advanstar Communications Inc. Advanstar Communications Inc. retains all rights to this article.)

Figure 4–10 shows how a CDMA base station transmitter uses a broad beam to provide universal pilot, sync, and paging channel coverage throughout a sector. Individual traffic channels are focused where they are required using downlink beamforming techniques as described in Section 3.11. The CDMA downlink beamforming system is illustrated in Figure 4–11, where power is allocated to each channel according to guidelines in Section 2.3.

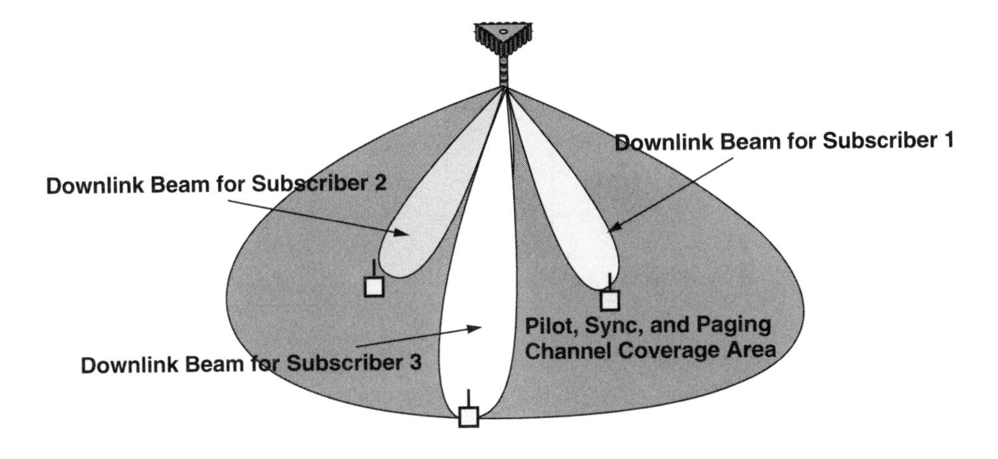

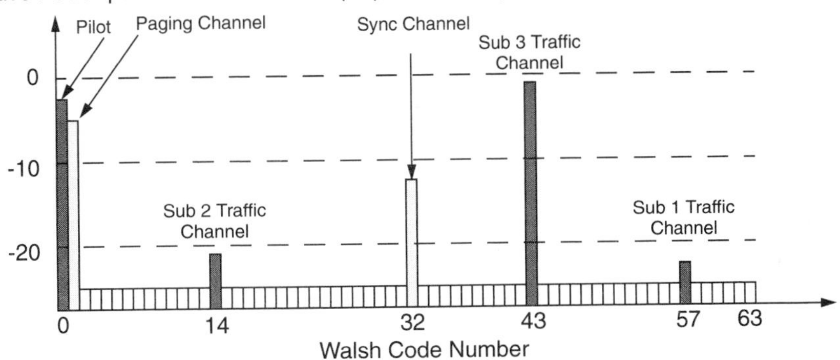

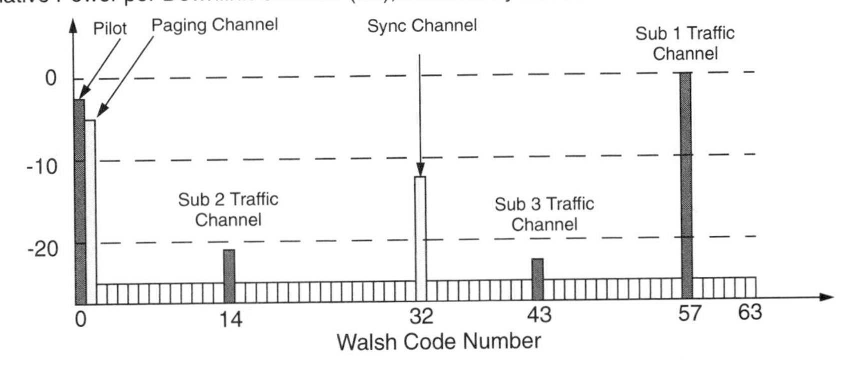

Figure 4–10 Downlink beamforming for IS-95 CDMA.

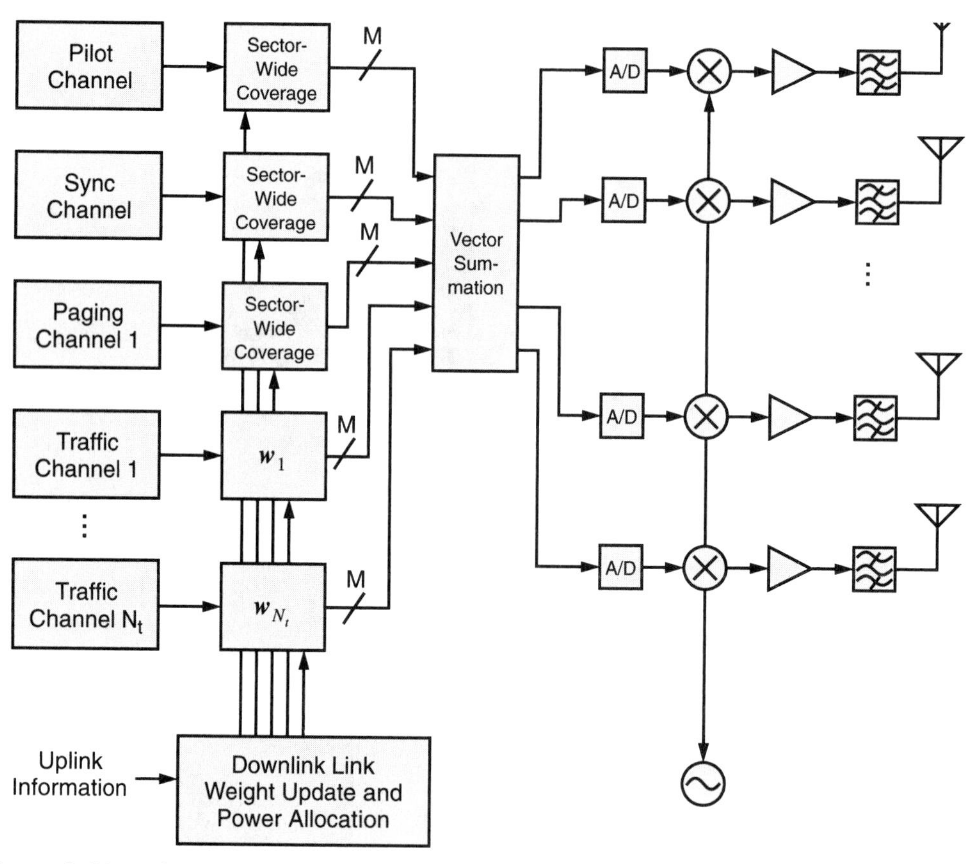

Figure 4–11 A CDMA downlink spatial power control system.

The key factor to note is that downlink beamforming is not used to separate signals with the cell. Rather, downlink beamforming is used to significantly reduce the overall transmitted power of the base station, since it focuses only sufficient power needed to meet FER requirements for a subscriber receiver, in the direction of that receiver.

Note that care must be taken whenever downlink beamforming is attempted. First, it is important that synthesized downlink beams are free of sidelobes and backlobes. Second, any change in the phase and amplitude response of an amplifier, filter, or antenna in Figure 4–11 will change the shape of any synthesized beam. This results in unwanted sidelobes that contribute interference to adjacent cells. Phase errors can also distort the beam to the point that the main lobe is no longer focused in the desired direction. Therefore, careful calibration must be performed periodically as discussed in Section 3.12.

The use of downlink beamforming in CDMA poses a potential problem. Recall from Chapter 2 that downlink traffic signals in IS-95 are phase-modulated relative to a pilot signal. In Figure 4–9 we note that if the traffic signal is transmitted using a narrowbeam, while the pilot

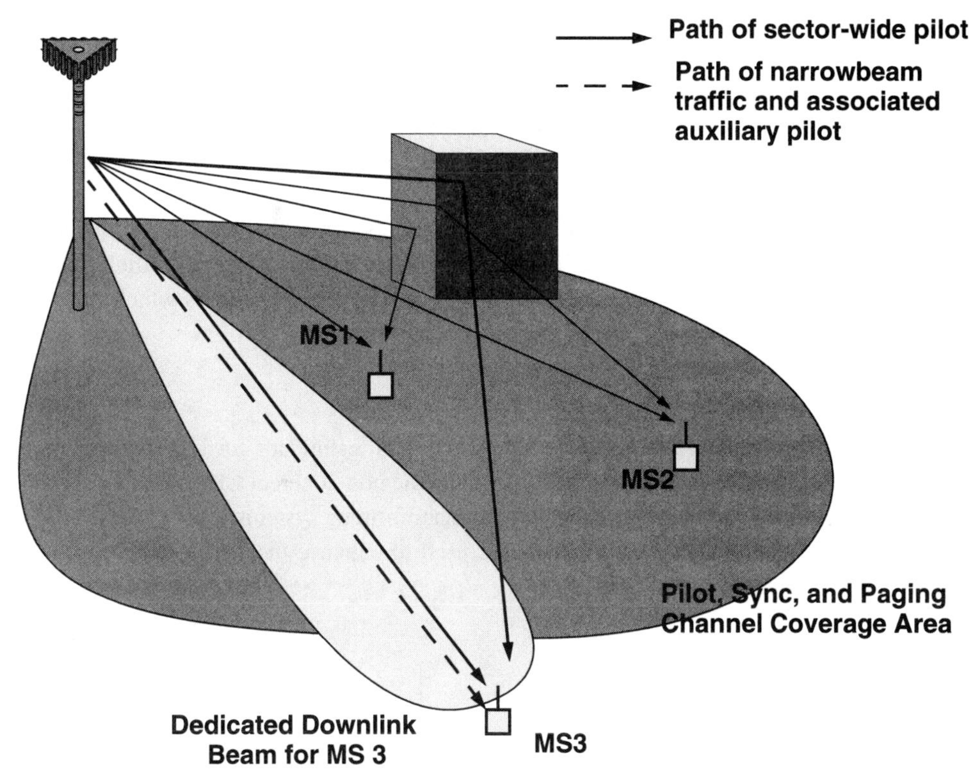

Figure 4–12 Multipath on the downlink can distort the phase relationship between a sector-wide pilot and a traffic channel transmitted in a narrowbeam. An auxiliary pilot, transmitted in the narrowbeam, can be used to assist in downlink demodulation, and help identify candidate Rake paths.

signal is transmitted using a broad beam that covers a sector, then the two signals do not encounter the same channel. As illustrated in Figure 4–12, the pilot signal, transmitted using a broad beam may encounter more scatterers than the signal transmitted using the narrow beam. There are also likely to be phase differences introduced between the two transmission antenna patterns. Both of these factors lead to distortion of the relationship between the phase of the narrowbeam traffic carrier and the phase of the sector-wide pilot signal.

Another consideration arises due to the Rake receiver used by the mobile station. The mobile station's Rake receiver may detect pilot energy from a multipath component; however, due to downlink beamforming, that multipath component may not contain useful traffic signal power for the mobile.

A way to address these problems, which is incorporated in the cdma2000 proposal, is the use of *auxiliary pilots*. Auxiliary pilots can be allocated in a number of ways. A fixed spot beam, that covers a high traffic area, may use an auxiliary pilot to serve many subscribers within that fixed beam. Several spot beams may share a paging channel and sync channel. For dynamic adaptive downlink beamforming, auxiliary pilots can either be assigned to clusters of subscribers falling within a single narrow beam, or an auxiliary pilot can be assigned to a single subscriber [Tie98][TIA98].

The auxiliary pilot passes through the same channel as the traffic channel, so coherent demodulation is not degraded. Furthermore, the mobile station can search for associated auxiliary pilots with its Rake receiver. Any auxiliary pilots that it finds associated with its traffic channel can be used for path diversity.

A potential disadvantage to the use of auxiliary pilots is that they do not exploit the efficiency that is inherent in using a single common sector-wide pilot for all downlink channels. This can be mitigated by assigning auxiliary pilots to clusters of subscribers which fall within the same beam.

4.6 Summary

In this chapter, we have discussed architectures and techniques for implementing smart antennas in CDMA systems. Key structures, including the non-coherent CDMA array combiner, the spatial processing Rake receiver, and downlink beamforming systems, were presented. In the next chapter, we will see how these systems are used to enhance the range and capacity of a CDMA wireless system.

CHAPTER 5

CDMA System Range and Capacity Improvement Using Spatial Filtering

In this chapter, analytical tools are developed to study the performance of CDMA cellular radio systems which use spatial filtering, as well as to determine practical limits on the number of simultaneous active users that can be supported in these systems. In later chapters, we discuss spatial signal processing techniques used to implement CDMA spatial filtering. However, before delving into details on how to implement smart antennas for CDMA, it is useful to examine the potential performance improvements expected. Because this is vital for determining system benefits and cost justification for smart antennas, we take several approaches to analyze these systems. First, we examine a single cell to develop simple rules and general design guidelines for uplink capacity improvement using CDMA technology. Next, we consider the impact of out-of-cell interference on CDMA performance and discuss the impact of spatial filtering on these systems. We then look at fixed wireless systems to discuss the potential for smart antennas at the subscriber end of the link. We conclude by formulating an alternative approach to CDMA capacity improvement calculations.

In this chapter, we present computer simulations and analytical results which demonstrate that spatial filtering at the base station significantly improves the reverse channel performance of multi-cell mobile radio systems, and we derive analytical techniques for characterizing mobile radio systems which employ frequency reuse. Finally, we also discuss the effects of using spatial filtering at the portable unit.

5.1 Range Extension In CDMA

An important benefit of smart antennas is *range extension*. Range extension allows the mobile to operate farther from the base station without increasing the uplink power transmitted by the mobile unit or the downlink power required from the base station transmitter.

First, consider the reverse link case of a single plane wave from a subscriber transmitter incident on the array. As described in Section 3.3, the data vector is equal to

$$\boldsymbol{u}(t) = s(t)\boldsymbol{a}(\theta, \phi) + \boldsymbol{n}(t) \tag{5.1}$$

The vector $\boldsymbol{n}(t)$ contains the noise contributed at each antenna element. Each element of $\boldsymbol{n}(t)$ is a complex Gaussian random variable with variance σ_n^2.

We can use (3.32) to find an optimal solution for the weight vector to extract $s(t)$.

$$\begin{aligned} \boldsymbol{w} &= \boldsymbol{R}^{-1}\boldsymbol{p} \\ &= (E[|s(t)|^2]\boldsymbol{a}(\theta, \phi)\boldsymbol{a}^H(\theta, \phi) + \sigma_n^2\boldsymbol{I})^{-1}E[|s(t)|^2]\boldsymbol{a}(\theta, \phi) \end{aligned} \tag{5.2}$$

With some rearranging, we have

$$\boldsymbol{w} = \frac{E[|s(t)|^2]}{\sigma_n^2}\boldsymbol{a}(\theta, \phi)(1 - \boldsymbol{a}^H(\theta, \phi)\boldsymbol{w}) \tag{5.3}$$

so $\boldsymbol{w} \propto \boldsymbol{a}(\theta, \phi)$, where the proportionality constant is unimportant. For simplicity, we set

$$\boldsymbol{w} = \boldsymbol{a}(\theta, \phi) \tag{5.4}$$

Then the output of the antenna combiner, illustrated in Figure 3–3, is

$$\begin{aligned} z(t) &= \boldsymbol{w}^H\boldsymbol{u}(t) \\ &= s(t)\boldsymbol{a}^H(\theta, \phi)\boldsymbol{a}(\theta, \phi) + \boldsymbol{a}^H(\theta, \phi)\boldsymbol{n}(t) \\ &= Ms(t) + \boldsymbol{a}^H(\theta, \phi)\boldsymbol{n}(t) \end{aligned} \tag{5.5}$$

The power of the desired signal component of $z(t)$ is

$$P_z = M^2E[|s(t)|^2] \tag{5.6}$$

The power of the noise component of $z(t)$ is

$$\begin{aligned} N_z &= E[\boldsymbol{a}^H(\theta, \phi)\boldsymbol{n}(t)\boldsymbol{n}^H(t)\boldsymbol{a}(\theta, \phi)] \\ &= \boldsymbol{a}^H(\theta, \phi)E[\boldsymbol{n}(t)\boldsymbol{n}^H(t)]\boldsymbol{a}(\theta, \phi) \\ &= \boldsymbol{a}^H(\theta, \phi)\sigma_n^2\boldsymbol{I}\boldsymbol{a}(\theta, \phi) \\ &= \sigma_n^2\boldsymbol{a}^H(\theta, \phi)\boldsymbol{a}(\theta, \phi) \\ &= M\sigma_n^2 \end{aligned} \tag{5.7}$$

Therefore, the signal to noise ratio of $z(t)$ is

$$\gamma_z = M\frac{E[|s(t)|^2]}{\sigma_n^2} \tag{5.8}$$

If, instead of using the array of M elements, only the signal at a single array element, $u_o(t)$, is used, then

$$u_o(t) = s(t) + n(t) \tag{5.9}$$

The signal-to-noise ratio of $u_o(t)$ is

$$\gamma_u = \frac{E[|s(t)|^2]}{\sigma_n^2} \tag{5.10}$$

Comparing (5.8) and (5.10), we see that an M-element array can achieve an SNR improvement of

$$G = 10\log(M)\ \text{dB} \tag{5.11}$$

in Additive White Gaussian Noise (AWGN) with no interference or multipath. This is the signal-to-noise ratio improvement provided by the array relative to the signal-to-noise ratio at a single element.

Next, consider the CDMA reverse link budget in Table 5–1, which shows both the smart antenna uplink and the uplink using a conventional base station antenna. For this example, we assume that the standard deviation, σ, of the shadowing component, χ_σ, in (1.58) is 8dB. Using (1.63), with a single base station receiver, a shadowing margin of 1.28σ, or 10.2 dB, is received to achieve a 10% probability of outage due to shadowing. As shown in Section 2.2.1, using two-way soft handoff, the required link margin is reduced to 0.77σ, or 6.2 dB, while maintaining the same outage probability.

We use a loading factor χ to account for multiple access interference from both in-cell and out-of-cell users [Vit95]. This will be discussed at greater length in Section 5.5. As shown in Table 5–1, the array is able to provide a 60% range improvement with no increase in transmitter power, or per-channel noise performance at the receiver, because $M = 8$ elements provide 9 dB more link margin, compared with conventional sectoring.

Table 5–1 Example of the uplink range extension achieved in a CDMA system using an 8 element array.

	Conventional Sectoring	Smart Antenna System
Portable Tx Power (Class II):........(a).	23.0 dBm	23.0 dBm
Portable Tx Antenna Gain:........(b).	0.0 dBi	0.0 dBi
EIRP:........(c=a+b)	23.0 dBm	23.0 dBm
BS. Antenna Element Gain (120° sector):........(d)	18.0 dBi	18.0 dBi
Ideal Spatial Processing Gain for M=8:........(e)		9.0 dB
Total Receiver Antenna Gain:........(f=d+e)	18.0 dBi	27.0 dBi
Required CINR at correlator input:........(g)	-12.3 dB	-12.3 dB
Receiver noise floor at NF=5 dB:........(h)	-108.1 dBm	-108.1 dBm
Total noise and interference at $\chi = 0.9$ ....(i)	-98.1 dBm	-98.1 dBm
Shadowing margin (10% outage probability, for two-way Soft Handoff, σ=8 dB)........(j)	6.2 dB	6.2 dB
Req'd Median Received Power:........(k=g+i+j)	-104.2 dBm	-104.2 dBm
Tolerable Median Path Loss:........(l=c+f-k)	145.2 dB	154.2 dB
Range, using a d^n model with n=2 up to 0.5 km, n=4.5 for greater distances, with f_c=1920 MHz:........	7.7 km	12.2 km

As a general rule-of-thumb, for a constant path loss exponent of $n \geq 2$, the range of a cell using smart antennas, R_s, is greater than the range using conventional antennas, R_c, by

$$\frac{R_s}{R_c} = M^{1/n} \tag{5.12}$$

and the ratio of the area of a cell covered with smart antennas, A_s, to the area of a cell covered using conventional antennas, A_c, is

$$\frac{A_s}{A_c} = M^{2/n} \tag{5.13}$$

Therefore, if N_c conventional base stations are required to provide coverage to an area, then the same area can be served using $N_s = N_c M^{(-2/n)}$ base stations. For a path loss exponent of $n = 4.5$, using eight-element arrays, this means that the area can be covered using 60% fewer base stations than would be required using conventional antenna systems. This is a powerful economic argument for service providers to consider smart antennas.

Table 5–1 may be overly optimistic because the gain of each individual array element will most likely be less than the 18 dBi afforded by a large (1.5 m) sector antenna. It is often not feasible to mount eight large sector panel antennas on a cell face. If the gain of each array element is reduced to 15 dBi, then the range of the smart antenna system in Table 5–1 is reduced to 10.3 km. Despite this decreased range, 44% fewer base stations are required relative to conventional sectored antenna systems.

This example does not take into account the fact that the *tele-density*, or erlangs of traffic supported per square kilometer, will be lower using a sparse smart antenna-based layout compared with a dense deployment using conventional base stations. However, in early deployments, or in rural areas, tele-density may not be a concern. In the next several sections, we show how smart antennas can also be used to increase the number of simultaneous subscribers that can be supported in each cell. Combining these capacity improvement techniques with the range extension discussed in this section, and given the high cost of base station towers, and difficulty in accessing tower sites, fewer base stations with more channels per base station could provide comparable tele-density to conventional deployments. However, one should be careful not to overestimate the ability of adaptive arrays to fulfill simultaneous range extension and capacity improvement goals. The practical limitations of the adaptive antennas technology along with the air interface standard must be taken into account. As we will see in the next several sections, smart antennas can also significantly improve system capacity in dense deployments where high tele-density is required.

5.2 Single Cell Systems with Spatial Filtering at the IS-95 Base Station

Consider a single cell CDMA system in which K mobile units simultaneously transmit to a single base station. This is a fundamental step toward developing concepts that will allow the multi-cell analysis required to predict CDMA capacity. As shown in Section 1.3.1, for interference limited, asynchronous, BPSK-CDMA over an *Additive White Gaussian Noise* (AWGN) channel, operating with perfect power control and with omni-directional antennas used at the base station, the bit error rate (BER), P_b, on the reverse link is approximated by [Pur77]

$$P_b \approx Q\left(\sqrt{\frac{3N}{K-1}}\right) \tag{5.14}$$

where K is the number of users in a cell and N is the spreading factor given in (1.35). Equation (5.14) assumes that the signature sequences are random and that K is sufficiently large to allow the Gaussian Approximation (GA) described in [Pur77] to be applied. Appendix A provides a thorough treatment of the Gaussian Approximation, as well as extensions that support analysis when users are not power-controlled.

In IS-95-based CDMA systems, (5.14) is not directly applicable because the uplink bit error rate is a complex function of different error control coding methods, orthogonal 64-ary modulation, multiple spreading codes, power control, and other factors. In IS-95, as described in Chapter 2, signals use modified M-sequences for short code and long code spreading. Offset-QPSK is used for spreading with BPSK data, and Data Burst Randomization helps to further randomize interference.

In IS-95 based systems, a critical parameter to measure link performance is the Carrier-to-Interference-and-Noise-Ratio (CINR) available for each subscriber. The CINR for each subscriber is measured after despreading. On the reverse channel, the CINR for a particular subscriber is measured at the input to the Walsh Chip Matched Filter, at the base station, as shown in Figure 4–3.

We define the CINR after despreading, as the ratio of the desired signal to the sum of interference and noise.

$$CINR = \frac{P_0}{\frac{1}{N}\sum_{k=1}^{K-1} P_k + \sigma_n^2} \tag{5.15}$$

This is essentially the argument of the Q-function in (1.36).

In (5.15), P_0 is the power of the desired signal at the input to the despreader at the base station, and P_k is the power from every other user for $k = 1 \ldots K-1$. The *spreading factor* is given by N, which was defined in (1.35) as

$$N = \frac{\text{Chip Rate}}{\text{Information Symbol Rate}} \tag{5.16}$$

Equation (5.15) reflects the fact that spreading reduces the impact of multiple access interference, P_k. The noise variance, σ_n^2, as in Section 1.3.1, represents the noise contribution to the decision variable after despreading.

Multiplying the numerator and denominator of (5.15) by the bit duration, T_b, we have

$$CINR = \frac{P_0 T_b}{\frac{1}{N}\sum_{k=1}^{K-1} P_k T_b + \sigma_n^2 T_b} \qquad (5.17)$$

The term $P_0 T_0$ is the energy per bit for the desired subscriber signal. After despreading, the noise bandwidth is approximately $1/T_b$. If the thermal noise has a power spectral density of N_n, then we can write the CINR as

$$CINR = \frac{E_b}{\frac{1}{N}\sum_{k=1}^{K-1} P_k T_b + N_n} = \frac{E_b}{\frac{I_0 \cdot T_b}{N} + N_n} = \frac{E_b}{N_i + N_n} = \frac{E_b}{N_o} \qquad (5.18)$$

The term N_i represents the power spectral density of the total multiple access interference after despreading.

Rather than using the spreading factor, as defined in (5.16), to compute the CINR in complex systems such as IS-95, it is more appropriate to compute the CINR using the *processing gain*. The processing gain for IS-95 systems results from a combination of PN-spreading and convolutional coding. This is because, as with PN-sequence spreading, convolutional coding increases the number of channel symbols compared with the number of information bits, and provides protection from the effects of MAI.

For the IS-95 uplink, the chip rate is 1.2288 Mcps. For both Rate Set 1 and Rate Set 2, the maximum symbol rate out of the convolutional encoder is 28.8 ksps. Therefore, the spreading factor is

$$\text{Spreading Factor} = \frac{1.2288 \times 10^6 \text{ cps}}{28.8 \times 10^3 \text{ sps}} = 42.667 \qquad (5.19)$$

or 16.3 dB. A common practice in IS-95 systems is to include also the impact of convolutional encoding in the processing gain. In Rate Set 1, which uses a 1/3 rate convolutional encoder on the reverse link, the processing gain is

$$\text{Processing Gain} = \frac{1.2288 \times 10^6 \text{ cps}}{9.6 \times 10^3 \text{ sps}} = 128 \qquad (5.20)$$

or 21.1 dB. For Rate Set 2, using the 1/2 rate coder, the processing gain is $42.667 \times 2 = 85.333$ or 19.3 dB. These rates also correspond to the ratio between the chip rate and the basic information rates of 9600 bps and 14400 bps, respectively, for each Rate Set, as shown in Tables 2–9 and 2–10. Therefore, we will use the term, N, to refer to the processing gain in (5.18).

As discussed in Chapter 2, the power control scheme in IS-95 uses Outer Loop Power Control to adjust the Frame Error Rate (FER) to a target value, typically 1%. The CINR required to deliver this FER is a function of rate set, vehicle speed, multipath conditions, and soft/softer handoff state. As described in Chapter 2, when linking to a single base station, for mobile subscribers, an E_b/N_o of 7-9 dB is required. Using soft handoff, the value of E_b/N_o required at each receiving base station drops to 4-6 dB due to macro-diversity that reduces the impact of shadow fading [Ket96][Wal94].

It is also important to account for the fact that IS-95-based CDMA systems take advantage of voice inactivity as described in Section 2.3.1. Because the vocoder reduces its output rate when the speaker is silent, the subscriber unit does not transmit continuously, but is gated on and off with a duty cycle as low as 1/8 during silent periods. This is captured in the *voice activity factor*, ν. Typically, the voice activity factor reduces the average Multiple Access Interference level seen by the base station receiver by 50-60% (ν=0.4 to 0.5) relative to the case where all subscribers are transmitting continuously.

We can modify (5.18) to take into account the CINR improvement due to the voice activity factor:

$$CINR = \frac{E_b}{\frac{\nu}{N}\sum_{k=1}^{K-1} P_k T_b + N_n} = \frac{E_b}{\nu\frac{I_o \cdot T_b}{N} + N_n} = \frac{E_b}{N_o} \tag{5.21}$$

To illustrate how directional base station antennas can improve the reverse link in a single cell CDMA system, consider the case in which each portable unit has an omni-directional antenna and the base station tracks each user in the cell, using a directive beam. It is assumed that a beam pattern, $F(\theta, \varphi)$, is formed at the base station so that the pattern has a steerable maximum in the direction of the desired user. For this analysis, we will assume that $F(\theta, \phi)$ is a *keyhole* pattern, as shown in Figure 5–1. This pattern, which resembles a keyhole when its cross-section is viewed in the *x-y*-plane, is not physically realizable, but it is useful for analysis.

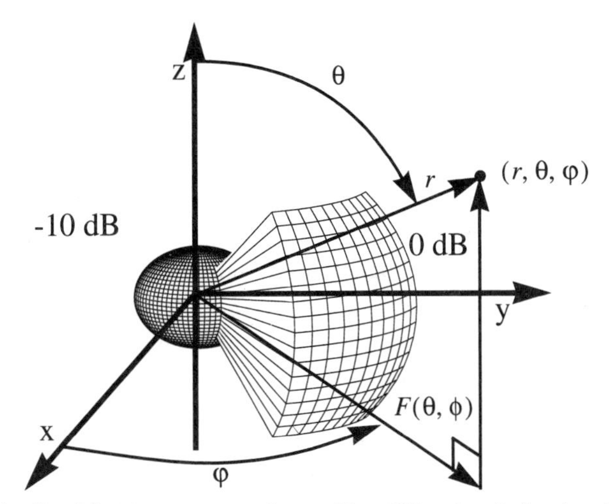

Figure 5–1 An idealized flat-top power pattern with a 60° azimuth (horizontal) beamwidth and a -10 dB sidelobe level. The pattern has a 90° elevation beamwidth.

We will assume that the pattern is separable in the θ (elevation) and ϕ (azimuth) dimensions, so that we can express the pattern as

$$F(\theta, \varphi) = F_e(\theta)F_a(\phi) \tag{5.22}$$

The elevation pattern, $F_e(\theta)$, is fixed with a maximum value on the horizon, at $\theta = \pi/2$. For simplicity, we will assume that

$$\max_{\phi}\{F_a(\phi)\} = 1 \qquad \max_{\theta}\{F_e(\theta)\} = 1 \tag{5.23}$$

The maximum gain of the overall pattern, given by (1.44), is

$$\begin{aligned} G_m &= \frac{4\pi}{\int_{\theta=0}^{\pi}\int_{\phi=0}^{2\pi} F(\theta, \phi)\sin\theta d\theta d\phi} \\ &= \frac{4\pi}{\int_{\theta=0}^{\pi}\int_{\phi=0}^{2\pi} F_e(\theta)F_a(\phi)\sin\theta d\theta d\phi} \\ &= \frac{4\pi}{\int_{\theta=0}^{\pi} F_e(\theta)\sin\theta d\theta \cdot \int_{\phi=0}^{2\pi} F_a(\phi)d\phi} \\ &= \left(\frac{2}{\int_{\theta=0}^{\pi} F_e(\theta)\sin\theta d\theta}\right)\left(\frac{2\pi}{\int_{\phi=0}^{2\pi} F_a(\phi)d\phi}\right) = G_eG_a \end{aligned} \tag{5.24}$$

When considered by itself, the maximum gain of the elevation pattern portion of the separable antenna pattern, $F_e(\theta)$, is

$$G_e = \frac{4\pi}{\int_{\theta=0}^{\pi}\int_{\phi=0}^{2\pi} F_e(\theta)\sin\theta d\theta d\phi} = \frac{2}{\int_{\theta=0}^{\pi} F_e(\theta)\sin\theta d\theta} \tag{5.25}$$

The azimuth pattern, $F_a(\varphi)$, can be steered through 360 degrees in the horizontal (φ) plane so that the desired user (user 0) is always in the main beam of the pattern. The maximum value of the horizontal antenna pattern occurs in the direction toward user 0.

The maximum gain of the horizontal portion of the pattern, $F_a(\phi)$, is

$$G_a = \frac{4\pi}{\int_{\theta=0}^{\pi}\int_{\phi=0}^{2\pi} F_a(\phi)\sin\theta d\theta d\phi} = \frac{2\pi}{\int_{\phi=0}^{2\pi} F_a(\phi)d\phi} \tag{5.26}$$

We assume that K users in the single cell CDMA system are uniformly distributed throughout a two-dimensional cell (in the horizontal plane, $\theta = \pi/2$). On the reverse link, the received power from the desired mobile is $P_{r;0}$. The powers of the signals incident at the base station antenna from the $K-1$ interfering users are given by $G_m F_\alpha(\phi_i) P_{r;i}$ where φ_i is the direction of the ith user in the horizontal plane, measured from the x-axis. Then the average total interference power, I_o, seen by the base station receiver for user 0, measured at the output of the receiving array, is given by

$$I_0 = G_m E\left\{\sum_{i=1}^{K-1} F_a(\phi_i) P_{r;i}\right\} \tag{5.27}$$

If perfect power control is applied so that the power incident at the base station antenna from each user is the same, then $P_{r;i} = P_c$ for each of the K users, and the average interference power seen by user 0 is given by

$$I_0 = G_m P_c E\left\{\sum_{i=1}^{K-1} F_a(\phi_i)\right\} \tag{5.28}$$

Assuming that users are independently and identically distributed throughout a circular region with radius R, the average total interference power received at the central base station may be expressed as

$$I_0 = G_m P_c (K-1) \int_0^R \int_0^{2\pi} f_{r,\phi}(r, \varphi) F_a(\varphi) d\varphi dr \tag{5.29}$$

where $f_{r,\phi}(r, \varphi)$ is the probability density function describing the geographic distribution of users throughout the cell. Assume users are uniformly distributed in the cell, then we have

$$f_{r,\phi}(r, \varphi) = \frac{r}{\pi R^2} \tag{5.30}$$

and

$$I_0 = G_m P_c \frac{(K-1)}{2\pi} \int_0^{2\pi} F_a(\varphi) d\varphi \tag{5.31}$$

Substituting (5.26) into (5.31) we have

$$I_0 = \frac{G_m P_c (K-1)}{G_a} \tag{5.32}$$

Assuming that the system is set to operate with $P_c \gg \sigma_n^2$, so that the system is not thermal noise-limited, then the mean CINR is found from (5.21)

$$CINR = \frac{G_m P_{r;0}}{\frac{\nu G_m P_c (K-1)}{NG_a}} = \frac{NG_a P_{r;0}}{\nu P_c (K-1)} = \frac{NG_a}{\nu(K-1)} \tag{5.33}$$

In other words, the CINR is proportional to the gain associated with the horizontal portion of the antenna pattern, $F_a(\phi)$. Note that (5.33) holds for a single cell CDMA system, with perfect power control applied to all subscribers, when the base station antenna pattern may be steered toward the desired subscriber.

With the canonical single-cell CDMA system with perfect power control as a backdrop, we now explore capacity limitations of multi-cell CDMA systems. Section 5.3 describes analytical techniques used to determine bit error rates in multi-cell CDMA systems employing spatial filtering. These techniques are applied to compare the performance of omni, sectorized, and smart antenna systems at the base stations. In Section 5.4, the effects of spatial filtering at the subscriber unit are examined, using several different base station configurations.

5.3 Reverse Channel Performance of Multi-cell Systems with Spatial Filtering at the Base Station

In this section, we investigate how smart antennas are used on the reverse link to improve CDMA system capacity in multi-cell systems. Note that equation (5.33) is valid only when a single cell is considered. To study the effects of spatial filtering on the reverse link when CDMA users are simultaneously active in *several* adjacent cells, we must first define the geometry of the cell region. For simplicity, we consider the geometry proposed in [Rap92][Rap96a] with a single layer of surrounding cells, as illustrated in Figures 5–2 and 5–3.

To develop a framework for studying CDMA systems, first consider the simple case where omni-directional antennas are used at each base station. Each antenna pattern is assumed to be of a separable form, as shown in (5.22), so that it can be expressed as $F(\theta, \phi) = F_a(\phi)F_e(\theta)$. For an omni-directional antenna, $F_a(\phi) = 1$ for all ϕ. Using (5.26), the azimuthal gain is $G_a = 1$, so that the overall antenna gain is $G_m = G_e$.

Let $d_{i,j}$ represent the distance from the ith user to base j as illustrated in Figure 5–3. Let $d_{i,0}$ represent the distance from the ith user to base station 0, the center base station. If Cell 0 contains the desired user, user 0, then the ith user shown in Figure 5–3 is an in-cell interferer if it is in Cell 0, and it is an out-of-cell interferer if it is outside of Cell 0. If user i is outside of Cell 0, then that user is power controlled to a base station in another cell j, and it represents a source of uplink interference that is not under control of the center cell.

Assume that the subscriber unit uses an isotropic antenna with a gain of $G_{\text{sub}} = 1$. The path loss between user *i* and base *j* is given by the log-distance path loss model described in Section 1.7.1, so that the power received at base station *j*, from the transmitter of user *i*, $P_{r;i,j}$, is given by

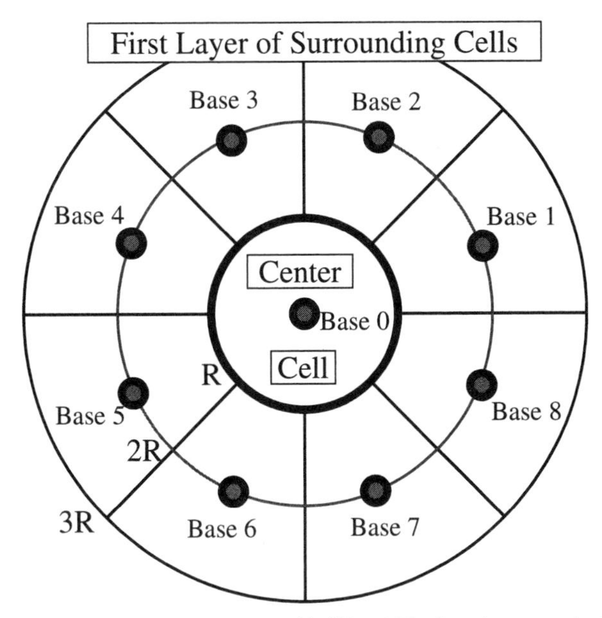

Figure 5–2 The wedge cell geometry proposed in [Rap92], showing a central cell surrounded by eight wedge-shaped cells that have the same area as the central cell. (©1992 IEEE.)

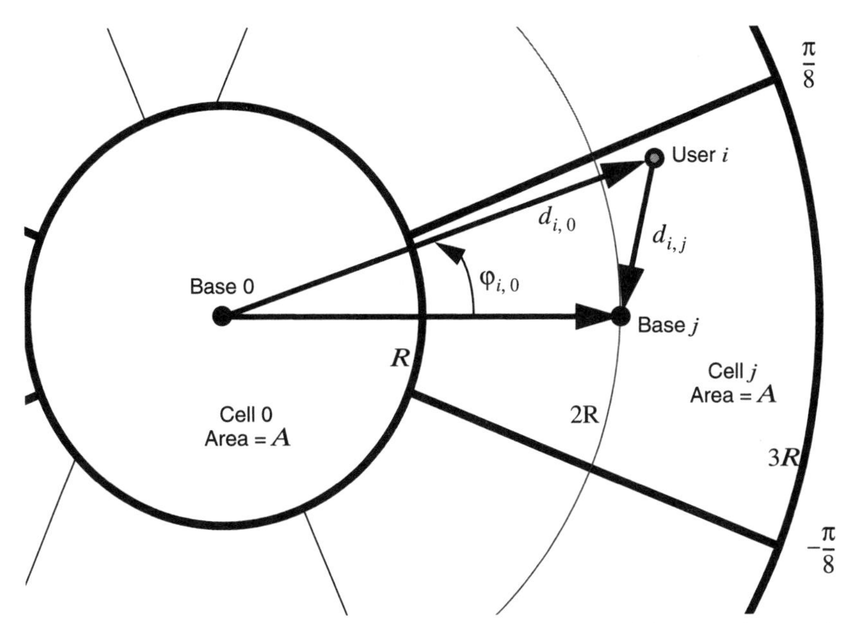

Figure 5–3 Geometry for determining $d_{i,j}$ as a function of $d_{i,0}$, the distance between out-of-cell user i and the central base station. The angle of user i relative to the line between the central base station and base station j is $\varphi_{i,0}$. (©1992 IEEE.)

$$P_{r;i,j} = G_{\text{sub}} G_m P_{t;i} \left(\frac{\lambda}{4\pi d_{ref}}\right)^2 \left(\frac{d_{ref}}{d_{i,j}}\right)^n L_{i,j} \tag{5.34}$$

where n is the path loss exponent, and d_{ref} is a close-in reference distance [Rap92]. The term $L_{i,j}$ represents a log-normally distributed shadowing loss on the link between subscriber transmitter i and base station receiver j.

Assume that base station j applies perfect power control to all users in cell j, so that power $P_{c;j}$ is received at base j, for each of the users in cell j, then the required power transmitted by user i, $P_{t;i}$, is

$$P_{t;i} = P_{c;j} \left(\frac{4\pi d_{ref}}{\lambda}\right)^2 \left(\frac{d_{i,j}}{d_{ref}}\right)^n \frac{1}{G_m L_{i,j}} \tag{5.35}$$

The power received at base station 0 from user i, $P_{r;i,0}$ is given by

$$P_{r;i,0} = L_{i,0} G_m P_{t;i} \left(\frac{\lambda}{4\pi d_{ref}}\right)^2 \left(\frac{d_{ref}}{d_{i,0}}\right)^n \tag{5.36}$$

Substituting (5.35) into (5.36), the reverse link power received at base 0 from user i, in adjacent cell j, is

$$P_{r;i,0} = \frac{L_{i,0}}{L_{i,j}} P_{c;j} \left(\frac{d_{i,j}}{d_{i,0}}\right)^n \tag{5.37}$$

To analyze (5.37), we consider the geometry shown in Figure 5–3. From the law of cosines,

$$d_{i,j}^2 = (2R)^2 + (d_{i,0})^2 - 2(2R d_{i,0}) \cos \varphi_{i,0} \tag{5.38}$$

Substituting (5.39) into (5.38), the power received at base 0 from user i is

$$P_{r;i,0} = \frac{L_{i,0}}{L_{i,j}} P_{c;j} \left(1 + \left(\frac{2R}{d_{i,0}}\right)^2 - \frac{4R}{d_{i,o}} \cos \varphi_{i,0}\right)^{n/2} \tag{5.39}$$

To determine the average out-of-cell interference power incident on the central base station, we assume that users are uniformly distributed in a typical adjacent cell from r=R to r=3R and from $\varphi = -\pi/8$ to $\pi/8$. Thus, we use a modified geometry from [Rap92] where 8 equal area cells surround the center cell. The probability density function (pdf) for the spatial distribution of users in a single adjacent cell is given by

$$f_{r,\phi}(r, \varphi) = \frac{r}{\pi R^2} \qquad R < r < 3R; -\pi/8 < \varphi < \pi/8 \tag{5.40}$$

Note that [Rap92] also considers the case in which users are not uniformly distributed throughout the cell. By allowing more users in an adjacent cell to concentrate closer to the central cell (in the smaller portion of the wedge), system performance is degraded [Rap92]. This is because out-of-cell users must transmit more power to reach their own base station, while generating more interference to the central base station due to their proximity. In this way, a worst-case analysis may be performed. For the present analysis, we consider only the case in which all users are uniformly distributed throughout the available coverage region.

Let ξ represent the expected value of the interference power from a single user in one of the adjacent cells when omni-directional base station antennas are used.

$$\xi = E\left[\frac{L_{i,0}}{L_{i,j}}\right] \cdot P_{c;j} \int_{R}^{3R} \int_{-\pi/8}^{\pi/8} f_{r,\phi}(r,\varphi)\left(\left(1+\left(\frac{2R}{r}\right)^2 - \frac{4R}{r}\cos\varphi\right)\right)^{n/2} dr d\varphi \tag{5.41}$$

The terms $L_{i,0}$ and $L_{i,j}$ are independent and log-normally distributed, with $E[L_{i,0}] = E[L_{i,j}]$, so $E[L_{i,0}/L_{i,j}] = 1$. If it is assumed that the center base station and all 8 surrounding base stations control power such that $P_{c;j} = P_c$, then, given a path loss exponent of n, we can express the expected value of central cell interference power from a single adjacent cell user as

$$\xi = \beta P_c \tag{5.42}$$

where

$$\beta = \int_{R}^{3R} \int_{-\pi/8}^{\pi/8} f_{r,\phi}(r,\varphi)\left(\left(1+\left(\frac{2R}{r}\right)^2 - \frac{4R}{r}\cos\varphi\right)\right)^{n/2} dr d\varphi \tag{5.43}$$

Table 5–2 lists values of β for several values of n.

Table 5–2 Values of β as a function of the path loss exponent, n as determined by (5.43).

n	β
2	0.1496
3	0.0824
4	0.0551

When omni-directional antennas are used at both the base station and the portable unit, β is related to the reuse factor, f, where $f \le 1$. The reuse factor is defined in [Rap92] as the ratio of received power from all in-cell interferers to the total interference from all users, with the assumption that users are in perfect power control to their closest base station.

$$f = \frac{\text{Multiple Access Interference from In-Cell Users}}{\text{Total Multiple Access Interference}} \tag{5.44}$$

As described in Chapter 8 of [Rap96a], this reuse factor is equivalent to the reciprocal of the frequency reuse ratio used in FDMA and TDMA systems.

For a single layer of adjacent cells, as shown in Figure 5–2,

$$f = \frac{N_0}{N_0 + N_{a1}M_1} \tag{5.45}$$

The term N_0 in (5.45) is the total Multiple Access Interference, seen at the central base station on the reverse link due only to subscribers in the central cell. N_{a1} is the total interference affecting the desired central cell user from all users in a single adjacent cell. M_1 is the number of cells which are adjacent to the central cell, which is always 8 for the geometry considered in this chapter.

When power control is performed by each base station on its own users, so that the power received from each mobile unit in the base station controlling that unit is P_c, then (5.45) may be expressed as

$$f = \frac{(K-1)P_c}{(K-1)P_c + 8K\beta P_c} \approx \frac{1}{1+8\beta} \qquad \text{for} \quad K \gg 1 \tag{5.46}$$

where we assume that there are K users in each of the nine cells. For $n = 4$, from Table 5–2, $\beta = 0.055$, and, from (5.46), $f = 0.69$, meaning that 31% of the interference power received at the central base station is due to users in adjacent cells. The reuse factor is summarized in Table 5–3 for different path loss exponents and different cell geometries. The table illustrates the impact of out-of-cell interference in CDMA systems.

Table 5–3 Reuse factor as a function of path loss exponent. The reuse factor is the fraction of Multiple Access Interference, seen at the base station, that is due to in-cell subscribers.

n	Reuse Factor, computed using concentric circle cell geometry with a single tier of adjacent cells. (5.46)	Reuse Factor, computed using concentric circle cell geometry with three tiers of adjacent cells [Rap92]	Reuse Factor, computed using 36 hexagonal cells [Lib96c]
2	0.46	0.19	0.17
3	0.60	0.37	0.34
4	0.69	0.49	0.48

When omni-directional antennas are used at both the base station and the portable unit, the total interference seen on the reverse link by the central base station is the sum of the interference from users within the central cell, $(K-1)P_c$, and from users in adjacent cells, $8K\beta P_c$.

$$I_o = \underbrace{(K-1)P_c}_{\text{In-Cell Interference}} + \underbrace{8K\beta P_c}_{\text{Out of Cell Interference}} \tag{5.47}$$

Now consider what happens as directional antennas are added at each base station. A horizontal beam pattern, $F_{a;i,j}(\phi)$, is applied for each subscriber and steered so that the desired subscriber is centered in the main beam of the pattern. As in the previous section, we assume the keyhole pattern of Figure 5–1 with the same beamwidth for each subscriber, so that the overall base station antenna gain is equal to $G_m = G_e G_a$ for all links. We assume that the same value of elevation gain G_e is used from the previous example.

Now assume that all adjacent cell subscribers are power controlled to the same level $P_{c,j}$ at base station j; however, the output power for each subscriber may be lower than the case using omni-directional antennas, due to increased gain, G_m for each subscriber, relative to G_e.

Let us assume that for the ith user in the central cell, an antenna beam from the central base station with pattern $F_{i,0}(\theta, \varphi) = F_e(\theta)F_{a;i,0}(\phi)$ may be formed with maximum gain in the direction of user i. The pattern for the desired user (user 0) is $F_{0,0}(\theta, \varphi) = F_e(\theta)F_{a;0,0}(\phi)$. For simplicity, we write this pattern for user 0 as $F(\theta, \varphi) = F_e(\theta)F_a(\phi)$.

The average interference power contributed by a single user in the central cell is thus given by

$$E[P_{r;i,0}|0 < r < R] = P_c \int_0^R \int_0^{2\pi} \frac{r}{\pi R^2} F_a(\phi) dr d\varphi = \frac{P_c}{G_a} \tag{5.48}$$

The average interference power at the array output of the antenna array at the base station due to a single user in an adjacent cell is given by

$$E[P_{r;i,0}|R < r < 3R] = \tag{5.49}$$

$$\frac{1}{8}\sum_{p=0}^{7} \int_R^{3R} \int_{-\pi/8}^{\pi/8} F_a\left(\varphi + \frac{p\pi}{4}\right) \frac{r}{\pi R^2} P_c \left(1 + \left(\frac{2R}{r}\right)^2 - \frac{4R}{r}\cos(\varphi)\right)^{n/2} dr d\varphi$$

If $F_a(\varphi)$ is piece-wise constant over the region $(2p-1)(\pi/8) < \varphi < (2p+1)(\pi/8)$ for $p = 0 \ldots 7$, then the antenna pattern may be expressed as

$$F_a(\varphi) = \sum_{p=0}^{7} F_{a,p} U\left(\varphi - \frac{p\pi}{4}\right) \tag{5.50}$$

where

$$U(\varphi) = \begin{cases} 1 & |\varphi| < \pi/8 \\ 0 & |\varphi| \geq \pi/8 \end{cases} \tag{5.51}$$

Substituting (5.50) into (5.49), we obtain

$$E[P_{r;i,0}|R < r < 3R] = \frac{P_c}{8}\sum_{p=0}^{7} F_{a,p} \int_R^{3R} \int_{-\pi/8}^{\pi/8} \frac{r}{\pi R^2}\left(1 + \left(\frac{2R}{r}\right)^2 - \frac{4R}{r}\cos(\varphi)\right)^{n/2} dr d\varphi \tag{5.52}$$

The gain, G_a of $F_a(\phi)$, as described by (5.50) is

$$G_a = \frac{8}{\sum_{p=0}^{7} F_{a,p}} \tag{5.53}$$

Therefore, (5.52) may be rewritten, using (5.53) and (5.43), as

$$E[P_{r;i,0}|R < r < 3R] = \frac{P_c\beta}{G_a} \tag{5.54}$$

It can be shown that (5.54) remains valid when the beam pattern, $G(\varphi)$, is rotated in the φ plane. Therefore, (5.54) is appropriate when $G(\varphi)$ is piece-wise constant over $(2p-1)(\pi/8) < \varphi - \varphi_d < (2p+1)(\pi/8)$ for any angle φ_d between $-\pi/8$ and $\pi/8$.

Using (5.54) with (5.48), the total interference power at the array output of the center base station receiver is

$$I_o = \frac{(K-1)P_c + 8KP_c\beta}{G_a} \tag{5.55}$$

which is $1/G_a$ times the interference received at an omni-directional base station, which was given by (5.47).

We can substitute (5.55) into (5.18) to obtain

$$CINR = \frac{NG_aP_c}{\nu(K-1)P_c + 8KP_c\beta} = \frac{NG_a}{\nu K(1+8\beta) - 1} \approx \frac{NG_a}{\nu K(1+8\beta)} = \frac{fNG_a}{\nu K} \tag{5.56}$$

Equation (5.56) shows how the azimuthal antenna pattern and the reuse factor combine to determine the CINR for a desired subscriber. The reuse factor is determined through the value of β, given in Table 5–2. It is assumed that perfect power control is applied as described in Section 5.2, with all base stations controlling reverse link received power to the same level, P_c.

To explore how (5.56) can be used to determine the capacity of a CDMA system, consider the three hypothetical base station antenna patterns illustrated in Figure 5–4. Descriptions of these patterns are summarized in Table 5–4. These antenna patterns are assumed to be directed so that maximum gain is in the direction of the desired mobile users. The first base station antenna pattern is an omni-directional pattern which models antennas used in traditional cellular systems. This configuration, shown in Figure 5–4(a) is used as a model for standard omni-directional systems without spatial filtering. This is Case A in Table 5–4.

The second configuration, given as Case B in Table 5–4 and illustrated in Figure 5–4(b), uses $120°$ sectorization at the base station. In this case, each base station uses three sectors, one covering the region from $30°$ to $150°$, the second covering the region from $150°$ to $270°$, and the third covering the region from $-90°$ to $30°$. The first sector is illustrated in Figure 5–4(b). This sector is active when the desired user is at an angle of $30°$ to $150°$. The azimuth gain, G_a, of this antenna is 4.8 dB.

Table 5–4 Summary of the spatial filters used in this analysis.

Name	G_a	Description
Case A: Omni-directional	0 dB	Pattern is illustrated in Figure 5–4(a).
Case B: Sectorized	4.8 dB	Perfect 120° sectorization with zero sidelobes, shown in Figure 5–4(b).
Case C: Flat-topped	9.3 dB	Uniform gain across the 30° main beam and uniform sidelobes which are 10 dB below the main beam, as shown in Figure 5–4(c).

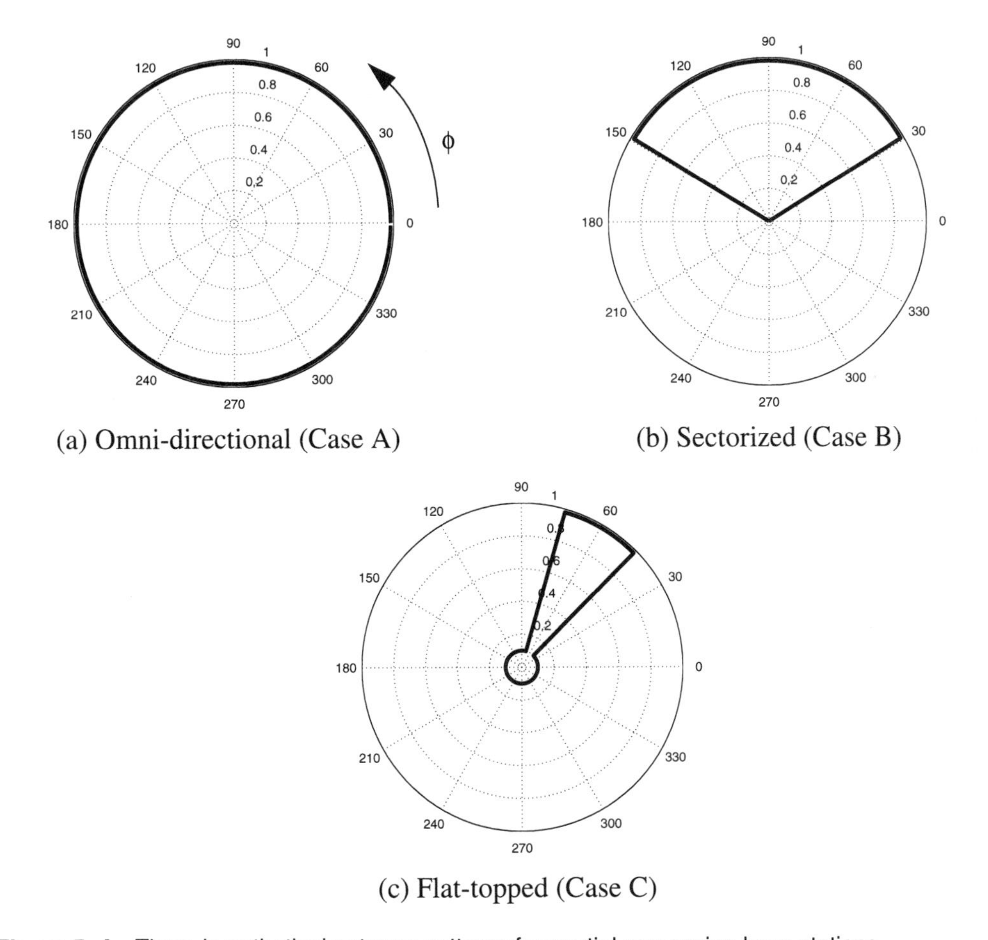

(a) Omni-directional (Case A)

(b) Sectorized (Case B)

(c) Flat-topped (Case C)

Figure 5–4 Three hypothetical antenna patterns for spatial processing base stations.

Table 5–5 Number of users supported in each cell, as limited by the reverse link for each spatial pattern. It is assumed that an E_b/N_0 of 9 dB is required, and a voice activity factor of 0.6 is used.

		Rate Set 1 (N=19.3 dB)			Rate Set 1 (N=21.1 dB)		
Name	G_a	**n=2**	**n=3**	**n=4**	**n=2**	**n=3**	**n=4**
Case A: Omni-directional	0 dB	12	16	18	8	10	12
Case B: Sectorized	4.8 dB	36	48	55	24	32	37
Case C: Flat-topped	9.3 dB	104	137	158	69	91	105

The third base station antenna pattern, shown in Figure 5–4(c), uses a flat-topped beam pattern similar to that shown in Figure 5–1. The main beam is 30 degrees wide with uniform gain in the main lobe. Side lobes are simulated by assuming a uniform side lobe gain which was 10 dB below the main beam gain. From (5.26), the azimuth gain of this beam is $G_a = 9.3$ dB. This is Case C in Table 5–4.

To evaluate the performance of each antenna pattern, assume that an E_b/N_0 of 9 dB is required. A voice activity factor of $\nu = 0.6$ is assumed. Then the number of users supported is found from (5.56).

$$K \le \frac{fNG_a}{\nu(E_b/N_0)} \tag{5.57}$$

For example, using Rate Set 1, the processing gain, N, is 128. For a path loss exponent of $n = 3$, from Table 5–3, the reuse factor is $f = 0.60$. For the case of an omni-directional base station antenna, the number of users that can be active at any one time is

$$K \le \frac{(0.60)(128)(10^{0/10})}{0.6 \cdot (10^{9/10})} = 16.2 \tag{5.58}$$

Therefore, a system using omni-directional antennas on the reverse link can support 16 simultaneous subscribers per cell. Using 120° sectorization,

$$K \le \frac{(0.603)(128)(10^{4.8/10})}{0.6 \cdot (10^{9/10})} = 48.9 \tag{5.59}$$

Therefore, 48 users can be supported in each cell. This is the reason that most CDMA systems are deployed using sectorization. By using three sectors at each base station, as shown by (5.58) and (5.59), CDMA base stations support three times the uplink traffic in each cell, compared with omni-directional systems. Since each sector covers one third of the cell, each 120° sector, or cell face, can support up to 48/3 or 16 subscribers.

When sectors are used on the uplink, they are also used on the downlink to provide three independent sets of forward link channels, one set on each cell face. In the IS-95 downlink, depending on the configuration, each sector can support a maximum of 61 downlink traffic channels. Thus, each sector transmits a different CDMA carrier (all on the same frequency, but with different pilot offsets). As described in Section 2.2, each carrier contains a complete set of forward link channels, including its own pilot, sync, and paging channels, which are spatially separated using the sector antennas. Because of soft and softer handoff, the actual number of subscribers that can be supported on each forward link carrier is considerably less than 61.

Finally, considering the steerable keyhole antenna pattern of Figure 5–4 (c), we have

$$K \leq \frac{(0.603)(128)(10^{9.3/10})}{0.6 \cdot (10^{9/10})} = 137.8 \tag{5.60}$$

Thus, up to 137 subscribers can be supported on the reverse link while meeting the required E_b/N_0 target.

To balance these capacity improvements on the forward link, multiple forward link sector-wide transmitters can be used with different PN offsets, or a scheme similar to that shown in Figure 4–9 can be employed. Together, the reverse channel result described in (5.60) and downlink spatial processing technologies can be combined to allow the current and future CDMA technologies to support tremendous capacity in a minimal amount of spectrum.

5.4 Reverse Channel Spatial Filtering at the WLL Subscriber Unit

In this section, we examine how the reverse channel is affected by using spatial filtering at the subscriber unit. In Section 1.5, we showed that the maximum gain that can be achieved with an antenna is limited by its size or aperture. As portable devices shrink, making them more convenient, the possibilities for simple spatial processing at the subscriber unit, beyond the use of two or three antenna elements, is diminished. However, for mobile systems, it is certainly possible to use arrays at the subscriber unit. Even more importantly, in WLL systems, described in Section 1.2.3, the benefits of spatial processing at the subscriber terminal are well-justified. In WLL systems, the goal is often to provide wireline quality voice and data services to residential and business customers, where capacity and link quality are at a premium. Many of these systems use antennas mounted on houses and buildings, where there is plenty of space for antenna arrays and other directional antenna systems. WLL subscriber terminals also typically draw power from a wired supply rather than batteries, so there is power to run digital signal processors required for spatial processing. Most of the CDMA WLL technologies discussed in Chapter 1 use directional antennas for precisely this reason. In this section, we show how subscriber-based spatial processing for the uplink can substantially improve system capacity.

To demonstrate the impact of subscriber-based uplink beamforming, a flat-topped beam shape, as illustrated in Figure 5–1, is used to model an adaptive antenna at the subscriber terminal. For analysis and simulation, let us assume that the subscriber terminal can achieve a beamwidth of 60 degrees with a side lobe level that was 6 dB down from the main beam. This corresponds to an antenna with a gain of 4.3 dB, which, using (5.11), can be achieved using only a few antenna elements. The pattern is similar to that shown in Figure 5–4(c) except that the beamwidth is wider in this case, representing the fact that subscriber-based equipment is lower in cost and has fewer antenna elements than the base station.

If each subscriber unit is capable of perfectly aligning the boresight of its antenna pattern with its serving base station, units then radiate maximum power in the direction of the desired base station, while reducing transmitter power in proportion to the gain of the subscriber antenna. In a fixed WLL or *Wireless Local Area Network* (WLAN) system, fixed directional antennas can be given the proper orientation when the system is installed. In future systems, steerable transmit arrays, as in Figure 3–23, may allow fixed subscriber terminals to select from multiple serving base stations.

Simulations were conducted, as described in [Lib94] and [Lib95] using this subscriber antenna configuration, along with the three base station patterns shown in Figure 5–4. In these simulations, subscribers were uniformly placed throughout a multi-cell coverage area, using the concentric circle cell geometry of Figure 5–2 with a central cell and a single tier of adjacent cells. For comparison, cases were run using both directional antennas and omni-directional antennas at the subscriber unit. The results are illustrated in Tables 5–6 and 5–7, respectively.

Table 5–6 Ratio of in-cell interference to total interference, f, as a function of path loss exponent, for the three base station antenna patterns with omni-directional antennas at the subscriber terminal.

Base Station Antenna Pattern	n=2	n=3	n=4
Case A: Omni-directional	0.454	0.601	0.693
Case B: Sectorized	0.453	0.601	0.692
Case C: Flat-topped	0.453	0.601	0.693
$\frac{1}{1+8\beta}$ (5.46) (values of β from Table 5–2)	0.455	0.603	0.694

Comparing Tables 5–6 and 5–7, we conclude that the use of spatial filtering at the base station does nothing to improve the reuse factor, f; however, the use of spatial filtering at the portable unit does allow f to be improved.

Spatial processing at the base station improves capacity by reducing the impact of both in-cell and out-of-cell interference, but it does so at the same rate on average. In other words, the gain G_a in (5.55) acts to reduce both in-cell and out-of-cell interference. On the other hand, when uplink spatial processing is used at the subscriber terminal, the effect is to increase the

Table 5–7 Ratio of in-cell interference to total interference, f, as a function of path loss exponent, for each base station antenna pattern with spatial filtering at the subscriber terminal.

Base Station Antenna Pattern	n=2	n=3	n=4
Case A: Omni-directional	0.675	0.816	0.883
Case B: Sectorized	0.675	0.815	0.882
Case C: Flat-topped	0.675	0.815	0.882

cell-to-cell isolation, so that every base station continues to see the same level of Multiple Access Interference from within the cell, but sees less interference from outside the cell. As shown in Tables 5–6 and 5–7, using directional antennas at the subscriber unit reduces the reuse factor, f, from 0.60 to 0.82 for $n = 3$. Using these values for the reuse factor in (5.57) for a given base station azimuth gain, G_a,

$$K_{\text{omni}} \le \frac{0.60NG_a}{\nu(E_b/N_0)} \tag{5.61}$$

using omni-directional antennas at the subscriber, compared with

$$K_{\text{dir}} \le \frac{0.82NG_a}{\nu(E_b/N_0)} \tag{5.62}$$

using directional subscriber antennas. So, using relatively low gain antennas at the subscriber unit, each base station receiver can support $K_{\text{dir}} = (0.82/0.60)K_{\text{omni}} = 1.37K_{\text{omni}}$ subscribers, for a 37% improvement in capacity. More directional subscriber units, with narrower beamwidths, provide greater capacity improvements.

When omni-directional antennas are used at the subscriber terminal, the reuse factor, f, is determined primarily by the path loss exponent, n, which is a function of environment and is not controlled by system designers. Using spatial filtering at the subscriber unit, it is possible to tailor f to a desired value which is greater than the reuse factor obtained using omni-directional antennas at the portable unit. Ideally, driving f to unity would allow system design to be virtually independent of inter-cell propagation environment, when perfect power control is assumed. This result is very important for Wireless Local Loop systems, where the highest possible system capacity is needed in a limited spectrum allocation to support advanced broadband data and video services.

5.5 Range and Capacity Analysis Using Smart Antennas - A Vector-Based Approach

The analysis and simulations in the previous sections provide a powerful means of quantifying how narrowbeam systems can be used to improve CDMA uplink capacity. This analysis is useful for basic research as well as back-of-the-envelope calculations, and it provides useful insight into the impact of spatial processing in CDMA systems. However, the approach assumes ideal antenna patterns and does not allow practical treatment of multipath, adaptive array weight misadjustment, or pattern degradation as a function of array geometry. Section 5.1 presented an analysis of CDMA uplink range extension when the Multiple Access Interference appears like noise. In this section, we reconsider the problem of CDMA uplink range and capacity improvements using a vector space approach and model MAI using a different approach.

Uplink range for CDMA is limited by the maximum power that can be transmitted by each subscriber unit, which is shown for each Station Class Mark in Table 5–8. Range also depends on the number of active users in the system. Even when the number of users in the system is small, the interference level will be significant relative to the thermal noise level and limits the CINR at the base station. Thus, CDMA systems are interference-limited rather than noise-limited. When the number of users decreases, the total interference level drops, and a subscriber can operate at an increased range while maintaining the same link margin. Similarly, if the number of active users in the system increases, the interference level rises and the maximum range of the cell decreases.

In the previous section, we ignored the impact of thermal noise on the CINR. In practical systems, if power control set points are established at high enough levels so that thermal noise, N_n is too much less than the interference power, N_i, then cell range will be extremely limited. Thus, set points for power control on the reverse link are set to be slightly higher than the thermal noise level.

Table 5–8 Maximum PCS CDMA portable station transmitter power for each station class.

Portable Station Class	Max Guaranteed EIRP (dBm)
I	28
II	23
III	18
IV	13
V	8

In general, it is desirable to strike a balance between the lowest power control set points possible, to allow maximum range to the edge of the cell, while not sacrificing system capacity to thermal noise. In practice, this is accomplished by using Outer Loop Power Control to adjust power control set points to achieve a 1% FER for each subscriber. In modeling, this effect is captured through the use of a *loading factor*, χ, which is the ratio of CDMA signal power from all users to the sum of CDMA power plus noise [Vit95].

$$\frac{\text{Received Signal Strength from All CDMA Sources}}{\text{Received Signal Strength from All CDMA Sources plus Noise}} \le \chi$$

The received signal strength from all CDMA sources, measured at the base station receiver, includes both in-cell and out-of-cell users. We can rewrite this as

$$\frac{\nu(P_0 + I_t)}{\nu(P_0 + I_t) + N\sigma_n^2} \le \chi \tag{5.63}$$

where P_0 is the power available at the base station receiver antenna from the desired subscriber, and I_t is the total sum of Multiple Access Interference power, measured at the same point.

Typical values of χ are between 0.50 and 0.75 [Wal94][Vit95]. If the spectrum containing the signal is viewed on a spectrum analyzer at a single receiving antenna element at the base station, then the total received signal and noise power will be greater than the power spectral density of the thermal noise by

$$\textit{Rise in Noise Level above Thermal Noise} = 10\log\left(\frac{1}{1-\chi}\right) \tag{5.64}$$

For example, using a loading factor of $\chi = 0.7$, the presence of the CDMA signals raises the thermal noise level measured at the base station by $5.2\ \text{dB}$. We will use the loading factor in this section to help analyze the combined impact of thermal noise and Multiple Access Interference.

The useful range of any PCS system must take into account both the uplink and downlink. Ideally, a balance is achieved so that when the desired uplink range is obtained, the downlink does not cause unwarranted interference into adjacent cells. The uplink range is limited by the MAI, the base station receiver noise figure, and the maximum mobile unit transmitter power.

To characterize the radio channel for an array-based receiver, the vector channel impulse response was introduced in Section 3.3. Let us consider the static case in which each user moves very slowly relative to the adaptation rate of the system and Doppler shift is not significant. The received signal vector for user k, from Section 3.3, is given by

$$\boldsymbol{u}_k(t) = \sum_{i=0}^{L_k-1} s_k(t - \tau_{k,i})\alpha_{k,i}\boldsymbol{a}(\phi_{k,i}) \tag{5.65}$$

While it is possible to analyze the general case of a multi-finger RAKE receiver, with L_k multipath components for user k, for simplicity we consider the case in which each subscriber in the system contributes a single component (i.e., no multipath) to the base station under consideration. Then the spatial signature for user k is simply

$$\boldsymbol{b}_k = \alpha_k \boldsymbol{a}(\phi_k) \tag{5.66}$$

Now assume that there are many active CDMA signals incident on the array, and that the combined effect of MAI appears spatially white. Then the optimal MMSE weight vector, $\boldsymbol{w}_k$,to extract spatial signature, $\boldsymbol{b}_k$, is proportional to $\boldsymbol{b}_k$, as shown in Section 5.1. Without loss of generality, we set

$$\boldsymbol{w}_k = \boldsymbol{b}_k / \|\boldsymbol{b}_k\| \tag{5.67}$$

where

$$\|\boldsymbol{b}_k\| = \sqrt{\boldsymbol{b}_k^H\boldsymbol{b}_k} \tag{5.68}$$

Then the total available power from the desired user at the base station despreader from the intended user, $k = 0$, is

$$\begin{aligned} P_0 &= |\boldsymbol{w}_0^H\boldsymbol{b}_0|^2 = \frac{|\boldsymbol{b}_0^H\boldsymbol{b}_0|^2}{\|\boldsymbol{b}_0\|^2} = \frac{|\boldsymbol{b}_0^H\boldsymbol{b}_0|^2}{\boldsymbol{b}_0^H\boldsymbol{b}_0} = \boldsymbol{b}_0^H\boldsymbol{b}_0 = \|\boldsymbol{b}_0\|^2 \\ &= \|\alpha_k\boldsymbol{a}(\phi_k)\|^2 = |\alpha_0|^2\boldsymbol{a}^H(\phi_k)\boldsymbol{a}(\phi_k) = M|\alpha_0|^2 \end{aligned} \tag{5.69}$$

where

$$\|\boldsymbol{b}_0\|^2 = \|\alpha_k\boldsymbol{a}(\phi_k)\|^2 = |\alpha_0|^2\boldsymbol{a}^H(\phi_k)\boldsymbol{a}(\phi_k) = M|\alpha_0|^2$$

The total MAI power at the input to the despreader is

$$I_o = \nu\sum_{k=1}^{K-1}|\boldsymbol{w}_0^H\boldsymbol{b}_k|^2 \tag{5.70}$$

where ν is the voice activity factor, which is assumed to be near 0.6. The CINR at the output of the despreader for user $k = 0$ is

$$\gamma_0 = \frac{\text{Desired Signal}}{\text{Interference + Noise}} = \frac{P_0}{\frac{1}{N}I_o + \sigma_n^2} \tag{5.71}$$

where P_0 is the desired signal power at the input to the despreader for subscriber 0, I_o is the total interference power, and σ_n^2 is the total thermal noise power. From the discussion in Section 5.2, N is the processing gain.

Based on the array geometry, the element pattern, and the distribution of users, we can determine an average value of $|\boldsymbol{w}_0^H\boldsymbol{b}_k|^2$, given by $\overline{I}_k$. The expected value of the interference component, $\overline{I}_o$, is

$$\overline{I}_o = E\left\{\nu\sum_{k=1}^{K-1}|\boldsymbol{w}_0^H\boldsymbol{b}_k|^2\right\} = \nu\sum_{k=1}^{K-1}E\{|\boldsymbol{w}_0^H\boldsymbol{b}_k|^2\} = \nu\sum_{k=1}^{K-1}\overline{I}_k \tag{5.72}$$

The value of each $\overline{I}_k$ is dependent on the array geometry. As a typical example, let us assume that the array is a linear, equally spaced array with half-wavelength spacing as shown in Figure 5–5. A reasonable assumption is that the Directions-Of-Arrival, $\{\phi_k\}$, of each of the signals incident on the array are independent and uniformly distributed on $\{0, \pi\}$.

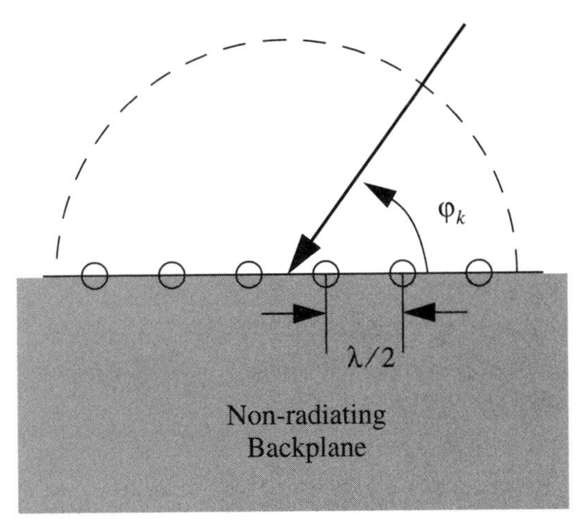

Figure 5–5 Geometry for a half-wavelength spaced linear array. It is assumed that subscriber signals arrive uniformly on $[0, \pi]$.

Using the value of $\boldsymbol{w}_0$ from (5.67), we express $\overline{I}_k$ as

$$\overline{I}_k = E\{|\boldsymbol{w}_0^H\boldsymbol{b}_k|^2\} = E\left\{\frac{|\boldsymbol{b}_0^H\boldsymbol{b}_k|^2}{\|\boldsymbol{b}_0^H\|^2}\right\} \tag{5.73}$$

$$= E\left\{\frac{|\alpha_0{}^*\boldsymbol{a}^H(\phi_0)\boldsymbol{a}(\phi_k)\alpha_k|^2}{|\alpha_0{}^*\boldsymbol{a}^H(\phi_0)\boldsymbol{a}(\phi_0)\alpha_0|}\right\} = \frac{|\alpha_k|^2}{M}E\{|\boldsymbol{a}^H(\phi_0)\boldsymbol{a}(\phi_k)|^2\}$$

Then we solve for the expected value of the inner product between two steering vectors:

$$E\{|\boldsymbol{a}^H(\phi_0)\boldsymbol{a}(\phi_k)|^2\} = E\left\{\sum_{n=0}^{M-1}\sum_{m=0}^{M-1} e^{-j\beta\Delta xn(\cos(\phi_0) - \cos(\phi_k))}e^{j\beta\Delta xm(\cos(\phi_0) - \cos(\phi_k))}\right\} \tag{5.74}$$

$$= \sum_{n=0}^{M-1}\sum_{m=0}^{M-1} E\{e^{-j\beta\Delta xn(\cos(\phi_0) - \cos(\phi_k))}e^{j\beta\Delta xm(\cos(\phi_0) - \cos(\phi_k))}\}$$

$$= \sum_{n=0}^{M-1}\sum_{m=0}^{M-1} E\{e^{-j\beta\Delta x(n-m)\cos(\phi_0)}\}E\{e^{j\beta\Delta x(n-m)\cos(\phi_k)}\}$$

The expectation in (5.65) results in a Bessel function expression:

$$E\{e^{-j\beta\Delta x(n-m)\cos(\phi_0)}\} = \int_0^{\pi}\frac{1}{\pi}e^{-j\beta\Delta x(n-m)\cos(\phi)}d\phi \tag{5.75}$$

$$= J_0(\beta\Delta x(n-m)) = J_0(\pi(n-m))$$

Therefore, the interference contribution from each subscriber may be expressed as

$$\overline{I_k} = \frac{|\alpha_k|^2}{M}\sum_{n=0}^{M-1}\sum_{m=0}^{M-1} J_0^2(\pi(n-m)) = |\alpha_k|^2 G_i(M) \tag{5.76}$$

The term $G_i(M)$ is called the *interference gain*. Substituting (5.76) into (5.72) and using the result in (5.71) along with (5.69), we obtain the CINR,

$$CINR = \gamma_0 \approx \frac{M|\alpha_0|^2}{\frac{\nu G_i(M)}{N}\sum_{k=1}^{K-1}|\alpha_k|^2 + \sigma_n^2} \tag{5.77}$$

The Interference Gain, $G_i(M)$, given in Table 5–9, is essentially a degradation from the ideal gain of M that can be achieved relative to noise due to the specific array geometry and the distribution of interferers. For the example discussed here, signals from interferers are uniformly distributed on $[0, \pi]$. Therefore, in a system which is entirely limited by thermal noise, the smart antenna can provide a gain of M, as shown in Section 5.1, whereas in a system limited by multiple access interference, a smaller gain, $M/G_i(M)$, can be obtained.

Table 5–9 Interference Gain for a linear half-wavelength spaced array with interferers distributed on $[0, \pi]$.

Number of Elements	$G_i(M)$
1	1.0
2	1.6
3	1.8
4	2.0
6	2.2
8	2.4
10	2.5

We can make a number of observations to simplify the expression for the CINR in (5.77). Let K_0 represent the number of active users in the same sector as the desired user (including the desired user), each of which is power controlled to a level P_0. Interference from adjacent sectors and other cells is modeled as I_a.

$$\gamma_0 \approx \frac{MP_0}{\frac{\nu G_i(M)(K_0-1)P_0 + \nu G_i(M)I_a}{N} + \sigma_n^2} \tag{5.78}$$

We may consider out-of-cell interference by using the reuse factor from Table 5–3,

$$f = \frac{\text{In-cell Interference}}{\text{Total Interference}} = \frac{(K_0 - 1)\nu P_0}{\nu I_a + (K_0 - 1)\nu P_0} \tag{5.79}$$

Using the reuse factor, we can express (5.79) as

$$CINR = \gamma_0 \approx \frac{MP_0}{\frac{\nu G_i(M)}{Nf}(K_0 - 1)P_0 + \sigma_n^2} \tag{5.80}$$

We can use the loading factor, χ, introduced in (5.63). We write

$$\frac{\nu(K_0 P_0 + I_a)}{\nu(K_0 P_0 + I_a) + N\sigma_n^2} = \chi \tag{5.81}$$

Rearranging (5.81)

$$\nu(K_0 P_0 + I_a) = \nu P_0\left(1 + \frac{K_0 - 1}{f}\right) = \left(\frac{\chi}{1 - \chi}\right)N\sigma_n^2 \tag{5.82}$$

When operating at the loading level, χ, we have

$$N\sigma_n^2 = \left(\frac{1 - \chi}{\chi}\right)\nu P_0\left(1 + \frac{K_0 - 1}{f}\right) \tag{5.83}$$

Substituting (5.83) into (5.80), we have

$$\gamma_0 = \frac{fNM/\nu}{G_i(M)(K_0 - 1) + \left(\frac{1 - \chi}{\chi}\right)(f + (K_0 - 1))} \tag{5.84}$$

which can be rearranged,

$$\gamma_0 = \frac{\chi NMf/\nu}{(\chi G_i(M) + 1 - \chi)(K_0 - 1) + (1 - \chi)f} \tag{5.85}$$

Since $(1 - \chi) < 1$ and $f < 1$, we approximate γ_0 using

$$\gamma_0 = \frac{\chi NMf}{\nu(\chi(G_i(M) - 1) + 1)(K_0 - 1)} \tag{5.86}$$

For a particular required value of CINR, $\bar{\gamma}$, the maximum number of simultaneous users supported is given by

$$K_{max} \leq \frac{\chi NMf}{(\chi(G_i(M) - 1) + 1)\nu\bar{\gamma}} + 1 \tag{5.87}$$

The maximum number of users that can be supported occurs as the loading factor is allowed to approach its maximum value, $\chi = 1$. In this case,

$$K_{max} \leq \frac{NMf}{G_i(M)\nu\bar{\gamma}} + 1 \tag{5.88}$$

Therefore, this provides an upper bound on the number of users supported in each cell of a CDMA system where a uniform path loss exponent applies throughout the system.

For example, using $n = 3$, which, from Table 5–6, is $f = 0.603$, with a processing gain of $N = 128$, $\nu = 0.6$, and a required CINR of 9 dB , we have

$$K_{max} \leq \frac{(128)M(0.603)}{G_i(M)(0.6)(10^{9/10})} + 1 = 16.2\frac{M}{G_i(M)} + 1 \tag{5.89}$$

For $M = 1$, $G_i(M) = 1$ so $K_{max} \leq 17$. This is approximately equivalent to the result obtained in (5.58) for omni-directional antennas.

Under the same conditions, with $M = 4$, $G_i(M) = 2.0$, we have

$$K_{max} \leq \frac{(128)4(0.603)}{(2)(0.6)(10^{9/10})} + 1 = 33 \tag{5.90}$$

For an 8-element array, $M = 8$, $G_i(M) = 2.4$, so

$$K_{max} \leq \frac{(128)8(0.603)}{(2.4)(0.6)(10^{9/10})} + 1 = 54 \tag{5.91}$$

Note that the 9.0 dB gain of the $M = 8$ array is roughly equivalent to the 9.3 dB keyhole antenna used in (5.60); however, while the analysis of the keyhole antenna showed that 137 users could be supported, the array system can support only 54 simultaneous subscribers. The difference is due to the Interference Gain and the fact that we did not assume that interference was distributed on $[0, \pi]$ in (5.91).

Table 5–10 shows the number of subscribers supported per sector as a function of the number of antenna elements used at the base station. Because it is assumed that the Direction-Of-Arrival is uniformly distributed on $[0, \pi]$, the results expressed here are applicable to sectorized antennas, such as those illustrated in Figure 4–7. These results are similar and comparable to those presented in Table 5–5, and they justify the analysis shown in earlier sections. Note, however, that the number of subscribers supported in each sector does not rise as rapidly as suggested in Table 5–5.

Table 5–10 Number of subscribers supported by each CDMA sector as a function of the path loss exponent, n, and the array size M.

			K_{max}		
M	$10\log(M)$	$10\log(G_i(M))$	**n=2**	**n=3**	**n=4**
1	0.0 dB	0.0 dB	13	17	19
4	6.0 dB	3.0 dB	25	33	38
8	9.0 dB	3.8 dB	41	54	62

5.6 Summary

In this chapter, we developed several tools to analyze the range and capacity improvements that can be achieved when smart antennas are used in CDMA wireless networks. Current IS-95 CDMA deployments use omni-directional or simple sectorized antennas, as discussed in Sections 5.3 and 5.5. Future systems that use arrays or switched beam systems will have much greater capacity, as shown in Tables 5–5 and 5–10. We also demonstrated how smart antennas and directional antennas at the subscriber unit can increase cell-to-cell isolation, increasing the reuse factor, f. In the next two chapters, we provide models and measurements of the spatial radio channel needed to better quantify smart antenna system performance.

CHAPTER 6

Characterization of Spatio-Temporal Radio Channels

The spatial properties of wireless communication channels are extremely important in determining the performance of smart antenna systems. In this chapter a range of spatial channel models are presented that have been developed to characterize the particular features of the radio channel that impact spatial processing receivers.

Classical channel models provide information on signal power level distributions and Doppler shifts of received signals. As shown subsequently, many emerging spatial models in the literature utilize the fundamental principles of the classical channel models. However, modern spatial channel models build upon the classical understanding of multipath fading and Doppler spread by incorporating additional concepts such as time delay spread, Angle-Of-Arrival, and adaptive array antenna geometries.

Fundamental channel models have led to the present-day theories of spatial diversity from both mobile user and base station perspectives. Modern channel models include both spatial and temporal features for design of wideband modems. Early channel models accounted only for the time-varying amplitude and phase of the channel. These models were then enhanced by adding time delay spread information, which is important when analyzing digital modulation over wireless communication links. Now, with the introduction of techniques and features that depend on the spatial distribution of the mobiles, wideband temporal and spatial information is required for relevant channel models.

The differentiation between the mobile and base station is important, too. Classical work has demonstrated that models must account for the physical geometry of scattering objects in the

---

Portions of this chapter have been adapted from R. Ertel, P. Cardieri, K. W. Sowerby, T. S. Rappaport, J. H. Reed, "Overview of Spatial Channel Models for Antenna Array Communication Systems," *IEEE Personal Communications*, Vol. 5, No. 1, pp. 10-22, February 1998, ©1998 IEEE.

vicinity of the antenna under study. The number and locations of these scattering objects are dependent upon the heights of the antennas relative to the local environment. In this chapter, we present some of the emerging models for spatial diversity and adaptive antennas and include the physical mechanisms and motivation behind the models.

6.1 Wireless Multipath Channel Models, Environment, and Signal Parameters

In analyzing the performance of smart antenna systems, it is important to understand the relationship between the response of the array and the multipath channel. In a wireless system, a signal transmitted through the channel interacts with the environment in a very complex way. There are reflections from large objects, diffraction of the electromagnetic waves around objects, and signal scattering. The result of these complex interactions is the presence of many signal components, or *multipath* signals, at the receiver. A simplified pictorial of the multipath environment with two mobile stations is shown in Figure 6–1.

The early classic models, which were developed for narrowband transmission systems, provide only information about signal amplitude level distributions and Doppler shifts of the received signals. These models have their origins in the early days of cellular radio [Cla68] [Gan72][Lee82][Tur72] when wideband digital modulation techniques were not readily available. As wireless systems became more complex and more accurate models were required, additional concepts, such as time delay spread, were incorporated into the channel models [Rap89][Sei91].

Each signal component experiences a different multipath environment which determines the amplitude ($\rho_{k,l}$), carrier phase shift ($\psi_{k,l}$), time delay ($\tau_{k,l}$), Direction-Of-Arrival ($\phi_{k,l}$), and Doppler shift ($f_{k,l}$) of the lth multipath component for the kth mobile. In general, each of these parameters will be time varying. It is notationally simpler to group the amplitude, phase shift, and Doppler frequency, using:

$$\alpha_{k,l}(t) = \rho_{k,l}e^{j(2\pi f_{k,l}t + \psi_{k,l})} \tag{6.1}$$

A discrete model, which is well suited to digital simulation, is used to characterize the channel where each multipath component is considered to be a plane wave, arriving from a discrete direction at a discrete time delay. For a particular cellular subscriber, the channel between the portable transmitter and the base station receiver is represented as a time-variant channel and modeled using the *Channel Impulse Response*,

$$h_k(t,\tau) = \sum_{l=0}^{L_k(t)-1} \alpha_{k,l}(t)\delta(t-\tau_{k,l}(t)) = \sum_{l=0}^{L_k(t)-1} \rho_{k,l}(t)e^{j\psi_{k,l}(t)}\delta(t-\tau_{k,l}(t)) \tag{6.2}$$

where $L_k(t)$ is the number of multipath components. The amplitude $\rho_{k,l}(t)$ of the multipath components is often modeled as a fixed, Rayleigh, Ricean, or log-normal distributed random variable while the phase shift $\psi_{k,l}$ is uniformly distributed. A time-varying discrete channel is illustrated in Figure 6–2.

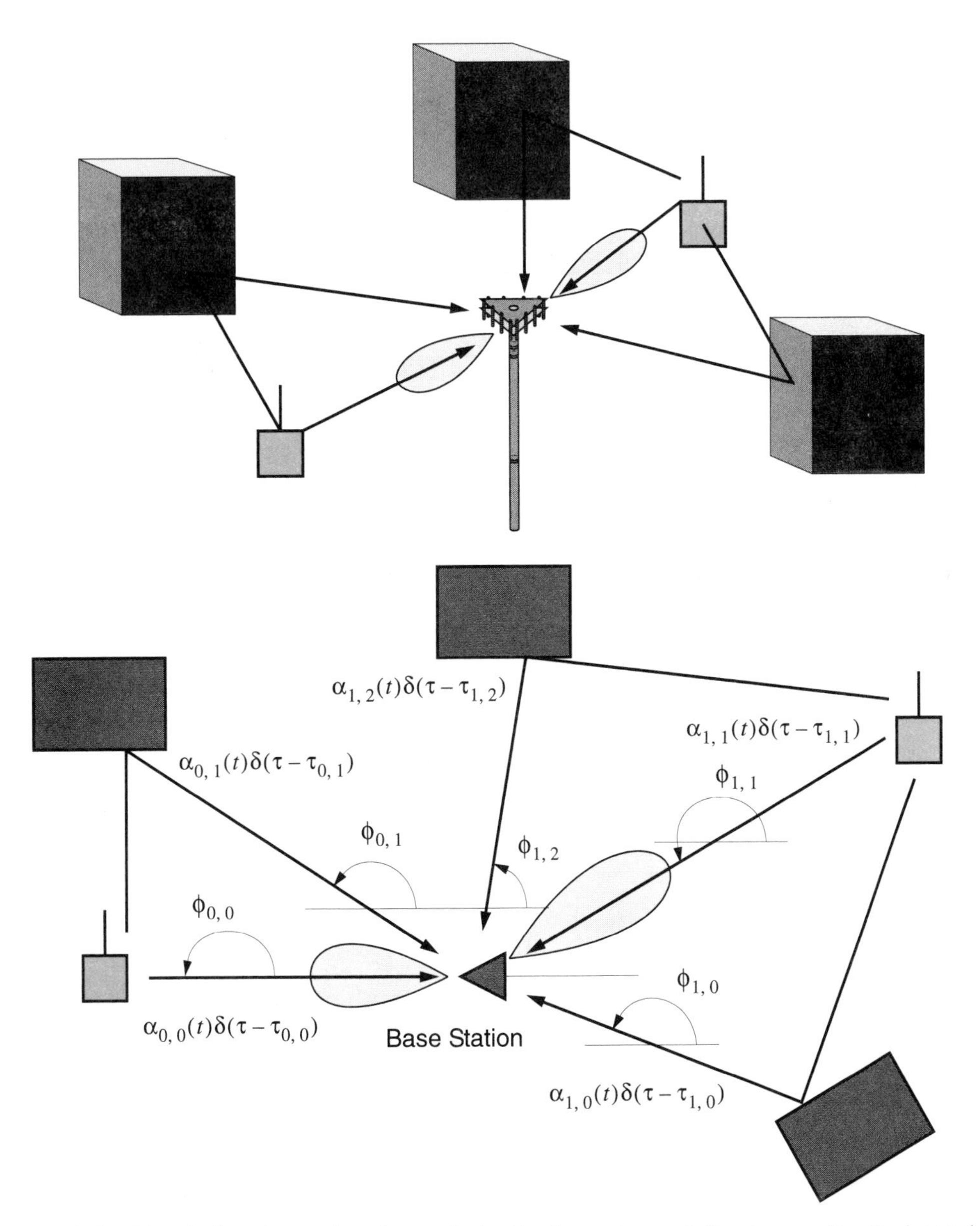

Figure 6–1 Signals from two subscriber units incident on a base station in a multipath channel.

The time varying nature of a wireless channel is caused by the motion of objects in the channel. A measure of the time rate of change of the channel is the *Doppler Power Spectrum*, introduced by M. J. Gans in 1972 [Gan72]. The Doppler Power Spectrum provides statistical information about the variation of the frequency of a tone received by a mobile traveling at

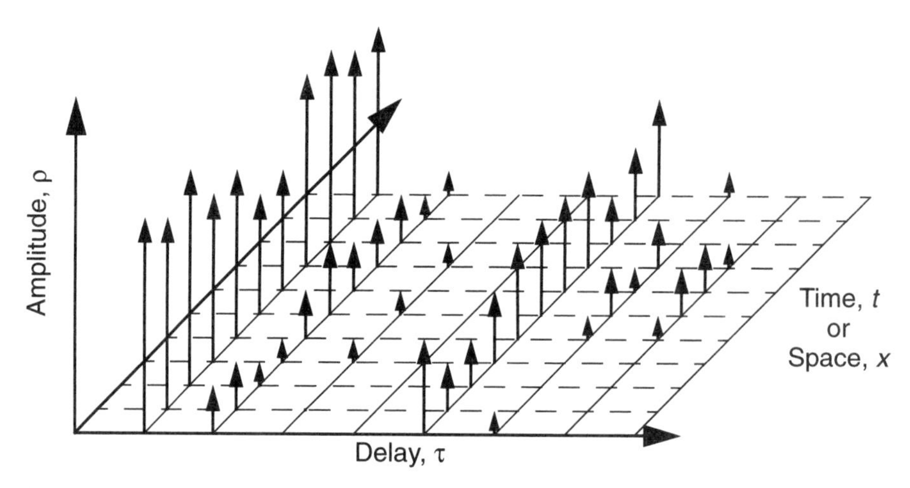

Figure 6–2 The time-varying channel impulse response, $h_k(t, \tau)$, is a function of both delay and time.

velocity, v. Based on the flat fading channel model developed by R. H. Clarke in 1968, Gans assumed that the received signal at the mobile station came from all directions and was uniformly distributed. Under these assumptions and for a $\lambda/4$ vertical antenna, the Doppler power spectrum is given by [Rap96a].

$$S(f) = \begin{cases} \dfrac{1.5}{\pi f_m \sqrt{1 - \left(\dfrac{f-f_c}{f_m}\right)^2}} & |f - f_c| < f_m \\ 0 & elsewhere \end{cases} \tag{6.3}$$

where f_m is the maximum Doppler shift given by v/λ, where λ is the wavelength of the transmitted signal at frequency f_c. Clarke's Doppler power spectrum is illustrated in Figure 6–3.

The channel impulse response in (6.2) can be modified to take into account the Direction-Of-Arrival of multipath components, yielding the Vector Channel Impulse Response introduced in Chapter 3,

$$\begin{aligned} \boldsymbol{h}(t, \tau) &= \sum_{l=0}^{L_k(t)-1} \boldsymbol{a}(\phi_{k,l}) \alpha_{k,l}(t) \delta(t - \tau_{k,l}(t)) \\ &= \sum_{l=0}^{L_k(t)-1} \boldsymbol{a}(\phi_{k,l}) \rho_{k,l}(t) e^{j\psi_{k,l}(t)} \delta(t - \tau_{k,l}(t)) \end{aligned} \tag{6.4}$$

where $\boldsymbol{a}(\phi_{k,l})$ is the *steering vector*. The steering vector is a function of the array geometry and Angle-Of-Arrival,

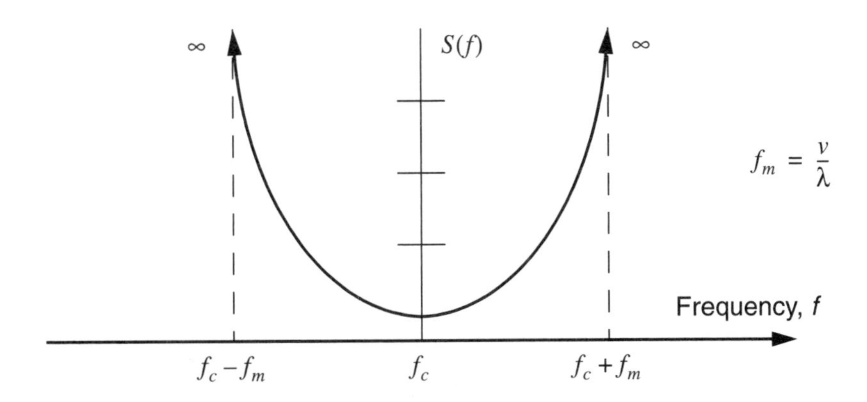

Figure 6–3 Clarke's model for the Doppler power spectrum where components arrive with equal probability from all directions.

$$\boldsymbol{a}(\theta, \phi) = \begin{bmatrix} 1 & a_1(\theta, \phi) & \ldots & a_{M-1}(\theta, \phi) \end{bmatrix}^H \tag{6.5}$$

where

$$a_m(\theta, \phi) = e^{j\beta(x_m \cos(\phi)\sin(\theta) + y_m \sin(\phi)\sin(\theta) + z_m \cos(\theta))} \tag{6.6}$$

where $\beta = 2\pi/\lambda$ is the *phase propagation factor.*

The Vector Channel Impulse Response provides a powerful tool which maps a transmitted signal $s(t)$ to the signal received at each array element. Each element of the VCIR is the channel impulse response between the portable terminal and each element of the receiving array. In the next section, several models that provide varying levels of information about the spatial channel are presented.

Given a transmitted signal, $s_k(t)$, the data vector is

$$\boldsymbol{u}_k(t) = \begin{bmatrix} u_{k,0}(t) & \ldots & u_{k,M-1}(t) \end{bmatrix}^T \tag{6.7}$$

$$= s_k(t) * \boldsymbol{h}_k(\tau, t) + \boldsymbol{n}(t) = \int_{-\infty}^{t} s_k(\lambda)\boldsymbol{h}_k(t-\lambda, t)d\lambda + \boldsymbol{n}(t)$$

$$= \sum_{l=0}^{L-1} \boldsymbol{a}(\phi_{k,l})\alpha_{k,l}(t)s(t-\tau_{k,l}) + \boldsymbol{n}(t)$$

where $\boldsymbol{n}(t)$ is the vector representing the noise introduced at each antenna element.

In channels where the differences between the path delays of multipath components are very small relative to the channel symbol or chip period, then the channel is called a *narrowband*

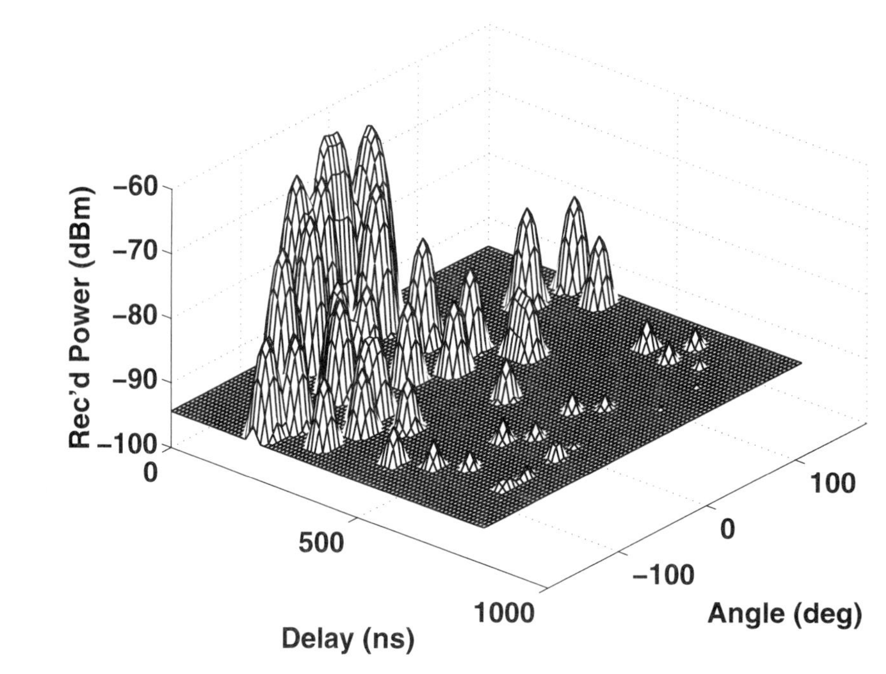

Figure 6–4 A plot of the magnitude of the channel impulse response, as a function of both delay and Angle-Of-Arrival [Lib98b].

channel and undergoes flat fading [Rap96b] whereby we make the approximation that $\tau_{k,l} = \tau_{k,0}$. Then the Vector Channel Impulse Response may be expressed as

$$\boldsymbol{u}_k(t) = s_k(t-\tau_0)\sum_{l=0}^{L-1}\boldsymbol{a}(\phi_{k,l})\alpha_{k,l}(t) + \boldsymbol{n}(t) = s(t-\tau_{k,l})\boldsymbol{b}(t) \tag{6.8}$$

where the term $\boldsymbol{b}(t)$ is called the *spatial signature* of the narrowband channel,

$$\boldsymbol{b}(t) = \sum_{l=0}^{L-1}\boldsymbol{a}(\phi_{k,l})\alpha_{k,l}(t) = \sum_{l=0}^{L-1}\boldsymbol{a}(\phi_{k,l})\rho_i e^{j(2\pi f_{k,l}t + \psi_{k,l})} \tag{6.9}$$

As noted in Chapter 3, $\boldsymbol{b}(t)$ represents a multiplicative channel rather than a convolutional channel. This means that the data vector, $\boldsymbol{u}_k(t)$ is obtained by multiplying $\boldsymbol{b}(t)$ by $s(t)$, rather than convolving. The spatial signature represents the instantaneous projection of the received signal from a particular user onto the array space.

The spatial channel impulse response given in (6.4) is a summation of several multipath components, each with its own amplitude, phase, and Direction-Of-Arrival. The distribution of these parameters is dependent upon the type of environment. For example, measurements have shown that the wireless channel between a base station and the ground-based subscriber terminal is a function of the richness of the local scattering environment and antenna heights [Rap89][Sei91][Feu94]. Measurements have been used to quantify the proper range and distribution of channel parameters. In the next section, we describe macrocell and microcell environments and discuss how the environment affects channel parameters.

6.1.1 The Macrocell Environment

Figure 6–5 shows the forward link channel for a macrocell environment. It is usually assumed that the scatterers surrounding the mobile station are approximately the same height as or higher than the mobile. This implies that the received signal at the mobile antenna arrives from all directions after bouncing from the surrounding scatterers as illustrated in Figure 6–5.

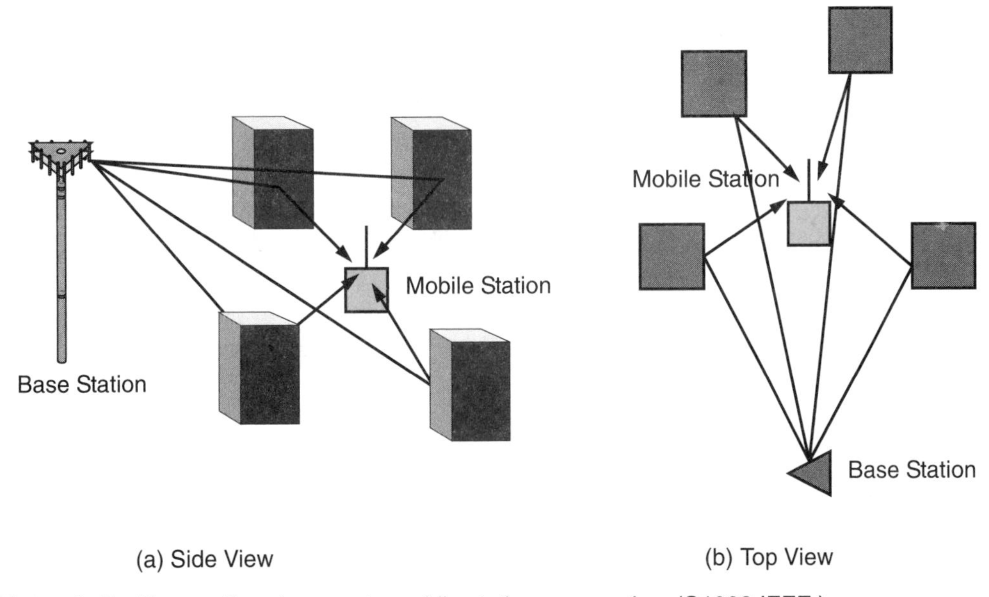

Figure 6–5 Macrocell environment - mobile station perspective. (©1998 IEEE.)

Under these conditions, Gans' assumption that the Angle-Of-Arrival is uniformly distributed over $[0, 2\pi]$ is valid. The classical Rayleigh fading envelope with deep fades approximately $\lambda/2$ apart results from this model [Rap96a].

However, the Angle-Of-Arrival of the received signal at the base station is quite different. In a macrocell environment, typically the base station is deployed above the surrounding scatterers. Hence, the received signals at the base station result from the scattering process in the vicinity of the mobile station, as shown in Figure 6–6. The multipath components at the base station are restricted to a smaller angular region, Φ_{BW}, and the distribution of the Direction-Of-Arrival is no longer uniform over $[0, 2\pi]$. Other Direction-Of-Arrival distributions are considered later in this chapter.

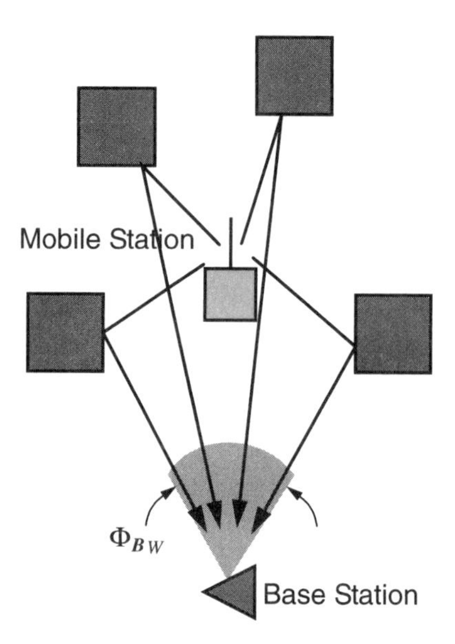

Figure 6–6 Macrocell environment - base station perspective, where Φ_{BW} is the angular spread of multipath components arriving at the base station. (©1998 IEEE.)

The base station model of Figure 6–6 was used to develop the theory and practice of base station diversity in today's cellular systems and has led to rules of thumb for the spacing of diversity antennas on cellular towers [Lee82].

6.1.2 The Microcell Environment

In the microcell environment, the base station antenna may be at the same height as the surrounding objects. In this case, the scattering spread of the DOA of the received signal at the base station is larger than in the macrocell case since the scattering process also occurs in the vicinity of the base station. Thus, as the base station antenna is lowered, the multipath DOA spread tends to increase. This change in the behavior of the received signal is very important for antenna array applications. Studies have shown that the angle spread increases as the antenna height is lowered into surrounding clutter.

6.1.3 Angle Spread

The *Root Mean Square (rms) angle spread* of the VCIR is a measure of the angular dispersiveness of the channel. For the signal received from subscriber, k, we define the central moment angle spread as

$$\sigma_\phi = \sqrt{E[\phi^2] - E^2[\phi]} \tag{6.10}$$

Since we typically do not have an expression for the Probability Density Function of the Direction-Of-Arrival, we often approximate the angle spread using

$$\sigma_\phi = \sqrt{\langle\phi^2\rangle - \langle\phi\rangle^2} \tag{6.11}$$

where

$$\langle\phi\rangle = \frac{\int_0^{2\pi} \phi P_r(\phi)d\phi}{\int_0^{2\pi} P_r(\phi)d\phi} \; ; \; \langle\phi^2\rangle = \frac{\int_0^{2\pi} \phi^2 P_r(\phi)d\phi}{\int_0^{2\pi} P_r(\phi)d\phi} \tag{6.12}$$

and $P_r(\phi)$ is the power received from angle ϕ. Using the discrete VCIR, we write

$$\sigma_\phi = \sqrt{\frac{\sum_{l=0}^{L-1} \alpha_{k,l}^2 \phi_{k,l}^2}{\sum_{l=0}^{L-1} \alpha_{k,l}^2} - \left(\frac{\sum_{l=0}^{L-1} \alpha_{k,l}^2 \phi_{k,l}}{\sum_{l=0}^{L-1} \alpha_{k,l}^2}\right)^2} \tag{6.13}$$

Equation (6.13) represents the rms angle spread for a narrowband channel. For a wideband channel with resolvable multipath signal components having time delay τ_i, each delay interval will have its own $\sigma_\phi(\tau_i)$.

Figures 6–7 and 6–8 show the rms angle spread as a function of antenna height and environment type for different environments in New Jersey. In these measurements, the impact of antenna height is clear. Figures 6–7 and 6–8 both show that angle spread increases as the antenna height is lowered. Figure 6–8 shows that the nature of the scattering environment also has an impact on angle spread, with angle spread increasing from rural to suburban and suburban to urban settings [Lib98a].

6.1.4 Measuring Time Variation of the Channel

There are several approaches that can be taken to characterize the time-variability of the spatial radio channel. A simple approach is to track the time-variability of the Direction-Of-Arrival of the strongest multipath component. This metric is closely tied to the relative position of the transmitter and receiver, as well as to the direction and speed of the mobile unit. Another metric, which is perhaps more appropriate for adaptive antenna systems, characterizes the time variation of the received spatial signature for a particular mobile subscriber.

This time-variability metric helps to determine the rate at which an adaptive spatial processor must update its weights to properly track a mobile unit. When co-channel interference is present in the system, the rates of change of the spatial signatures, $\boldsymbol{b}$, for all co-channel signa-

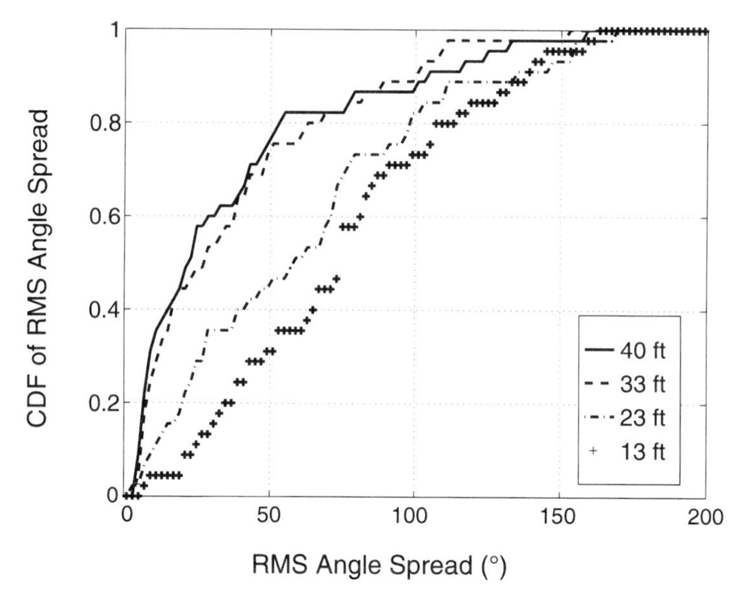

Figure 6–7 RMS angle spread as a function of receiver antenna height for locations in New Jersey. Measurements were taken in urban, suburban, and rural environments at 1900 MHz [Lib97] (Reprinted with permission of ATIRP). Transmitters are 5 ft. above ground level. (©1998 IEEE.)

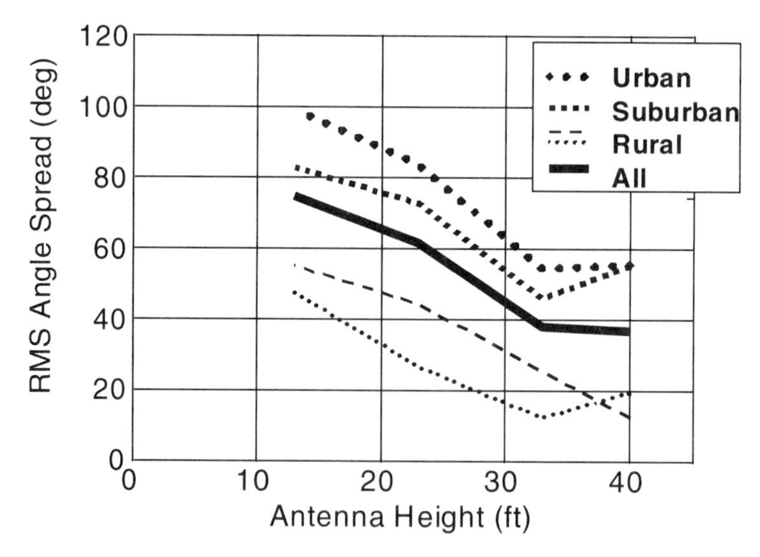

Figure 6–8 RMS angle spread in different environments at 1900 MHz, as seen by a receiver at varying heights [Lib97] (Reprinted with permission of ATIRP). Transmitters are 5 ft. above ground level. (©1998 IEEE.)

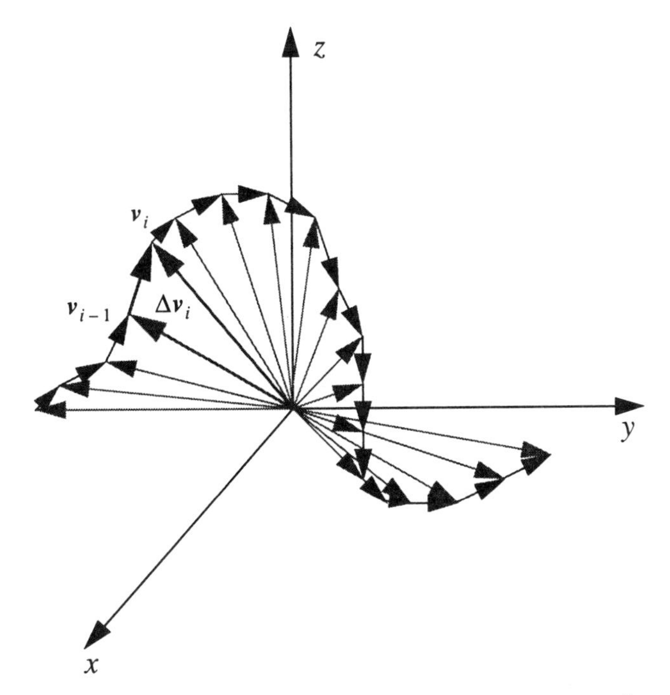

Figure 6–9 Representation of the time-variation of the spatial signature in $\mathcal{R}^3$ space [Lib98a] (Reprinted with permission of ATIRP) [Lib98b] (©1998 IEEE).

tures must be modeled. Figure 6–9 illustrates the time-variability of the spatial signature in 3-dimensional, real-valued ($\mathcal{R}^3$) space, although the actual spatial signature will vary in M-dimensional, complex-valued ($\mathcal{C}^M$) space.

One way to model the variation in spatial signature is to model the change in the "direction" of the spatial signature. If we define the spatial signature at times $i\Delta T$ and $(i+1)\Delta T$ as $\boldsymbol{b}_i$ and $\boldsymbol{b}_{i+1}$, the change in direction is defined as

$$\Delta \boldsymbol{b}_i = \boldsymbol{b}_{i+1} - \left(\frac{\boldsymbol{b}_i^H \boldsymbol{b}_{i+1}}{\boldsymbol{b}_i^H \boldsymbol{b}_i}\right)\boldsymbol{b}_i \tag{6.14}$$

Let us define the rate of change in direction of the spatial signature as

$$\delta_i = \frac{\|\Delta \boldsymbol{b}_i\|}{\|\boldsymbol{b}_{i+1}\|\Delta T} = \frac{1}{\Delta T}\sqrt{1 - \frac{|\boldsymbol{b}_i^H \boldsymbol{b}_{i+1}|^2}{\|\boldsymbol{b}_i\|^2 \|\boldsymbol{b}_{i+1}\|^2}} \tag{6.15}$$

The change in spatial signature is plotted as a function of vehicle speed in Figure 6–10, for a series of measurements along a highway.

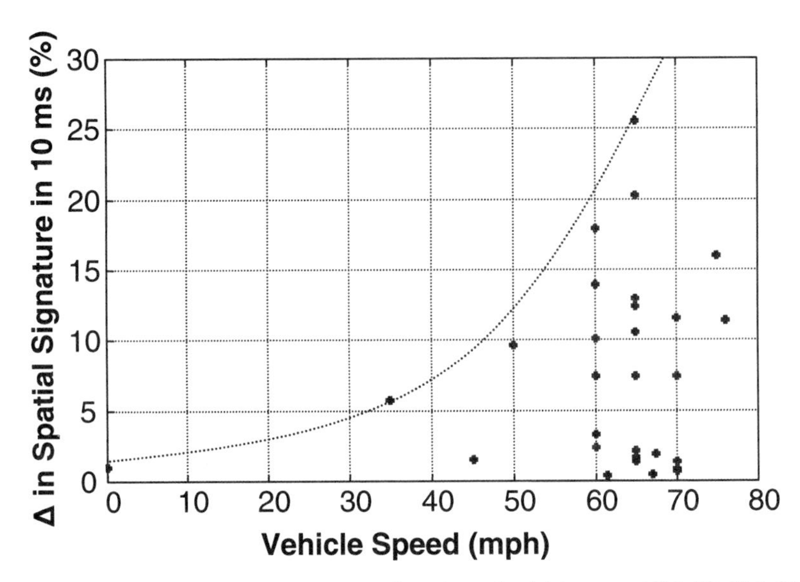

Figure 6–10 Change in spatial signature as a function of vehicle speed [Lib98a][Lib98b] (©1998 IEEE).

6.2 Spatio-Temporal Channel Models for Smart Antennas

In a radio communication system, multipath can limit performance either by introducing fading in narrowband systems or by causing intersymbol interference in wideband systems [Rap96a]. In CDMA systems, which exploit path diversity, it is also important to characterize multipath. In systems employing spatial filtering in the form of fixed directional antennas, switched beam systems, or adaptive antennas, the angular (or spatial) distribution of the multipath components is important in determining the performance of the radio link.

To simulate these systems without resorting to measured data or site-specific propagation prediction techniques, a model must be used to generate multipath channel parameters, including information about the Direction-Of-Arrival (DOA) of components. When no other information is available, researchers have often assumed that the azimuthal DOA of multipath components, ϕ, is uniformly distributed on $[0, 2\pi]$ [Par92]. However, this is inaccurate in general.

In this section, a number of realistic spatial channel models are introduced, and the defining equations (or geometry) and the key results are described. The various models were developed and used for different applications. Some of the models provide information about only one single channel characteristic, such as angle spread, while others attempt to capture all the properties of the wireless channel. In discussing the models, an effort is made to identify the original motive for the model and to convey the information the model is intended to provide.

6.2.1 Lee's Model

In Lee's model, scatterers are evenly spaced on a circular ring about the mobile as shown in Figure 6–11. Each of the scatters is intended to represent the effect of many scatterers within the region and are referred to as *effective scatterers*. The model was originally used to accurately predict the correlation between the signals received by two sensors as a function of element spacing at the base station or mobile. However, since the correlation matrix of the received signal vector of an antenna array can be determined by considering the correlation between each pair of elements, the model has application to any arbitrary array size.

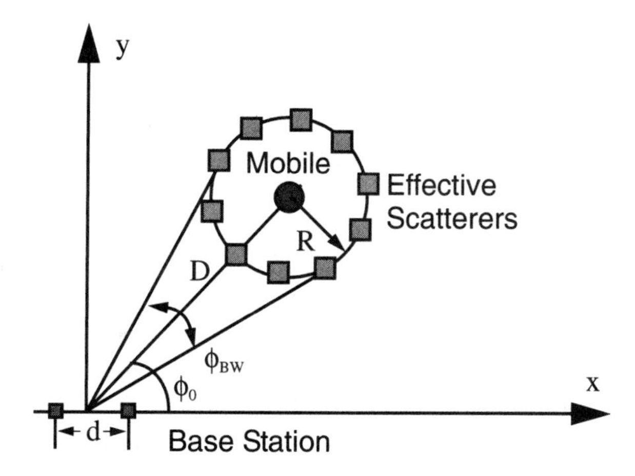

Figure 6–11 Lee's Model. (©1998 IEEE.)

The level of correlation will determine the performance of spatial diversity methods [Lee82], [Asz96]. In general, larger angle spreads and element spacings result in lower correlations, which provide an increased diversity gain for fixed physical element spacing. Measurements of the correlation observed at both the base station and the mobile are consistent with a narrow angle spread at the base station and a large angle spread at the mobile. Correlation measurements made at the base station indicate that the typical radius of scatterers is from 100 to 200 wavelengths [Lee82]. This model explains how a narrowband signal at a mobile receiver only requires about 0.2λ spacing for good diversity (low correlation), whereas good diversity requires a spacing of about 40λ at a tall base station.

Assuming that N scatterers are uniformly placed on the circle with radius R and oriented so that a scatterer is located on the line of sight, the discrete DOAs are [Asz96].

$$\phi_i \approx \frac{R}{D}\sin\left(\frac{2\pi}{N}i\right) \quad \text{for } i = 0, 1, \ldots, N-1 \tag{6.16}$$

From the discrete DOAs, the correlation of the narrowband signals between any two elements of the array can be found using [Asz96]

$$r(d, \phi_0, R, D) = \frac{1}{N}\sum_{i=0}^{N-1} \exp[-j2\pi d \cos(\phi + \phi_i)] \tag{6.17}$$

where d is the element spacing and ϕ_0 is measured with respect to the line between the two elements, as shown in Figure 6–11.

6.2.2 Stapleton's Extension of Lee's Model

The original model only provided information about the correlation of signals received at different elements. Motivated by the need to consider small-scale fading in diversity systems, Stapleton *et al.* proposed an extension to Lee's model that accounts for Doppler shift by imposing an angular velocity on the ring of scatterers [Sta94], [Sta96]. For the model to give the appropriate maximum Doppler shift, the angular velocity of the scatterers must equal v/R, where v is the vehicle velocity and R is the radius of the scatterer ring [Sta96]. Stapleton illustrated that this model could be used to simulate the BER for $\pi/4$-DQPSK systems in a manner which closely matches measured data.

When the model is used to provide joint DOA and TOA channel information, one finds that the resulting power delay profile is "U-shaped" [Ert98]. By considering the intersections of the effective scatterers by ellipses of constant delay, one finds that there is a high concentration of scatterers in the ellipses with minimum delay, a high concentration of scatterers in ellipses with maximum delay, and a lower concentration of scatterers in between. Higher concentrations of scatterers with a given delay correspond with larger powers, thus larger values on the power delay profile. The "U-shaped" power delay profile is not consistent with measurements. Therefore, an extension to Lee's model, in which additional scatterer rings are added to provide different power delay profiles, is proposed in [Sta96].

While the model is quite useful in predicting the correlation between any two elements of the array (thus the array correlation matrix), it is not well suited for simulations requiring a complete VCIR model of the wireless channel.

6.2.3 Discrete Uniform Distribution

A model similar to Lee's model in terms of both motivation and analysis was proposed in [Asz96]. The model (referred to here as the discrete uniform distribution) evenly spaces N scatterers within a narrow beamwidth centered about the bearing to the mobile as shown in Figure 6–12. The discrete possible DOAs, assuming N is odd, are given by [Asz96]

$$\phi_i = \frac{1}{N-1}\phi_{BW} i, \quad i = -\frac{N-1}{2}, \ldots, \frac{N-1}{2} \tag{6.18}$$

From this, the correlation of the signals present at two antenna elements with a separation of d is found to be

$$r(d, \phi_0, \phi_{BW}) = \frac{1}{N}\sum_{i=\frac{N-1}{2}}^{\frac{N-1}{2}} \exp[-j2\pi d \cos(\phi_0 + \phi_i)] \tag{6.19}$$

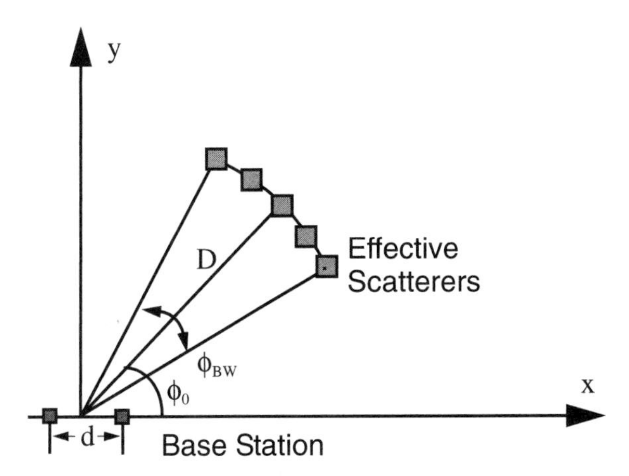

Figure 6–12 Discrete uniform geometry. (©1998 IEEE.)

Measurements reported in [Asz96] suggest that the DOA statistics in rural and suburban environments are Gaussian distributed (see the discussion of the GAA model later). However, in practice, the DOA will be discrete (i.e., a finite number of samples from a Gaussian distribution), and therefore it is not valid to use a continuous DOA distribution to estimate the correlation present between different antenna elements in the array. The correlation that results from a continuous DOA distribution decreases monotonically with element spacing, whereas the correlation that results from discrete DOA exhibits damped oscillations present. Therefore, a continuous DOA distribution will underestimate the correlation that exists between the elements in the array [Asz96].

In [Asz96], a comparison is made between the correlation obtained using the discrete uniform distribution model, Lee's model, and a continuous Gaussian AOA as a function of element spacing. The comparisons indicate that, for small element separations (two wavelengths), the three models have nearly identical correlation characteristics. For larger element separations (greater than two wavelengths), the correlation values using the continuous Gaussian DOA are close to zero, while the two discrete models have oscillation peaks with correlation as high as 0.2 even beyond four wavelengths. Additionally, it was found that the correlation of the discrete uniform distribution falls off more quickly than the correlation in Lee's model.

Again, while the model is useful for predicting the correlation between any pair of elements in the array (which can be used to calculate the array correlation matrix), it fails to include all the channel phenomena, such as delay spread and Doppler spread, required for certain types of simulations.

6.2.4 Geometrically Based Single Bounce Statistical Channel Models

Geometrically Based Single Bounce (GBSB) Statistical Channel Models are developed by defining a spatial scatterer density function. These models are useful for both simulation and analysis purposes. Use of the models for simulation involves randomly placing scatterers in the scatterer region according to the form of the spatial scatterer density function. From the location of each of the scatterers, the DOA, TOA, and signal amplitude are determined.

From the spatial scatterer density function, it is possible to derive the joint and marginal TOA and DOA probability density functions. Knowledge of these statistics can be used to predict the performance of an adaptive array. Furthermore, knowledge of the underlying structure of the resulting array response vector may be exploited by beamforming and position location algorithms.

The shape and size of the spatial scatterer density function required to provide an accurate model of the channel is subject to debate. Validation of these models through extensive measurements remains an active area of research.

Two variations of the GBSB model are considered. The *GBSB Circular Model* (GBSBCM) is applicable to high tier, macro-cellular scenarios where the base station is very high relative to scatterers, and is based on a circular distribution of scatterers near the mobile unit. A second model, a *GBSB Elliptical Model* (GBSBEM), assumes that scatterers are distributed throughout the transmitter/receiver path. The GBSBEM is more applicable to low tier microcell systems and cases where the base station antenna is at the same height as the surrounding clutter.

6.2.5 The Geometrically Based Single Bounce Circular Model (Macrocell Model)

The geometry of the Geometrically Based Single Bounce Circular Model (GBSBCM) is shown in Figure 6–13. It assumes that the scatterers lie within radius R_m about the mobile. Often the requirement that $R_m < D$ is imposed. The model is based on the assumption that in macrocell environments where antenna heights are relatively high, there will be no signal scattering from locations near the base station. The idea of a circular region of scatters centered about the mobile was originally proposed by Jakes [Jak74], and used in [Lee82] to derive theoretical results for the correlation observed between two antenna elements. Later, it was used to determine the effects of beamforming on the Doppler spectrum [Pet97a], [Pet97b] for narrowband signals. It was shown that the rate and the depth of the envelope fades are significantly reduced when a narrow-beam beamformer is used at the base station.

The joint TOA and DOA density function obtained from the model provides some insights into the properties of the model. Using a Jacobian transformation, it is possible to derive the joint TOA and DOA density function at both the base station and the mobile. The resulting joint probability density function (PDF) at the base station is

$$f_{\tau,\phi}(\tau,\phi) = \begin{cases} \dfrac{(D^2-\tau^2c^2)(D^2c-2D\tau c^2\cos\phi+\tau^2c^3)}{4\pi R_m^2(D\cos\phi-\tau c)^3} & \dfrac{(D^2-2\tau c\cos\phi+\tau^2c^2)}{\tau c-D\cos\phi} \le 2R_m \\ 0 & \text{otherwise} \end{cases} \quad (6.20)$$

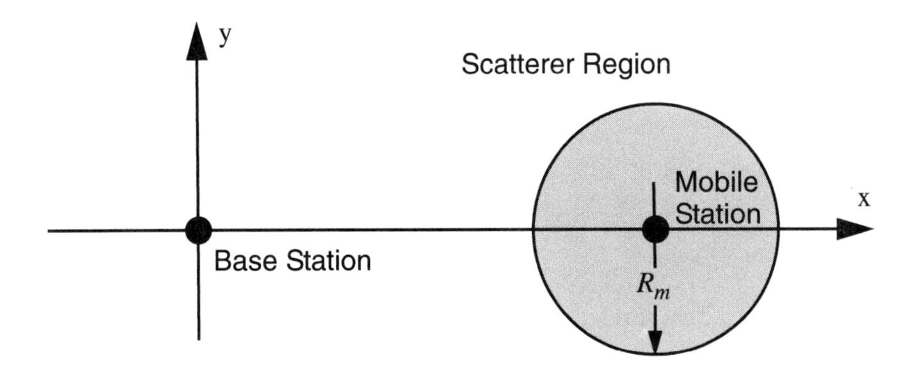

Figure 6–13 Circular scatterer density geometry. (©1998 IEEE.)

At the mobile unit, the pdf is

$$f_{\tau,\phi}(\tau,\phi) = \begin{cases} \dfrac{(D^2-\tau^2c^2)(D^2c-2D\tau c^2\cos\phi+\tau^2c^3)}{4\pi R_m^2(D\cos\phi-\tau c)^3} & \dfrac{D^2-\tau^2c^2}{D\cos(\phi)-\tau c} \le 2R_m \\ 0 & \text{otherwise} \end{cases} \tag{6.21}$$

The circular model predicts a relatively high probability of multipath components with small excess delays along the line of sight. From the base station perspective, all of the multipath components are restricted to lie within a small range of angles. The appropriate values for the radius of scatterers can be determined by equating the angle spread predicted by the model (which is a function of R_m) with measured values. Measurements reported in [Asz96] suggest that typical angle spreads for macrocell environments with T-R separation of 1 km are approximately two to six degrees.

The GBSBCM can be used to generate sample channels for simulation purposes. Generation of sample channel impulse responses from the GBSBCM is accomplished by uniformly placing scatterers in the circular scatter region about the mobile and then calculating the corresponding DOA, TOA, and power levels.

6.2.6 The Geometrically Based Single Bounce Elliptical Model (Microcell Model)

In the GBSBEM model, scatterers are uniformly distributed within an ellipse, as shown in Figure 6–16. The base station and the mobile unit are at the foci of the ellipse. This model was developed for microcell and picocell environments where antenna heights are low, so that multipath scattering is just as likely near the base station as it is near the mobile [Lib95][Lib96a].

An essential feature of the GBSBEM model is the physical interpretation that only multipath signals that arrive with an absolute delay of less than or equal to τ_m are considered.

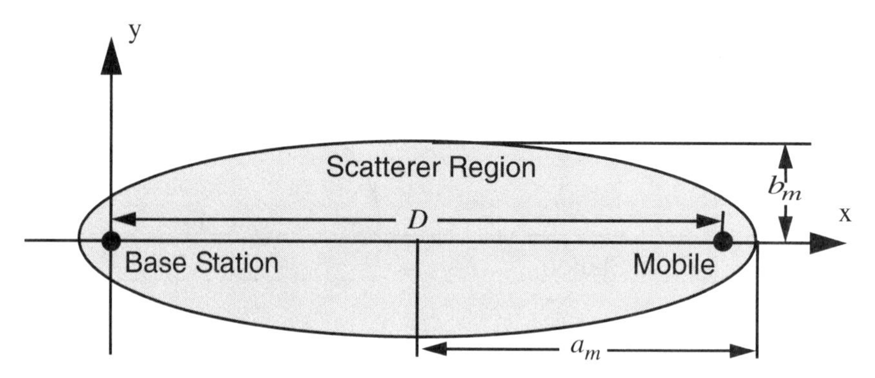

Figure 6–14 Elliptical scatterer density geometry.

Ignoring components with longer delays is possible because very long delays correspond to long paths, which experience higher path loss. Provided that τ_m is chosen sufficiently large, the model will account for nearly all power and DOA of multipath components. The parameters a_m and b_m are the major and minor axes of the of the ellipse containing the scatterers.

$$a_m = c\tau_m/2 \tag{6.22}$$

$$b_m = \sqrt{c^2\tau_m^2 - D^2} \tag{6.23}$$

where c is the speed of light, and τ_m is the maximum delay to be considered.

$$f_{\tau,\phi}(\tau,\phi) = \begin{cases} \dfrac{(D^2-\tau^2c^2)(D^2c-2D\tau c^2\cos\phi+\tau^2c^3)}{4\pi a_m b_m(D\cos\phi-\tau c)^3} & \dfrac{D}{c} < \tau \le \tau_m \\ 0 & \text{otherwise} \end{cases} \tag{6.24}$$

The power-delay-angle profile, also called a *pda-gram*, shown in Figure 6–4 was generated from the GBSBEM model, using $\tau_m = 5.5\ D/c$. In Chapter 7, the GBSBEM model is presented in detail and it is used to derive a wide range of channel characteristics, including characterization of the channel angle spread, power delay profile, and narrowband fading characteristics. Commercial simulation software that implements models like this is available from [Wir98].

6.2.7 The Gaussian Wide Sense Stationary Uncorrelated Scattering (GWSSUS) Model

The GWSSUS is a statistical channel model that makes assumptions about the form of the received signal vector [Zet94][Zet95][Zet96a][Zet96b]. The primary motivation of the model is to provide a general equation for the received signal correlation matrix. In the GWSSUS model, scatterers are grouped into clusters in space. The delay differences within each cluster are not resolvable within the transmission signal bandwidth. By including multiple clusters, resolvable

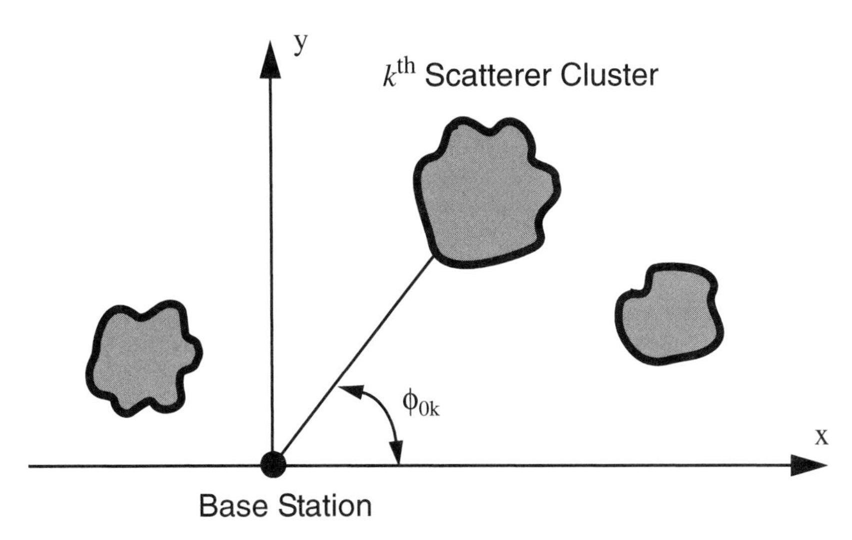

Figure 6–15 GWSSUS geometry. (©1998 IEEE.)

multipath which introduces frequency selective fading channels is modeled using the GWSSUS. Figure 6–15 shows the geometry assumed for the GWSSUS model for $D = 3$ clusters.

The mean DOA for the kth cluster is denoted ϕ_{0k}. It is assumed that the location and delay associated with each cluster remain constant over b data bursts. The form of the received signal vector is

$$\boldsymbol{x}_b(t) = \sum_{k=0}^{D-1} \boldsymbol{v}_{k,b} s(t - \tau_k) \tag{6.25}$$

where $\boldsymbol{v}_{k,b}$ is the superposition of the steering vectors during the bth data burst within the kth cluster, which may be expressed as

$$\boldsymbol{v}_{k,b} = \sum_{i=0}^{N_k - 1} \rho_{k,i} e^{j\psi_{k,i}} \mathbf{a}(\phi_{0k} - \phi_{k,i}) \tag{6.26}$$

where N_k denotes the number of scatterers in the kth cluster, $\rho_{k,i}$ is the amplitude, $\psi_{k,i}$ is the phase, $\phi_{k,i}$ is the Angle-Of-Arrival of the ith reflected scatterer of the kth cluster, and $\mathbf{a}(\phi)$ is the array response vector in the direction of ϕ [Asz96]. It is assumed that the steering vectors are independent for different k.

If N_k is sufficiently large (approximately 10 or more [Zet94]), for each cluster of scatterers, the central limit theorem may be applied to the elements of $\boldsymbol{v}_{k,b}$. Under this condition, the elements of $\boldsymbol{v}_{k,b}$ are Gaussian distributed. Additionally, it is assumed that $\boldsymbol{v}_{k,b}$ is wide sense stationary. The time delays τ_k, are assumed to be constant over several bursts, whereas the phases $\psi_{k,i}$ change much more rapidly. The vectors $\boldsymbol{v}_{k,b}$ are assumed to be zero mean, complex Gaussian wide sense, stationary random processes where b plays the role of the time argument. The vector $v_{k,b}$ is a multivariate Gaussian distribution, which is described by its mean and covariance matrix. When no line of sight component is present, the mean will be zero due to the random phase $\psi_{k,i}$, which is assumed to be uniformly distributed in the range 0 to 2π. When a direct path component is present, the mean becomes a scaled version of the corresponding array response vector $E\{v_{k,b}\} \propto \mathbf{a}(\phi_{0k})$ [Asz96]. The covariance matrix for the kth cluster is given by [Zet96b].

$$R_k = E\{\mathbf{v}_{k,b}\mathbf{v}_{k,b}^H\} \tag{6.27}$$

$$= \sum_{i=0}^{N_k-1} |\alpha_{k,i}|^2 E\{\mathbf{a}(\phi_{0k} - \phi_{k,i})\mathbf{a}^H(\phi_{0k} - \phi_{k,i})\}$$

The model provides a fairly general result for the form of the covariance matrix. However, it does not indicate the number or location of the scattering clusters, and hence the model requires some additional information for application to typical environments.

6.2.8 Gaussian Angle-Of-Arrival (GAA)

The Gaussian Angle-Of-Arrival (GAA) channel model is a special case of the GWSSUS model described above, where only a single cluster is considered ($D = 1$ in (6.25)), and the DOA statistics are assumed to be Gaussian distributed about some nominal angle, ϕ_0, as shown in Figure 6–16. Since only a single cluster is considered, the model is a narrowband channel model that is valid when the time spread of the channel is small compared to the inverse of the signal bandwidth; hence, time shifts may be modeled as simple phase shifts and the channel is flat fading [Ott95].

The statistics of the steering vector are distributed as a multivariate Gaussian random variable. Similar to the GWSSUS model, if no line of sight is present, then $E\{v_{k,b}\} = 0$; otherwise, the mean is proportional to the array response vector $\mathbf{a}(\phi_{0k})$. For the special case of uniform linear arrays, the covariance matrix may be described by

$$\mathbf{R}(\phi_0, \sigma_\phi) \approx p\mathbf{a}(\phi_0)\mathbf{a}^H(\phi_0) \otimes \mathbf{B}(\phi_0, \sigma_\phi) \tag{6.28}$$

where the (k,l) element of $\mathbf{B}(\phi_0,\sigma_\phi)$ is given by

$$B(\phi_0, \sigma_\phi)_{k,l} = \exp[-2(\pi\Delta(k-l))^2\sigma_\phi^2\cos^2\phi] \tag{6.29}$$

and p is the receiver signal power, Δ is the element spacing, and $\otimes$ denotes element-wise multiplication [Ott95].

6.2.9 Time-Varying Vector Channel Model (Raleigh's Model)

Raleigh's time varying vector channel model was developed to provide both small scale Rayleigh fading and theoretical spatial correlation properties [Ral95a]. The propagation envi-

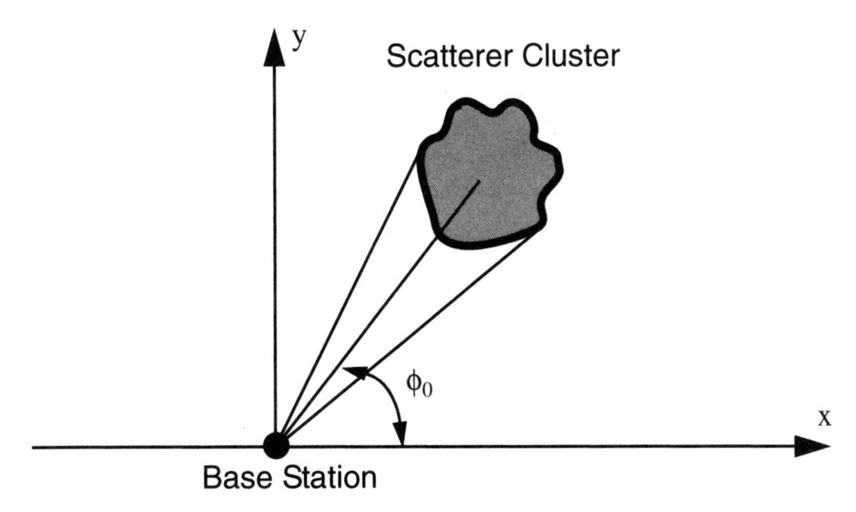

Figure 6–16 GAA geometry. (©1998 IEEE.)

ronment considered is densely populated with large dominant reflectors, as shown in Figure 6–17. It is assumed that at a particular instant of time, the channel is characterized by L dominant reflectors. The received signal vector is then modeled as

$$\mathbf{x}(t) = \sum_{l=0}^{L(t)-1} \mathbf{a}(\phi_l)\alpha_l(t)s(t-\tau) + \mathbf{n}(t) \tag{6.30}$$

where $\mathbf{a}$ is the steering vector, $\alpha_l(t)$ is the complex path amplitude, $s(t)$ is the modulated signal, and $\mathbf{n}(t)$ is additive noise. This is equivalent to the impulse response given in (6.4).

The unique feature of the model is in the calculation of the complex amplitude term, $\alpha_l(t)$, which is expressed as

$$\alpha_l(t) = \beta_l(t) \cdot \sqrt{\Gamma_l \cdot \psi(\tau_l)} \tag{6.31}$$

where Γ_l accounts for log-normal fading, $\psi(\tau_l)$ describes the total received power delay profile, and $\beta_l(t)$ is the complex intensity of the radiation pattern as a function of time. The complex intensity is described by

$$\beta_l(t) = K\sum_{n=0}^{N_t-1} C_n(\phi_l)\exp(j2nf_d\cos(\Omega_{n,l})t) \tag{6.32}$$

where N_t is the number of signal components contributing to the lth dominant reflector, K accounts for the antenna gain and transmit signal power, $C_n(\phi_l)$ is the complex radiation on the nth component of the lth dominant reflecting surface in the direction of ϕ_l, f_d is the maximum Doppler shift, and $\Omega_{n,l}$ is the angle toward the nth component of the lth dominant reflector with respect to the motion of the mobile. The resulting complex intensity, $\beta_l(t)$, exhibits a complex

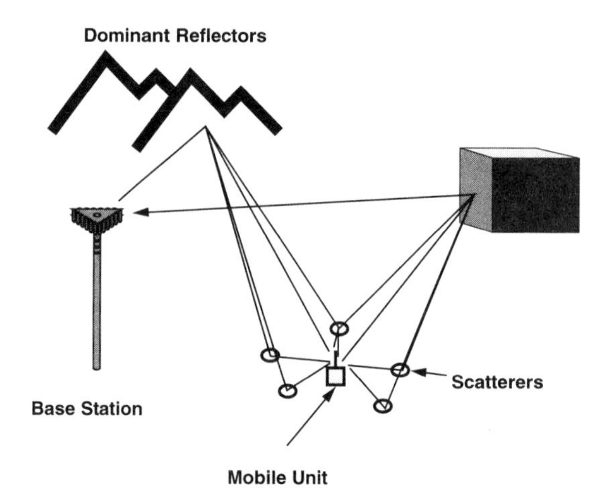

Figure 6–17 Raleigh's model signal environment. (©1998 IEEE.)

Gaussian distribution in all directions away from the mobile [Ral95a]. Both the time and spatial correlation properties of the model are compared to theoretical results in [Ral95a]. The comparison shows that there is good agreement between the two.

6.2.10 Two GSM Simulation Models (TU and BU)

Two spatial channel models have been developed for simulation purposes in the GSM standard [Ans95a]. The Typical Urban (TU) model is designed to have time properties appropriate for large towns in flat environments, while the Bad Urban (BU) model was developed to model large delay spread environments with large reflectors that are not in the vicinity of the mobile.

Both of these models obtain the received signal vector using

$$\mathbf{x}(t) = \sum_{n=0}^{N-1} \alpha_n(t) \exp\left(-j2\pi f_c \frac{d_n(t)}{c} + \psi\right) s\left(t - \frac{l_n(t)}{c} + \Delta_t\right) \mathbf{a}(\phi_n(t)) \tag{6.33}$$

where N is the number of scatterers, f_c is the carrier frequency, c is the speed of light, $l_n(t)$ is the path propagation distance, ψ is a random phase, and Δ_t is random delay for each multipath component. In general, the path propagation distance $d_n(t)$ will vary continuously with time; hence, Doppler fading occurs naturally in the model.

6.2.11 The Typical Urban (TU) Model

In the TU model, 120 scatterers are randomly placed within a 1 km radius about the mobile [Mog96], similar to the concept used in SIRCIM and SMRCIM developed by Seidel and Rappaport [Rap96a][Wir98]. The position of the scatterers is held fixed over the duration in which the mobile travels a distance of 5 m. At the end of the 5 m, the scatterers are returned to their original position with respect to the mobile. At each 5-m interval, random phases are assigned to the scatterers as well as randomized shadowing effects, which are modeled as log-normal with distance with a standard deviation of 5-10 dB [Mog96]. The received signal delay is determined by assuming a ray-traced path from the location of each scatterer. An exponential path loss law is also applied to account for large-scale fading [Zet96a]. Simulations have shown that the TU model and the GSM-TU model have nearly identical power delay profiles, Doppler spectrums, and delay spreads [Mog96]. Furthermore, the DOA statistics are approximately Gaussian and similar to those of the GAA model described in Section 6.2.8.

6.2.12 The Bad Urban (BU) Model

The BU model is identical to the TU model with the addition of a second scatterer cluster with another 120 scatterers offset 45° from the first, as shown in Figure 6–18. The scatterers in the second cluster are assigned 5 dB less average power than the original cluster [Mog96]. The presence of the second cluster results in an increased angle spread, which in turn reduces the off-diagonal elements of the array covariance matrix. The presence of the second cluster also causes an increase in the delay spread.

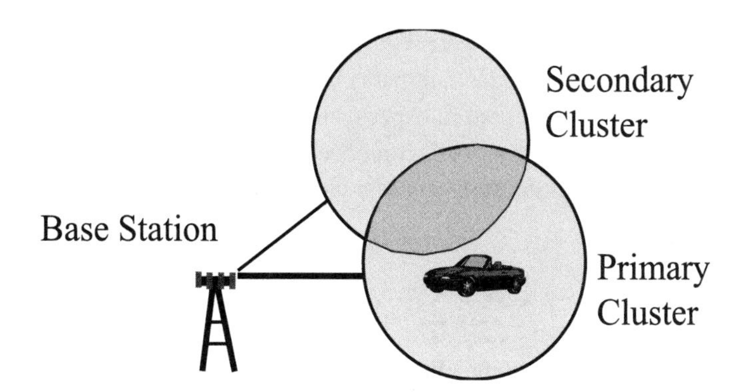

Figure 6–18 Bad Urban vector channel model geometry. (©1998 IEEE.)

6.2.13 The Uniform Sectored Distribution Model

Distribution (USD) is shown in Figure 6–19 [Nor94]. The model assumes that scatterers are uniformly distributed within an angle distribution of ϕ_{BW} and a radial range of Δ_R centered about the mobile. The magnitude and phase associated with each scatterer are selected at random from a uniform distribution of [0, 1] and [0, 2π], respectively. As the number of scatterers approaches infinity, the signal fading envelope becomes Rayleigh with uniform phase [Nor94]. In [Nor94], the model is used to study the effect of angle spread on spatial diversity techniques.

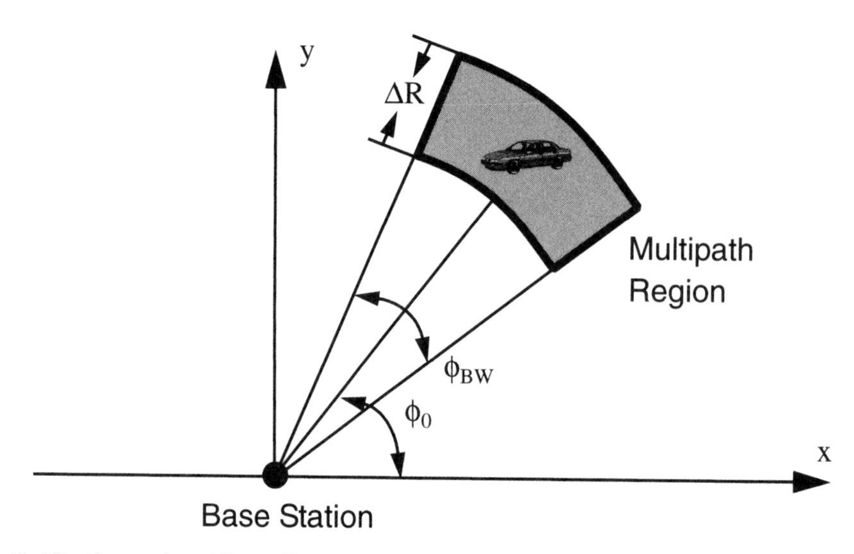

Figure 6–19 Geometry of the uniform sectored distribution. (©1998 IEEE.)

6.2.14 Modified Saleh-Valenzuela's Model

Saleh and Valenzuela developed a multipath channel model for indoor environments based on the clustering phenomenon observed in experimental data [Sal87]. The clustering phenomenon refers to the observation that multipath components arrive at the antenna in groups. It was found that both the clusters and the rays within a cluster decayed in amplitude with time. The impulse response of the model is given by

$$h(t) = \sum_{i=0}^{\infty} \sum_{j=0}^{\infty} \alpha_{ij}\delta(t - T_i - \tau_{ij}) \tag{6.34}$$

where the sum over i corresponds to the clusters and the sum over j represents the rays within a cluster. The variables $\{\alpha_{ij}\}$ are Rayleigh distributed with the mean square value described by a double-exponential decay given by

$$\overline{\alpha_{ij}^2} = \overline{\alpha_{00}^2}\exp(-T_i/\Gamma)\exp(\tau_{ij}/\gamma) \tag{6.35}$$

where Γ and γ are the cluster and ray time decay constants, respectively. Motivated by the need to include DOA in the channel mode, Spencer *et al.* proposed an extension to the Saleh-Valenzuela's model [Spe97a], assuming that time and the angle are statistically independent, or

$$h(t, \phi) = h(t)H(\phi) \tag{6.36}$$

Similar to the time impulse response in (6.34), the proposed *angular impulse response* is given by

$$f(\phi) = \frac{1}{\sqrt{2}\sigma}\exp\left(-\left|\frac{\sqrt{2}\phi}{\sigma}\right|\right) \tag{6.37}$$

where σ is the rms angle spread of each cluster. This model was proposed based on indoor measurements as discussed later in this chapter.

6.2.15 Extended Tap-Delay-Line Method

A wideband channel model that is an extension of the traditional statistical tap-delay-line model and includes DOA information was developed by Klein and Mohr [Kle96a]. The channel impulse response is represented by

$$h(\tau, t, \phi) = \sum_{w=0}^{W-1} \alpha_w(t)\delta(\tau - \tau_w)\delta(\phi - \phi_w). \tag{6.38}$$

This model is comprised of W taps, each with an associated time delay τ_w, complex amplitude α_w, and DOA ϕ_w. The joint density functions of the model parameters should be determined from measurements. As shown in [Kle96a], measurements can provide histograms of the joint distribution of $|\alpha|$, τ, and ϕ, and the density functions, which are proportional to these histograms, can be chosen.

6.2.16 Elliptical Subregions Model (Lu, Lo, and Litva's Model)

Lu *et al.* [Lu97] proposed a model of multipath propagation based on the distribution of the scatterers in elliptical subregions, as shown in Figure 6–20. Each subregion (shown in a different shade) corresponds to one range of the excess delay time.

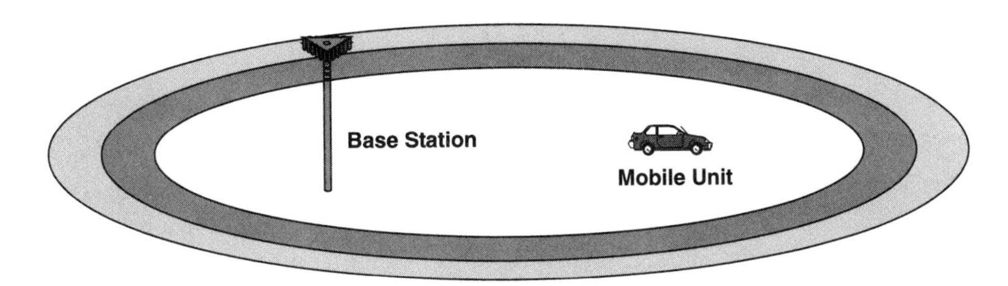

Figure 6–20 Elliptical subregions spatial scatterer density. (©1998 IEEE.)

This approach is essentially the same as the GBSBEM developed by Liberti and Rappaport [Lib96a] in that an ellipse of scatterers is considered. The primary difference between the two models is in the selection of the number of scatterers and the distribution within the entire ellipse. In Lu, Lo, and Litva's model, the ellipse is first subdivided into a number of subregions. The number of scatterers within each subregion is then selected from a Poisson random variable, whose mean is chosen to match the measured time delay profile data.

It was also assumed that the multipath components arrive in clusters due to the multipath reflecting points of the scatterers. Thus, assuming that there are L scatterers with K_l reflecting points, the model proposed is represented by

$$h(t, t_0) = \sum_{i=0}^{L-1} E_t(\phi_i^{(t)}) \times \sum_{k=0}^{K_i-1} \rho_{ik}\exp(-(2\pi f_{ik}t_0 + \gamma_{ik}))\delta(t - \tau_{ik})E_r(\phi_{ik}) \tag{6.39}$$

where ρ_{ik}, τ_{ik}, and γ_{ik} correspond to the amplitude, time delay, and phase of the signal component from the ik^{th} reflecting point, respectively. Here, f_{ik} is the Doppler frequency shift of each individual path, ϕ_{ik} is the angle between the ik^{th} path and the receiver-to-transmitter direction, and $\phi_i^{(t)}$ is the angle of the ith scatterer as seen from the transmitter. $E_t(\phi)$ and $E_r(\phi)$ are the radiation patterns of the transmit and receive antennas, respectively. The variable ϕ_{ik} is assumed to be Gaussian distributed.

Simulation results using this model were presented in [Lu97], showing that a 60° beamwidth antenna reduces the mean RMS delay spread by about 30-40 percent. These results are consistent with similar measurements made in Toronto using a sectorized antenna [Sou94].

6.2.17 Measurement Based Channel Model

A channel model in which the parameters are based on measurement was proposed by Blantz *et al.* [Bla96]. The idea behind this approach is to characterize the propagation environment, in terms of scattering points, based on measurement data. The time-variant impulse response takes the form

$$h(\tau, t) = \int_0^{2\pi} (\vartheta(\tau, t, \phi))^* \gamma(\tau, \phi)^* f(\tau) d\phi \tag{6.40}$$

where $f(\tau)$ is the impulse response representing the joint transfer characteristic of the transmission system components (modulator, demodulator, filters, etc.) and $\gamma(\tau, \phi)$ is the characteristic of the base station antenna. The term $\vartheta(\tau, t, \phi)$ is the time-variant directional distribution of channel impulse response seen from the base station. This distribution is time-variant due to mobile motion and depends on the location, orientation, and velocity of the mobile station antenna and the topographical and morphological properties of the propagation area. SIRCIM and SMRCIM, two measurement-based channel simulators, use the GBSBEM model to synthesize multipath data with realistic DOAs, delays, and amplitude variations [Rap96a][Wir98].

6.2.18 Ray Tracing Models

The models presented in Section 6.2 are based on statistical and/or geometrical analysis and measurements and provide the average path loss, delay spread, and Angle-Of-Arrival characteristics. In recent years, *ray tracing* techniques have been extensively developed, based on geometric theory, reflection, diffraction, and scattering propagation models [Rap96a].

By using site-specific information, such as building databases of architectural drawings, this technique deterministically models the propagation channel [Sou94], [Val93], [Sei91], including the path loss exponent, Angle-Of-Arrival, and delay spread. However, the high computational burden and the difficulty in obtaining detailed terrain and building databases can make ray tracing models difficult to use. Some progress has been made in overcoming the computational burden, and powerful commercial tools that perform site-specific ray tracing coverage analysis are just now becoming available for widespread use [Wir98]. A building database used for ray tracing analysis of smart antenna systems is shown in Figure 6–21.

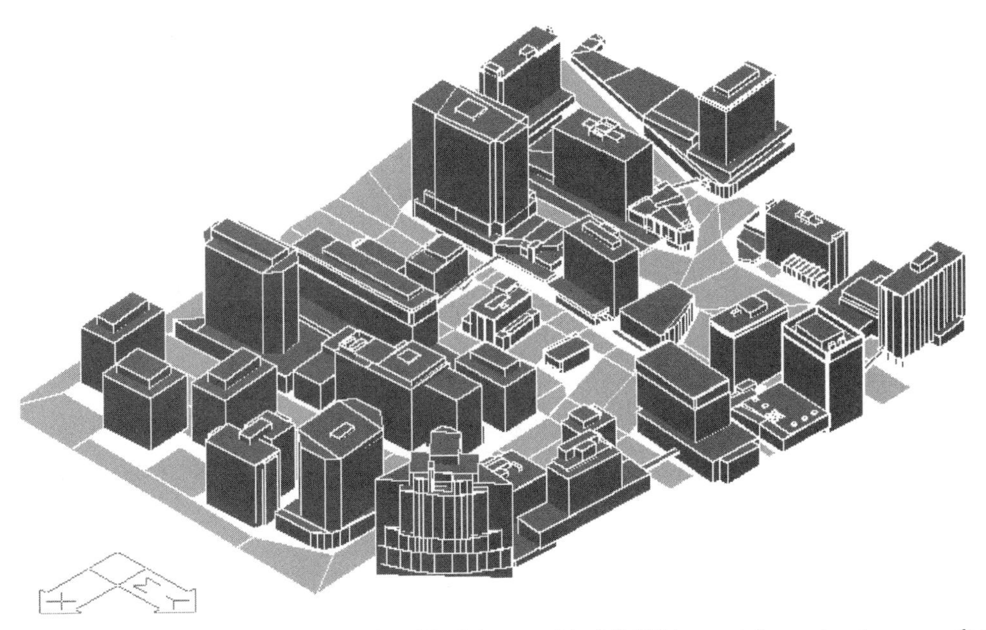

Figure 6–21 Digital model of Rosslyn, Virginia, used in [Lib95] to model smart antenna performance in real world radio channels, using ray-traced multipath channels.

6.3 Spatial Channel Measurements

To date, relatively few publications have described spatial channel measurements. In [Kle96b], TOA and DOA measurements are presented for outdoor macrocellular environments. The measurements were made using a rotating 9° azimuth beam directional receiver antenna with a 10 MHz bandwidth centered at 1840 MHz. Three environments near Munich were considered, including rural, suburban, and urban areas with base station antenna heights of 12.3 m, 25.8 m, and 37.5 m, respectively. The key observations include [Kle96b]:

- Most of the signal energy is concentrated in a small interval of delay and within a small range of DOA in rural, suburban, and even many urban environments.
- By using directional antennas, it is possible to reduce the time dispersion.

In 1996, measurements performed by Liberti *et al*, derived statistics of the rms angle spread in a wide variety of environments at 1900 MHz [Lib97]. In these measurements, as illustrated in Figures 6–22, a rotating dish antenna, with a 10° beamwidth was used to characterize the angle spread in rural, suburban, and urban environments. At each location, measurements were performed at four antenna heights, 13, 23, 33, and 40 feet above ground level, to characterize the impact of ground clutter on smart antenna performance. Results from these measurements are shown in Figures 6–7 and 6–8.

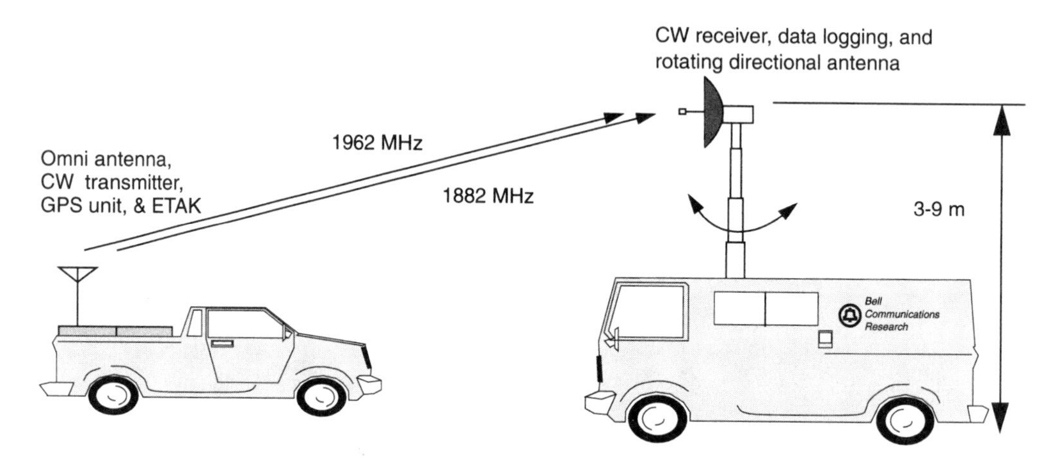

Figure 6–22 Measurements performed by Liberti, Murray, and Bergmann in 1996 [Lib97]. A photograph of the van is shown in Figure 6–23.

These results clearly show that

- Angle spread increases as the base station antenna height is lowered.
- Angle spread increases from rural to suburban environments, and from suburban to urban settings.

In 1997, array measurements were performed along a highway in New Jersey to characterize the time varying spatial signature [Lib98a]. These experiments used an 8-element circular array, shown in Figure 6–23. In these measurements, array data was captured from a high speed mobile transmitter traveling at speeds of 35 to 75 mph. The time-variation of the spatial signature for these measurements is shown in Figure 6–10. These measurements illustrated that

- Very small aperture arrays could provide highly accurate Direction-Of-Arrival information for high speed mobile transmitters, using very short 10 ms signal snapshots.
- The time-variation of the spatial signature tends to increase as mobile vehicle speed increases. As shown in Figure 6–10, the time-variation of the spatial signature is overbounded by a function which increases with vehicle speed.

Figure 6–23 Mast mounted, 8-element circular array used by Liberti and Murray to perform dynamic spatial signature measurements from high speed mobile transmitters, as described in [Lib98a][Lib98b] (©1998 IEEE).

Another set of TOA and DOA measurements is reported in [Tho92] for urban areas. The measurements were made using a two-element receiver mounted on a test vehicle with an elevation of 2.6 m. The transmitting antenna was placed 30 m high on the side of a building. A bandwidth of 10 MHz with a carrier frequency of 2.33 GHz was used. The delay-Doppler spectra observed at the mobile was used to obtain the delay-DOA spectra. The results indicate that it is possible to account for most of the major features of the delay-DOA spectra by using ray tracing and considering the large buildings in the environment.

Motivated by diversity combining methods, earlier measurements were concerned primarily with determining the correlation between the signals at two antenna elements as a function of the element separation distance. These studies found that, at the mobile, relatively small separation distances were required to obtain a low correlation between the elements, whereas at the base station very large spacing was needed. These findings indicate that there is a relatively small angle spread observed at the base station, following Lee's original work [Ada86].

Previously, an extension to Saleh-Valenzuela's indoor model, including DOA information, was proposed based on indoor measurements of delay spread and DOA made at Brigham Young University at 7 GHz [Spe97b]. The DOAs were measured using a 60 cm parabolic dish antenna with a 3 dB beamwidth of 6°. The results showed a clustering pattern in both time and angle domain, which led to the proposed channel model described in Section 6.2.14. Also, it was observed that the cluster mean Angle-Of-Arrival was uniformly distributed [0, 2π]. The distribution of the Angle-Of-Arrival of the rays within a cluster presented a sharp peak at the mean, leading to the use of the Laplacian distribution for modeling. The standard deviation found for this distribution was about 25°. Based on these measurements, a channel model including delay spread and DOA information was proposed, under the assumption that time delay and Angle-Of-Arrival were independent variables, which is not true in general.

In [Fuh97], two-dimensional DOA and delay spread measurement and estimation were presented. The measurements were made in downtown Paris using a channel sounder at 900 MHz and a horizontal rectangular planar array at the receiver. The estimation of DOA, including azimuth and elevation angle, was performed using 2D unitary ESPRIT [Zol96] with a time resolution of 0.1 μs and angle of resolution of 5°. The results presented confirmed assumptions made in urban propagation, such as the wave-guiding mechanism of streets and the exponential decay of the power delay profile.

Finally, in [Jen95], measurements are used to show the variation in the spatial signature with both time and frequency. Two measurements of change were given, the relative angle change given by

$$\text{Relative Angle Change } (\%) = 100 \times \sqrt{1 - \left|\frac{\mathbf{a}_i^*}{\|\mathbf{a}_i\|} \cdot \frac{\mathbf{a}_j^*}{\|\mathbf{a}_j\|}\right|^2} \tag{6.41}$$

and the relative amplitude change, found by using

$$\text{Relative Amplitude Change (dB)} = 20\log_{10}\frac{\|\mathbf{a}_j\|}{\|\mathbf{a}_i\|} \tag{6.42}$$

where $\mathbf{a}_i$ and $\mathbf{a}_j$ are the two spatial signatures (array response vectors) being compared. The measurements indicate that when the mobile and surroundings are stationary, there are relatively small changes to the spatial signature. Likewise, there are moderate changes when objects and the environment are in motion and large changes when the mobile itself is moving. Also, it was found that the spatial signature changes significantly with a change in carrier frequency. In particular, the measurements found that the relative amplitude change in the spatial signal could exceed 10 dB with a frequency change of only 10 MHz. This result indicates that the uplink spatial signature cannot be directly applied for the downlink beamforming in most of today's cellular and PCS systems that have 45 MHz and 80 MHz separation between the uplink and downlink frequencies, respectively.

It is instructive to note that the relative angle change is related to the change in spatial signature by:

$$\text{Relative Angle Change } (\%) = 100 \times \frac{\|\Delta \boldsymbol{b}_i\|}{\|\boldsymbol{b}_{i+1}\|} = 100 \times \sqrt{1 - \left|\frac{\boldsymbol{b}_i^H}{\|\boldsymbol{b}_i\|} \cdot \frac{\boldsymbol{b}_{i+1}}{\|\boldsymbol{b}_{i+1}\|}\right|^2} \tag{6.43}$$

This metric was used to derive the time-rate-of-change of the spatial signature given in (6.15).

6.4 Application of Spatial Channel Models

The effect that classical channel properties such as delay spread and Doppler spread have on system performance has been an active area of research for several years and is fairly well understood. Spatial channel models include the DOA properties of the channel, which are often characterized by the angle spread. The angle spread has a major impact on the correlation observed between the pairs of elements in the array. These correlation values specify the received signal vector correlation matrix, which in turn determines the performance of linear combining arrays [Mon80]. In general, the higher the angle spread, the lower the correlation observed between any pair of elements in the array. The various spatial channel models provide different angle spreads and hence will predict different levels of system performance.

The channel models presented here have various applications in the analysis of systems that utilize adaptive antenna arrays. Some of the models were developed to provide analytical models of the spatial correlation function, while others are intended primarily for simulation purposes.

Increasingly, wireless service providers and manufacturers are relying on detailed simulations to help design and develop today's wireless networks. The application of adaptive antennas is no exception. However, to obtain reliable results, accurate spatial channel models are needed. Fortunately, statistical channel models that include DOA and wideband time delay spread are available for modem and smart antenna development [Wir98]. With accurate simulations of adaptive antenna array systems, researchers will be able to predict the capacity improvement, range extension, and other performance measures of the system, which in turn will determine the cost of effectiveness of adaptive array technologies.

6.5 Summary

In this chapter, an overview of a wide range of modeling techniques used to characterize spatial radio channels has been provided. These models have varying complexity, and the set of assumptions underlying each model tends to be appropriate for particular environments or geometries. We also provided an overview of the key measurement campaigns that have endeavored to characterize spatial radio channels. In the next chapter, we present a detailed derivation of the Geometrically Based Single Bounce Elliptical Model (GBSBEM), and explore different ways that the model can be applied in investigating spatial radio channels.

CHAPTER 7

The Geometrically Based Single Bounce Elliptical Model

In the previous chapter, an overview of spatial channel models, metrics, and measurements was presented. Here we describe one particular model, the Geometrically Based Single Bounce Elliptical Model (GBSBEM) [Lib95][Lib96a]. The complete derivation of the model is provided. The model is used to simulate power-delay-angle profiles (*pda-grams*), power delay profiles, joint time-angle statistics, marginal characteristics of the Direction-Of-Arrival, and narrowband fading envelopes. The GBSBEM model is appropriate for low-tier systems, including microcell and picocell systems, where base station antennas are surrounded by clutter, and scatterers are distributed between and around both the transmitter and receiver. Other models, described in the previous chapter, may be applicable under different conditions and in different environments. However, the rich extensions and analytical potential of the GBSBEM model make it an ideal candidate for study, showing the inter-relationship among different aspects of the spatial radio channel.

When omni-directional antennas are used at the transmitter and receiver, a complex envelope model for the multipath channel impulse response is

$$h(t) = \sum_{i=0}^{L-1} \alpha_i \delta(t-\tau_i) = \sum_{i=0}^{L-1} \rho_i e^{j\psi_i} \delta(t-\tau_i) \tag{7.1}$$

where $\rho_i = |\alpha_i|$ is the magnitude of the ith multipath component, γ_i is the phase of the ith multipath component, and τ_i is the delay associated with that component. The parameter L is the total number of multipath components. In [Rap89], [Sei89], and [Sal87], the statistics of L, ρ_i, and τ_i are presented based on measured data.

To model system performance with directional antennas, it is also necessary to model the Direction-Of-Arrival (DOA), ϕ_i and the Direction-Of-Departure (DOD), θ_i, of each component. Strictly speaking, it is also necessary to model the polarization state of each component as well; however, rather than explicitly model the polarization state, polarization mismatches are absorbed into other aspects of the GBSB model.

Unlike previous statistical characterizations which are primarily based on measurements, the model presented here is based entirely on geometry. We begin by modeling a transmitter and receiver in free space with a separation d_0. We then add scatterers so that all scatterers lie in the plane which includes the transmitter and receiver. For simplicity, it is convenient to view this plane as horizontal, although the plane may, in fact, be slightly inclined to allow for differences between the transmitter and receiver antenna heights. If all scatterers lie in a plane, then the paths of all multipath components will also lie in the plane, so that, at the receiver, all plane waves will appear to arrive from the horizon ($\theta = \pi/2$ in Figure 3–2.). In macro-cellular systems, the vertical distribution of measured Directions-Of-Arrival supports the notion that rays arrive primarily from the horizontal direction [Rap96a].

If a line-of-sight (LOS) path exists between the transmitter and the receiver, then the first component will arrive at a time $\tau_0 = d_0/c$ where c is the speed of light, 3×10^8 m/s.

Let us consider a multipath component which arrives at time τ_i. If the component is the result of a single reflection (a single bounce path), then the scatterer causing the reflection must lie on an ellipse with major axis half length a, and minor axis half length b, as shown in Figure 7–1 [Sei89]. The quantities a, b, and f are given by

$$f = d_0/2 \tag{7.2}$$

$$a = c\tau_i/2 \tag{7.3}$$

$$b = \sqrt{a^2 - f^2} \tag{7.4}$$

The scatterer resulting in a single bounce multipath component arriving at time τ_i lies at coordinates (x_s, y_s) such that x_s and y_s satisfy

$$\frac{x_s^2}{a^2} + \frac{y_s^2}{b^2} = 1 \tag{7.5}$$

This is because the distance from the scatterer at (x_s, y_s) to the transmitter at $(-f, 0)$, and the distance from the scatterer to the receiver at $(f, 0)$, sum to $2a = c\tau_i$ provided that the scatterer lies on the ellipse $\{a, b\}$.

Let us assume that scatterers are uniformly distributed in space. All of the scatterers giving rise to single bounce multipath components arriving between time τ and time $\tau + \Delta\tau$ lie in the region bounded on the inner edge by an ellipse with parameters

$$a_1 = c\tau/2 \tag{7.6}$$

$$b_1 = \sqrt{a_1^2 - f^2} \tag{7.7}$$

and on the outer edge by the ellipse with parameters

$$a_2 = c(\tau + \Delta\tau)/2 \tag{7.8}$$

$$b_2 = \sqrt{a_2^2 - f^2} \tag{7.9}$$

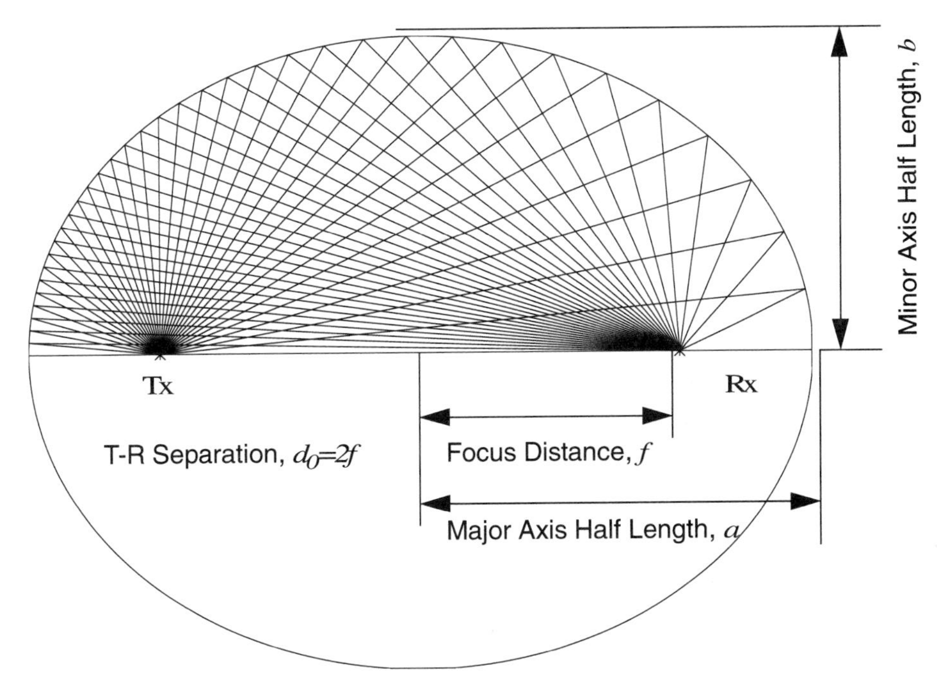

Figure 7–1 The locus of all points where a scatterer must lie, which results in a single bounce multipath component with delay τ_i, is an ellipse with $a = c\tau_i/2$, $f = d_0/2$, and $b = \sqrt{f^2 - a^2}$ [Sei89][Par89][Rap96a].

The region bounded by these two ellipses is illustrated in Figure 7–2. The area of this region is

$$A = \pi(a_2 b_2 - a_1 b_1) \tag{7.10}$$

The probability density function (pdf) for the x and y coordinates of scatterers, giving rise to single bounce multipath components arriving between times τ and $\tau + \Delta\tau$, is given by

$$f_{xy}(x, y) = \frac{1}{A} = \frac{1}{\pi(a_2 b_2 - a_1 b_1)} \qquad x, y \in \left\{ \frac{x^2}{a_1^2} + \frac{y^2}{b_1^2} \geq 1, \frac{x^2}{a_2^2} + \frac{y^2}{b_2^2} \leq 1 \right\} \tag{7.11}$$

To determine the cumulative distribution function (cdf) for the DOA conditioned on the TOA, we first find the probability that a single bounce multipath component arrives with a Direction-Of-Arrival, ϕ_i, between 0 and an angle Φ, at time τ_i prior to τ. The time τ determines an ellipse given by parameters $\{a, b\}$ from (7.3) and (7.4). To find this cdf, we first compute the area of the region in the ellipse $\{a, b\}$ for which $0 \leq \phi < \Phi$, where ϕ is measured clockwise from a line drawn between the transmitter and receiver as shown in Figure 7–3. This area is

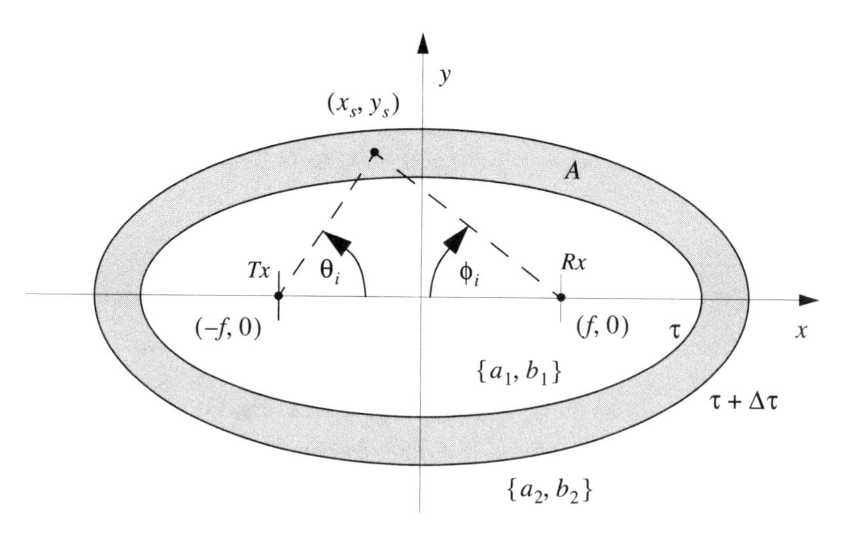

Figure 7–2 Geometry for determining the statistics of the Direction-Of-Arrival (DOA), ϕ_i, given that a multipath component arrives at time τ_i, where $\tau \le \tau_i < \tau + \Delta\tau$.

$$
\begin{aligned}
R_{a,b}(\Phi) &= \int_{-a}^{a\left(\frac{f - a \cdot \cos\Phi}{a - f\cos\Phi}\right)} b\sqrt{1 - \left(\frac{x}{a}\right)^2}\,dx + \frac{(f - x_\Phi)y_\Phi}{2} \\
&= -\frac{ab}{2}\left(\cos^{-1}\left(\frac{f - a \cdot \cos\Phi}{a - f\cos\Phi}\right) - \pi - \frac{(f - a \cdot \cos\Phi)b\sin\Phi}{(a - f\sin\Phi)^2}\right) + \frac{b^4\sin\Phi\cos\Phi}{2(a - f\cos\Phi)^2} \\
&= \frac{-ab}{2}\left(\cos^{-1}\left(\frac{f - a \cdot \cos\Phi}{a - f\cos\Phi}\right) - \pi\right) + \frac{b^2 f\sin\Phi}{2(a - f\cos\Phi)}
\end{aligned}
\tag{7.12}
$$

Thus, the probability that a single bounce multipath component, arriving with a Time-Of-Arrival τ_i, such that $\tau \le \tau_i < \tau + \Delta\tau$, has a Direction-Of-Arrival between 0 and Φ is given by

$$
\Pr(0 \le \phi_i < \Phi \mid 0 \le \phi_i \le \pi, \tau \le \tau_i \le \tau + \Delta\tau) = \frac{R_{a_2,b_2}(\Phi) - R_{a_1,b_1}(\Phi)}{A/2} \tag{7.13}
$$

Through symmetry, we may deduce that for $-\pi \le \phi_i < 0$,

$$
\Pr(-\pi \le \phi_i < \Phi \mid -\pi \le \phi_i \le 0, \tau \le \tau_i \le \tau + \Delta\tau) = \frac{A/2 - (R_{a_2,b_2}(-\Phi) - R_{a_1,b_1}(-\Phi))}{A/2} \tag{7.14}
$$

Then for $-\pi \le \phi_i < \pi$ we obtain the following cumulative distribution function (cdf) for Φ

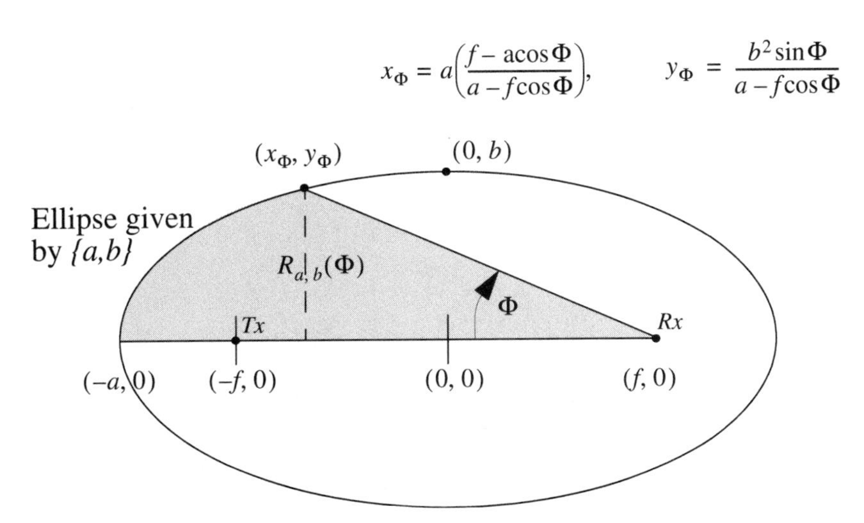

Figure 7–3 The region of the ellipse $\{a, b\}$ swept out by Φ.

$$F_{\phi|\tau}(\Phi \mid \tau \le \tau_i \le \tau + \Delta\tau) = \Pr(-\pi \le \phi_i < \Phi \mid \tau \le \tau_i \le \tau + \Delta\tau)$$

$$= \begin{cases} \dfrac{1}{2} - \dfrac{R_{a_2,b_2}(-\Phi) - R_{a_1,b_1}(-\Phi)}{A} & -\pi \le \Phi \le 0 \\[2ex] \dfrac{1}{2} + \dfrac{R_{a_2,b_2}(\Phi) - R_{a_1,b_1}(\Phi)}{A} & 0 \le \Phi < \pi \end{cases} \tag{7.15}$$

Our goal is to determine the distribution of the Direction-Of-Arrival, ϕ_i, for a particular multipath component as a function of Time-Of-Arrival. In order to simplify notation, it is convenient to introduce the normalized multipath delay $r_i = c\tau_i/d_0 = \tau_i/\tau_0$. To obtain the cdf for ϕ_i conditioned on the normalized multipath delay, r_i, we take the limit of (7.15) as $\Delta\tau$ goes to zero which gives:

$$F_{\phi|r}(\Phi \mid r_i) = \begin{cases} \dfrac{1}{2\pi}\cos^{-1}\left(\dfrac{1 - r_i\cos\Phi}{r_i - \cos\Phi}\right) - \dfrac{\sqrt{r_i^2 - 1}\,\sin(-\Phi)(1 - r_i\cos\Phi)}{2\pi(2r_i^2 - 1)(r_i - \cos\Phi)^2} & -\pi \le \Phi \le 0 \\[2ex] 1 - \dfrac{1}{2\pi}\cos^{-1}\left(\dfrac{1 - r_i\cos\Phi}{r_i - \cos\Phi}\right) + \dfrac{\sqrt{r_i^2 - 1}\,\sin(\Phi)(1 - r_i\cos\Phi)}{2\pi(2r_i^2 - 1)(r_i - \cos\Phi)^2} & 0 \le \Phi < \pi \end{cases} \tag{7.16}$$

The conditional pdf for the DOA, ϕ_i, may be found by differentiating (7.16) with respect to Φ. This gives

$$f_{\phi|r}(\phi \mid r_i) = \frac{(r_i^2 - 1)^{3/2}(r_i^2 - 2r_i\cos\phi + 1)}{\pi(2r_i^2 - 1)(r_i - \cos\phi)^3} \qquad -\pi \le \phi \le \pi \tag{7.17}$$

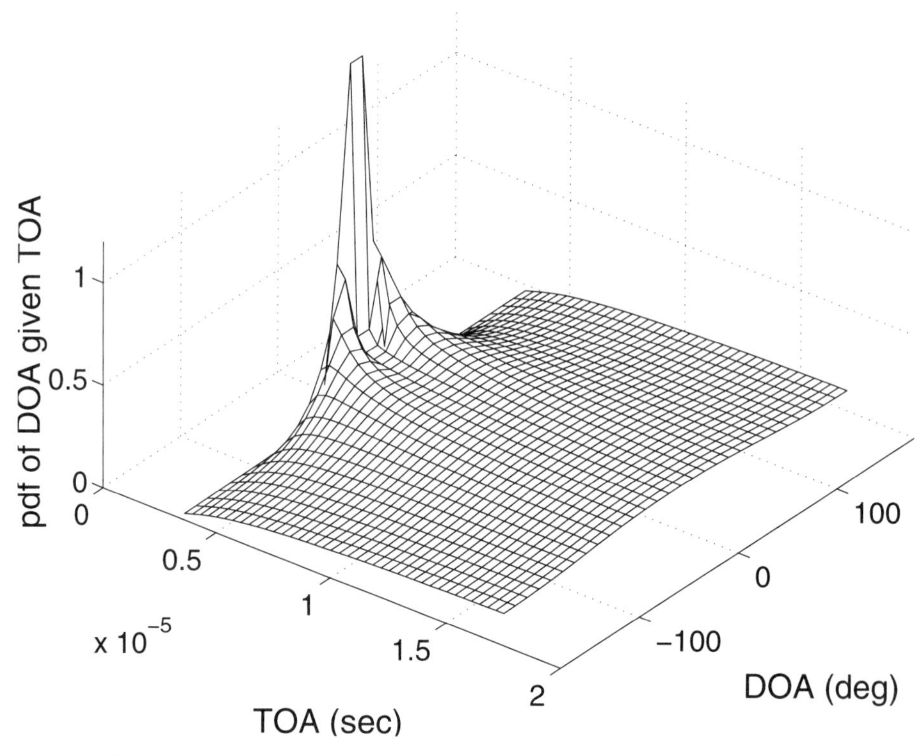

Figure 7–4 Probability density function of Direction-Of-Arrival conditioned on Time-Of-Arrival for single bounce multipath components for a T-R separation of 1 km.

The probability density function for ϕ_i, conditioned on τ_i, $f_{\phi \mid r}(\phi, \tau_i/\tau_0)$, is illustrated in Figure 7–4.

Using the same geometrical model used to develop the conditional statistics for Direction-Of-Arrival, we may derive a pdf for the distribution of Time-Of-Arrival (TOA), τ_i. We begin by introducing a maximum observation time, τ_m. The maximum observation time defines a window such that all observed multipath components have delays so that $\tau_0 \le \tau_i < \tau_m$. All single bounce scatterers that give rise to multipath components, which arrive prior to τ_m, must lie within the ellipse with area

$$A_m = \pi\left(\frac{c\tau_m}{2}\right)\sqrt{\left(\frac{c\tau_m}{2}\right)^2 - \left(\frac{d_0}{2}\right)^2} \tag{7.18}$$

as shown in Figure 7–2. Similarly, all multipath components arriving before time τ, such that $\tau < \tau_m$, must result from scatterers located in an ellipse with area

$$A_\tau = \pi\left(\frac{c\tau}{2}\right)\sqrt{\left(\frac{c\tau}{2}\right)^2 - \left(\frac{d_0}{2}\right)^2} \qquad \tau_0 \le \tau < \tau_m \tag{7.19}$$

The probability that a component arrives before time τ is equal to the probability that a scatterer lies within the ellipse corresponding to time τ. Since scatterers are uniformly distributed in space, and all scatterers are contained within the ellipse corresponding to time τ_m, the probability that a component arrives before time τ is

$$F_\tau(\tau) = \Pr(\tau_0 \le \tau_i \le \tau) = \frac{A_\tau}{A_m} = \frac{\left(\frac{c\tau}{d_0}\right)\sqrt{\left(\frac{c\tau}{d_0}\right)^2 - 1}}{\beta} \qquad \tau_0 \le \tau < \tau_m \tag{7.20}$$

where

$$\beta = \left(\frac{c\tau_m}{d_0}\right)\sqrt{\left(\frac{c\tau_m}{d_0}\right)^2 - 1} \tag{7.21}$$

We may differentiate (7.20) with respect to τ in order to obtain the pdf

$$f_\tau(\tau) = \frac{c}{\beta d_0}\frac{2(c\tau/d_0)^2 - 1}{\sqrt{(c\tau/d_0)^2 - 1}} \qquad \tau_0 \le \tau < \tau_m \tag{7.22}$$

It is more convenient to generate the distribution of the normalized multipath delay, r_i directly. The distribution for the normalized multipath delay is

$$f_r(r) = \frac{2r^2 - 1}{\beta\sqrt{r^2 - 1}} \qquad 1 \le r < r_m \tag{7.23}$$

where $\beta = r_m\sqrt{r_m^2 - 1}$ and $r_m = \tau_m/\tau_0$ is the maximum value of the normalized multipath component delay. Combining (7.17) and (7.23), we obtain the joint pdf for ϕ_i and r_i for single bounce multipath components

$$f_{\phi,r}(\phi, r) = \frac{(r^2 - 1)(r^2 - 2r\cos\phi + 1)}{\pi\beta(r - \cos\phi)^3} \qquad \begin{array}{c} -\pi \le \phi \le \pi \\ 1 \le r < r_m \end{array} \tag{7.24}$$

By integrating over r, we obtain the marginal pdf for distribution of Direction-Of-Arrival for single bounce multipath components:

$$f_\phi(\phi) = \frac{1}{2\pi\beta}\left(\frac{(1 + r_m^4) - 4r_m\cos\phi + (2 - 3r_m^2)\cos^2\phi + 6r_m\cos^3\phi - 3\cos^4\phi}{(r_m - \cos\phi)^2}\right) - \frac{1}{4\pi\beta}(1 - 3\cos 2\phi) \tag{7.25}$$

which may be simplified to obtain

$$f_\phi(\phi) = \frac{1}{2\pi\beta}\frac{(r_m^2 - 1)^2}{(r_m - \cos\phi)^2} \tag{7.26}$$

Equation (7.26) is useful for characterizing multipath components when power, P_i, and Time-Of-Arrival, τ_i, are not required.

We conclude this section by summarizing the assumptions required to apply the geometrical model.

1. All scatterers lie in the same plane as the transmitter and receiver. This plane is approximately parallel with the ground so that all multipath components appear to arrive from the horizon.
2. A line-of-sight direct path exists between the transmitter and receiver.
3. Single bounce multipath is the dominant mode of propagation (in addition to direct path propagation). Rough surface scattering, diffraction, multiple bounce multipath, and other mechanisms are not included in the model.
4. In the next section, the model will be extended to include ground reflections, in which case we will add the assumption that the T-R separation, d_0, is much larger than the transmitter antenna height, h_t, and the receiver antenna height, h_r.

7.1 Simulation of Multipath Component Parameters in GBSBEM

In this section, an algorithm is developed for generating values of the Time-Of-Arrival (TOA), τ_i, Direction-Of-Arrival (DOA), ϕ_i, and power P_i for multipath components using the statistics developed in the previous section. This procedure is useful in performing Monte Carlo simulations of the multipath channel.

In simulating the DOA, TOA, and power levels for multipath components, it is necessary to generate samples of random variables from specified distributions. Using the GBSBEM approach, the simplest way to create sample spatio-temporal channels is to generate a set of L scatterers which are uniformly distributed in x and y and fall within the ellipse described by (7.2)-(7.5) for τ_i=τ_m. The DOA, DOD, and delay can then be computed from the coordinates of the transmitter, receiver, and scatterer. Alternatively, it is sometimes useful to generate the DOA, DOD, and delay directly. An approach to do this is described in this section.

Typically, in simulations, a uniform random number generator is available which may be used to generate instances, x_i, of a random variable which is uniformly distributed on $[0, 1]$. To generate instances of a random variable from an arbitrary distribution, $f_z(z)$, we require a smooth monotonic function $z = g(x)$ such that [Sta86]

$$f_z(z) = \left.\frac{f_x(x)}{\left|\frac{dz}{dx}\right|}\right|_{x = g^{-1}(z)} = \left|\frac{dx}{dz}\right| \qquad 0 \le x \le 1 \tag{7.27}$$

since $f_x(x) = 1$ over the indicated range.

A relationship between x and z which satisfies this is $x = F_z(z)$ where $F_z(z)$ is the desired cumulative distribution function for z. Since $F_z(z)$ is monotonically increasing, its functional inverse, $z = g(x)$, is also monotonically increasing so that (7.27) may be applied. Also, the CDF satisfies $0 \le F_z(z) \le 1$ so the support region given in (7.27) is satisfied. Therefore, provided that $F_z(z)$ is a smooth function of z, the functional inverse of $F_z(z)$, provides a function, $g(x)$, which may be used to generate samples from the desired distribution, given samples from a uniform distribution.

To generate a multipath channel impulse response using the GBSB model, we generate the multipath delay, τ_i, the Direction-Of-Arrival, ϕ_i, the component power, P_i, and the phase, γ_i, for each of the L multipath components. It is assumed that the number of multipath components, L, and the T-R separation d_0 are known.

The first step is to generate the Time-Of-Arrival, τ_i of each multipath component. The distribution of the normalized multipath component delay, r_i is given by (7.23):

$$f_r(r) = \frac{2r^2 - 1}{\beta\sqrt{r^2 - 1}} \qquad 1 \le r < r_m \tag{7.28}$$

Note that $\beta = r_m\sqrt{r_m^2 - 1}$ depends on the maximum value of the normalized multipath delay. For now we will simply choose a value of r_m corresponding to the maximum observation time. In the next section a procedure will be outlined for determining r_m based on the path loss exponent, n, and the reflection loss, L_r. Later, in Section 7.4, a technique will be presented for selecting r_m based on the delay spread of the channel.

In order to generate instances of the normalized multipath delay with the distribution given by (7.28), it is necessary to find a function $r_i = g(x_i)$ such that r_i has the desired distribution when x_i is drawn from a uniform distribution on $[0, 1]$. To find $g(x)$, we first find the cdf of r,

$$F_r(r) = \frac{r\sqrt{r^2 - 1}}{\beta} \qquad 1 \le r < r_m \tag{7.29}$$

Solving $x = F_r(r)$ for r we obtain a function, $r = g(x)$, which transforms a uniform random variable x into a random variable r with the pdf given by (7.28),

$$r_i = \sqrt{\frac{1}{2} + \frac{1}{2}\sqrt{1 + 4\beta^2 x_i^2}} \tag{7.30}$$

Having generated r_i, we obtain the multipath component delay from $\tau_i = r_i d_0/c$. Thus we may generate an instance of the normalized multipath delay which gives the Time-Of-Arrival (TOA) of a given multipath component.

The next step is to generate a Direction-Of-Arrival from the distribution given in (7.17). To generate samples of the DOA, given a Time-Of-Arrival, we determine a function, $\phi = g(y)$, which transforms samples, y_i, from a uniform distribution into samples from the distribution given in (7.17). From the previous discussion, $g(y_i)$ is the functional inverse of $y_i = F_{\phi|r}(\phi_i | r_i) = h(\phi_i)$ where

$$h(\phi) = \begin{cases} \dfrac{1}{2\pi}\operatorname{acos}\left(\dfrac{1 - r_i\cos\phi}{r_i - \cos\phi}\right) - \dfrac{\sqrt{r_i^2 - 1}\sin(-\phi)(1 - r_i\cos\phi)}{2\pi(2r_i^2 - 1)(r_i - \cos\phi)^2} & -\pi \le \phi \le 0 \\[2ex] 1 - \dfrac{1}{2\pi}\operatorname{acos}\left(\dfrac{1 - r_i\cos\phi}{r_i - \cos\phi}\right) + \dfrac{\sqrt{r_i^2 - 1}\sin(\phi)(1 - r_i\cos\phi)}{2\pi(2r_i^2 - 1)(r_i - \cos\phi)^2} & 0 \le \phi < \pi \end{cases} \tag{7.31}$$

```
% a=DOA(y,r,M)
% ----------------------
% Returns an instance of the random variable phi which is the
% Direction-Of-Arrival corresponding to normalized delay r. The input,
% x, is a vector of uniformly distributed random variables on [0,1].
% The returned vector of DOA's is the same length as x. M is an optional
% parameter giving the number of points to generate for interpolation.

function a=DOA(y,r,M);

if (nargin<3)
 M=361;
end;

% Generate values of phi
phi=((0:M-1)*2*pi-pi)/(M-1);

% Separate into positive and negative angles:
phip=phi(phi>=0);
phin=phi(phi<0);
h=zeros(size(phi));

% Compute h(phi)
h(phi<0)= 1/(2*pi)*acos((1-r*cos(phin))./(r-cos(phin)))...
                    -(r^2-1)^(1/2)*sin(-phin).*(1-r*cos(phin))./...
                    (2*pi*(r-cos(phin)).^2)/(2*r^2-1);
h(phi>=0)= 1-1/(2*pi)*acos((1-r*cos(phip))./(r-cos(phip)))...
                    +(r^2-1)^(1/2)*sin(phip).*(1-r*cos(phip))./...
                    (2*pi*(r-cos(phip)).^2)/(2*r^2-1);

% Find phi that gives y=h(phi) given y
a=interp1(h,phi,y);
```

Figure 7–5 A MATLAB function, DOA.m, which generates values of ϕ_i with the distribution given in Equation (7.17).

Since (7.31) is a complicated expression, we resort to numerical techniques to find its functional inverse.

Using MATLAB, we generate a vector containing samples of $h(\phi)$ at M points in ϕ such that $-\pi \le \phi < \pi$. We then generate a *single value*, y_i, from a uniform distribution, $U(0, 1)$, using the randomization function in MATLAB. The Direction-Of-Arrival is determined by finding the value of ϕ_i which gives $y_i = h(\phi_i)$ using MATLAB's interpolation function. A short MATLAB function which transforms y_i into a value of ϕ_i is shown in Figure 7–5.

Using this technique, 10 000 values of ϕ_i were generated and a normalized histogram was constructed. This normalized histogram is plotted along with the ideal pdf, given by (7.17), in Figure 7–6.

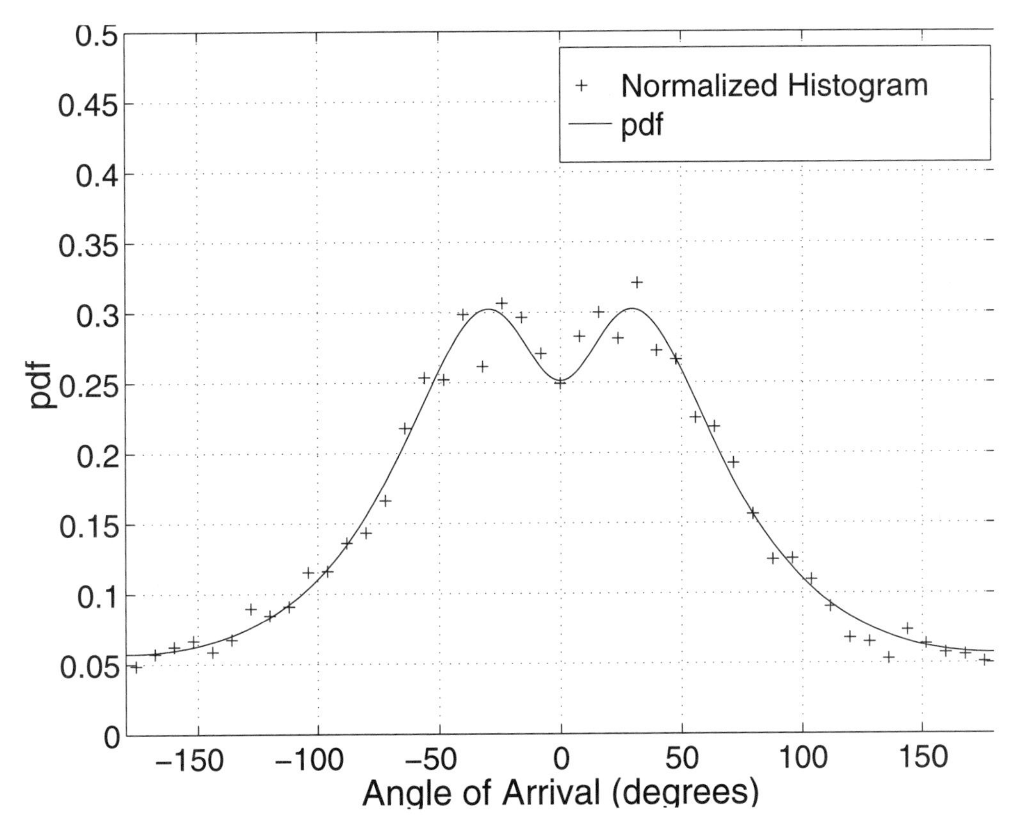

Figure 7–6 A plot of the normalized histogram for 10,000 values of ϕ_i generated using the MATLAB function shown in Figure 7–5, for a multipath normalized delay of $r_i = 1.6$. Also shown is the ideal pdf for ϕ given in Equation (7.17).

To account for the effects of a directional transmitting antenna, we require knowledge of the Direction-Of-Departure θ_i for the ith multipath component. The Direction-Of-Departure, θ, is defined, as shown in Figure 7–2, such that $\theta = 0$ is in the direction of the receiver as seen from the transmitter. Given a Direction-Of-Arrival at the receiver, ϕ_i, it can be shown that the Direction-Of-Departure at the transmitter is

$$\theta_i = 2\,\mathrm{atan}\left(\frac{\sin(\phi_i)}{\frac{r_i+1}{r_i-1} - \cos(\phi_i)}\right) \tag{7.32}$$

It is assumed that both the direct path component and multipath components experience d^n path loss such that the received power for the direct path component is [Rap89][Feu94]

$$P_0 = P_{ref} - 10n\log\left(\frac{c\tau_0}{d_{ref}}\right) + G_r(0) + G_t(0) \tag{7.33}$$

where P_0 and P_{ref} are in dBm. P_{ref} is a reference power that is measured at a distance d_{ref} from the transmitting antenna when omni-directional antennas are used at both the transmitter and receiver. This may be either a measured value or, using Friis' free space link equation with omni-directional antennas, we obtain

$$P_{ref} = P_T - 20\log\left(\frac{4\pi d_{ref} f}{c}\right) \tag{7.34}$$

where P_T is the transmitted power in dBm and f is the carrier frequency in Hz. The path loss exponent, n, determines the rate at which path loss increases as a function of distance between the transmitter and receiver for each multipath component. In free space the path loss exponent is $n = 2$, as described in Chapter 1. Typical values of n for microcell environments are between 2 and 4 [Feu94]. $G_r(\phi)$ is the receiver antenna gain as a function of the Direction-Of-Arrival, ϕ. Similarly, $G_t(\theta)$ is the gain of the transmitting antenna in the Direction-Of-Departure θ. It is assumed that the direct path component is co-polarized with the receiving antenna so that there is no polarization mismatch for the direct path component.

The received power for the ith multipath component is

$$P_i = P_{ref} - 10n\log\left(\frac{c\tau_i}{d_{ref}}\right) - L_r + G_r(\theta_i) + G_t(\phi_i) \tag{7.35}$$

where L_r is the loss, in dB, due to reflection from the scatterer. The loss, L_r, includes the effects of polarization mismatch between the received multipath component and the receiving antenna. Assume that the complex amplitudes of the multipath components are normalized such that $\alpha_0 = 1$. Then the power in each RF multipath component, P_i, is related to the term α_i in (7.1) by $P_i = P_0 + 20\log|\alpha_i|$. The power of the multipath component relative to the power of the direct path component is

$$P_i - P_0 = -10n\log(r_i) - L_r + G_r(\phi_i) - G_r(0) + G_t(\theta_i) - G_t(0) \tag{7.36}$$

Thus, it is possible to generate Time-Of-Arrival, τ_i, Direction-Of-Arrival, ϕ_i, and power P_i, for single bounce multipath components for simulations.

The geometrical model may be easily extended to accommodate the ground reflection by using a plane earth model to generate a ground reflection for each multipath component [Rap96a]. In this case, received power from the combination of each multipath component with its ground reflection is

$$P_i = P_T - 20\log\left(\frac{(r_i d_0)^2}{h_t h_r}\right) - L_r + G_r(\phi_i) + G_t(\theta_i) \qquad r_i d_0 \gg h_t, h_r \tag{7.37}$$

where h_t is the transmitter antenna height, h_r is the receiver antenna height, and P_T is the transmitted power in dBm [Rap96a]. Then the received power of the ith multipath component (combined with its ground reflection) relative to the direct path (combined with its ground reflection) is

$$P_i - P_0 = -40\log(r_i) - L_r + G_r(\phi_i) - G_r(0) + G_t(\theta_i) - G_t(0) \tag{7.38}$$

which is simply (7.36) with $n = 4$. Note that the d^n path loss model is applied here to each multipath component individually. Typically, the d^n path loss model is applied to the total received power which is the result of many multipath components. It is, however, reasonable to apply the d^n path loss model to account for ground reflections along each path. Care should be taken in applying path loss exponents derived from total received power measurements to the GBSB model.

Note that the geometrical model gives no indication of the number of multipath components L. In [Sei89] the measured statistics for L are presented for indoor UHF radio channels in manufacturing environments. In that study, it was found that L could be modeled as a discrete uniform random variable between 9 and 35. The steps presented in this section may be used to generate any number of multipath components using the procedure summarized in Figure 7–7.

We conclude this section by generating multipath profiles for $n = 4.0$, $L_r = 6.0$ dB, and a T-R separation of $d_0 = 500$ m. Simulated power delay profiles and power-angle-delay profiles are shown in Figure 7–8 for the cases of $L = 5$, 20, and 50 multipath components. For these simulations, it was assumed that omni-directional antennas were used at both the transmitter and receiver. A threshold was set at 20 dB below the power of the direct path component and all multipath components arriving below the threshold were ignored. The maximum normalized delay was set to $r_m = 2.0$. To help visualize the multipath components, the profiles are plotted assuming that the channel was sounded using a 100 ns Gaussian pulse and a Gaussian beam antenna with a beamwidth of 20 degrees.

7.2 Marginal Distribution of the Direction-Of-Arrival in the GBSBEM Model

In the previous section, techniques were demonstrated for simulating the power, DOA, and TOA of multipath components. In a narrowband communication system it may be desirable to determine the distribution of ϕ without separately generating τ_i and P_i.

Note that the marginal pdf for ϕ, given by (7.26), requires knowledge of β which requires a maximum value of r_m. In a narrowband system, the maximum observation time may have no obvious intuitive interpretation, so another technique for developing r_m is required.

One method which may be used to determine a maximum value of r_i is to consider only multipath components arriving with a power level that is within a margin, T dB, of the direct path. Using (7.33), with omni-directional antennas, the received power for the direct path component is

$$P_0 = P_{ref} - 10n\log\left(\frac{c\tau_0}{d_{ref}}\right) \tag{7.39}$$

The received power for the multipath component is given by

$$P_i = P_{ref} - 10n\log\left(\frac{c\tau_i}{d_{ref}}\right) - L_r \geq P_{ref} - 10n\log\left(\frac{c\tau_m}{d_{ref}}\right) - L_r \tag{7.40}$$

Given: T-R separation d_0, path loss exponent n, reference power P_{ref} in dBm, and reflection loss L_r.

Either assume a value of r_m, or calculate an appropriate value of r_m based on the maximum multipath delay, a power threshold, the delay spread, or the maximum excess delay, as summarized in Table 7–1.

Compute $\beta = r_m\sqrt{r_m^2 - 1}$.

Create a direct path component with $P_0 = P_{ref} - 10n\log\left(\frac{d_0}{d_{ref}}\right) + G_r(0) + G_t(0)$ with $\theta_0 = 0$, $\phi_0 = 0$, and $\tau_0 = d_0/c$. Assume $\alpha_0 = 1$.

For i = 1 to $L - 1$,

Generate a value of x_i from $U(0, 1)$.

Compute the normalized multipath delay, $r_i = \sqrt{\frac{1}{2} + \frac{1}{2}\sqrt{1 + 4\beta^2 x_i^2}}$.

Generate a value of y_i from $U(0, 1)$.

Compute the Direction-Of-Arrival ϕ_i from y_i using DOA.m shown in Figure 7–5.

Compute the Direction-Of-Departure,

$$\theta_i = 2\,\mathrm{atan}\left(\sin(\phi_i) \cdot \left(\frac{r_i + 1}{r_i - 1} - \cos(\phi_i)\right)^{-1}\right)$$

Determine the power of the multipath component,

$$P_i = P_0 - 10n\log(r_i) - L_r + G_r(\phi_i) - G_r(0) + G_t(\theta_i) - G_t(0)$$

Generate a phase angle, γ_i, from $U(0, 2\pi)$.

Compute $\alpha_i = 10^{(P_i - P_0)/20} e^{j\gamma_i}$

end

Figure 7–7 Pseudo code to generate all of the necessary parameters to characterize L multipath components [Lib95].

where τ_m is the maximum observation time. We are interested only in multipath components arriving with a received power greater than $P_0 - T$. For this case all P_i will satisfy $P_0 - P_i \le T$. If

$$T = -10\log\left(\frac{c\tau_0}{d_{ref}}\right) + 10n\log\left(\frac{c\tau_m}{d_{ref}}\right) + L_r = 10n\log(r_m) + L_r \tag{7.41}$$

(a) L=5 Multipath Components

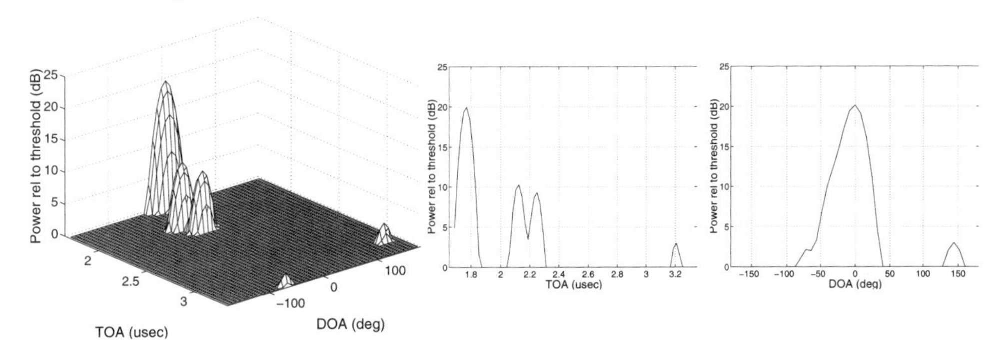

(b) L=20 Multipath Components

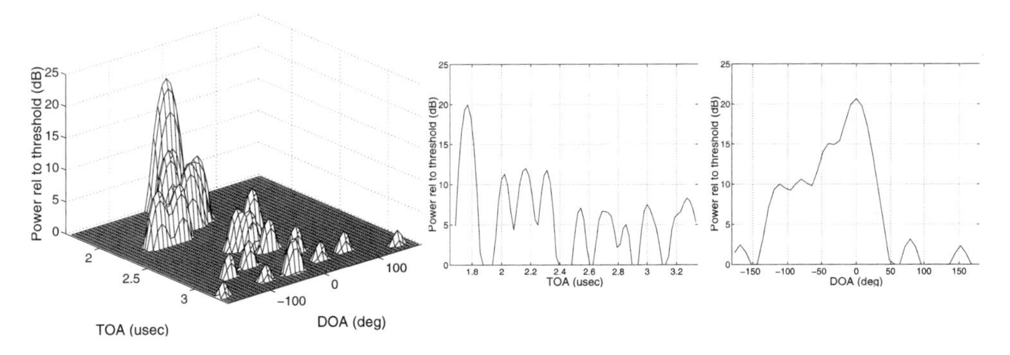

(c) L=50 Multipath Components

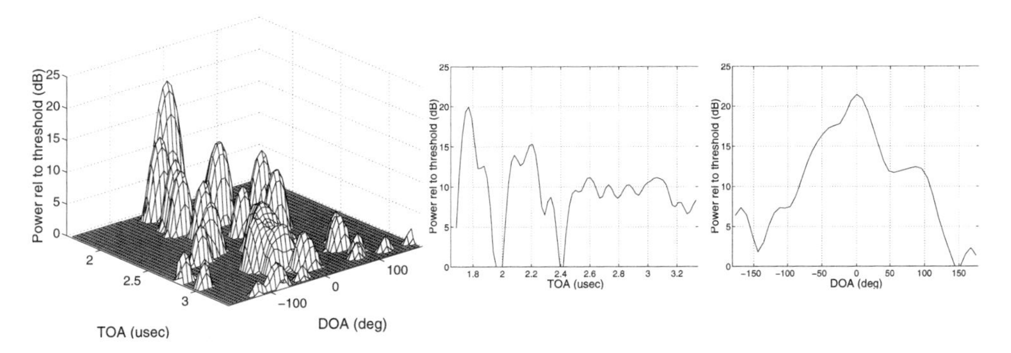

Figure 7–8 Power-delay-angle profiles, power-delay profiles, and power-angle profiles, generated using $n = 4.0$, $L_r = 6.0$ dB, and $d_0 = 500$ m for 5, 20, and 50 multipath components.

then r_m is given by

$$r_m = 10^{\frac{T - L_r}{10n}} \tag{7.42}$$

Alternative techniques for selecting r_m, based on the delay spread and maximum excess delay of the channel, are presented in Section 7.4.

Given r_m, we determine the pdf for ϕ_i using (7.26).

$$f_\phi(\phi) = \frac{1}{2\pi\beta} \frac{(r_m^2 - 1)^2}{(r_m - \cos\phi)^2} \tag{7.43}$$

To illustrate how (7.43) might be useful, let us examine the effectiveness of using a directional antenna at the receiver to mitigate multipath. Note that we assumed that omni-directional antennas were used at both ends of the link to derive (7.42). If we now use a directional antenna at the receiver, then the power levels given in (7.39) and (7.40) may be interpreted as the power levels incident on the receiving antenna. As an example, we assume that we have a directional receiver antenna with a beamwidth of 60°, which we model as having a gain of 7.8 dB within $\pm 30°$ of the boresight direction and a gain of zero ($-\infty$ dB) in other directions. It is assumed that the boresight is aligned in the direction of the transmitter.

To determine the number of multipath components rejected by the directional antenna, we note that the expected value of the number of multipath components received in the main beam, L_m, divided by the total number of multipath components, L, is equal to the probability that any particular multipath component arrives within $\pm 30°$ of the boresight direction of the antenna. This quantity is found by integrating from $-\pi/6 \le \phi \le \pi/6$. Assume that the reflection loss, L_r, is 6 dB, and the path loss exponent is $n = 2.0$.

For $T = 10$ dB, the directional antenna rejects only 51% of the multipath components. If we extend the margin to $T = 20$ dB to include lower level multipath components (effectively increasing our observation interval), we find that our directional antenna has rejected 76% of the multipath components. Note that using the traditional uniform model for Direction-Of-Arrival we would, rather optimistically, predict that 83% of the multipath components was rejected by the directional antenna. This is due to the fact that the GBSB model predicts that paths with low excess delays, and therefore higher power levels, tend to be clustered about the direct path. In the uniform model, all multipath components, regardless of power level, have the same angular distribution.

In (6.10), the angle spread was defined as

$$\sigma_\phi = \sqrt{E[\phi_i^2] - E[\phi_i]^2} \tag{7.44}$$

Here, we consider only scattered components, not the direct path component. Using (7.43), we find that the mean of ϕ is zero, so the ensemble average angle spread is

$$\sigma_\phi = \sqrt{\int_{-\pi}^{\pi} \frac{\phi^2}{2\pi\beta} \frac{(r_m^2 - 1)^2}{(r_m - \cos\phi)^2} d\phi} \tag{7.45}$$

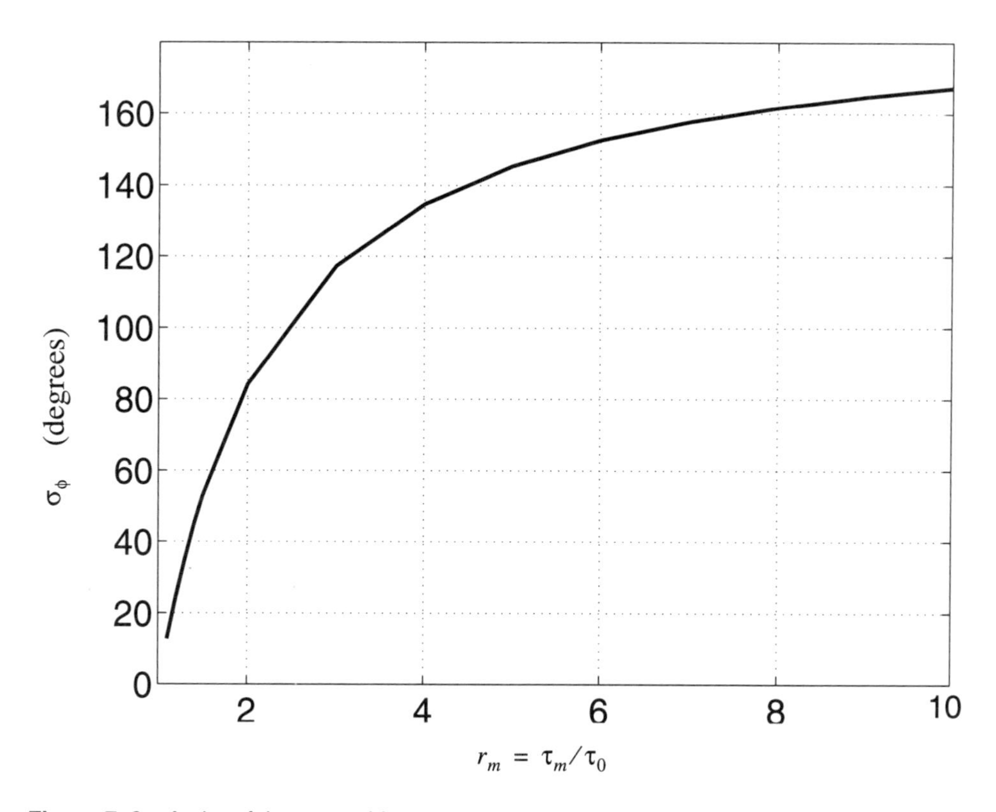

Figure 7–9 A plot of the ensemble average angle spread as a function of r_m.

This expression can be evaluated numerically. A plot of σ_ϕ as a function of r_m is illustrated in Figure 7–9.

7.3 Doppler Spectra and the Fading Envelope

Consider a mobile user, travelling at speed, v, which is receiving a CW signal at a carrier frequency $f_c = c/\lambda$. Because of multipath, several replicas of the transmitted signal arrive at the receiver. Each multipath replica of the received signal at the mobile unit will experience a Doppler shift due to motion of the user. Assume that the mobile unit is travelling in a direction ϕ_v measured clockwise with respect to the direct path between the transmitter and receiver as shown in Figure 7–10. The ith multipath component plane wave is incident from a direction ϕ_i measured with respect to the direct path between the transmitter and receiver.

The Doppler shift, υ_i, of the ith multipath component is dependent on the DOA, ϕ_i of the plane wave relative to the direction of motion of the user, ϕ_v:

$$\upsilon_i = f_m \cos(\phi_i - \phi_v) \tag{7.46}$$

where f_m is the maximum possible Doppler shift which is given by $f_m = v/\lambda$ [Rap96a].

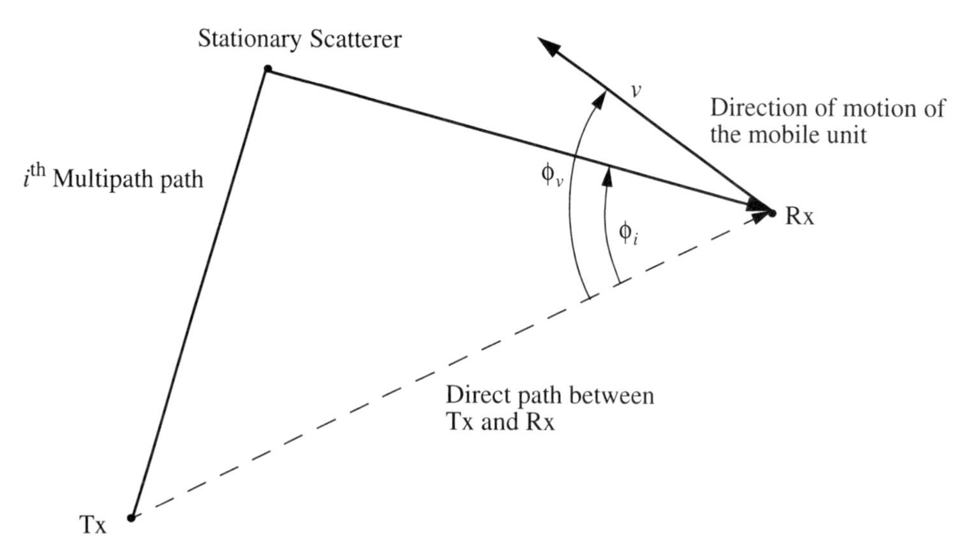

Figure 7–10 The mobile unit is traveling in a direction ϕ_v, measured with respect to the direct path between the transmitter and receiver.

The baseband complex equivalent representation of the received signal at the mobile unit is

$$r(t) = E_0 \sum_{i=0}^{L-1} \alpha_i e^{j2\pi \upsilon_i t} \tag{7.47}$$

where α_i is the complex multipath amplitude for multipath component i, given in (7.1).

The power spectral density (PSD) of the received signal is given by [Sta86]

$$S_r(f) = \lim_{T \to \infty} \frac{1}{2T} \mathrm{E}[R_T(f) R_T^*(f)] \tag{7.48}$$

where $R_T(f)$ is the finite time Fourier transform of $r(t)$. The PSD for the received signal, which shows the spreading of the signal in the frequency domain due to the motion of the mobile unit, is referred to as the *Doppler spectrum*. The finite time Fourier transform of $r(t)$ is

$$\begin{aligned} R_T(f) &= \int_{-T}^{T} \left(E_0 \sum_{i=0}^{L-1} \alpha_i e^{j2\pi \upsilon_i t} \right) e^{-2\pi f t} dt = E_0 \sum_{i=0}^{L-1} \alpha_i \int_{-T}^{T} e^{j2\pi(\upsilon_i - f)t} dt \\ &= E_0 \sum_{i=0}^{L-1} \alpha_i T \frac{\sin(2\pi(\upsilon_i - f)T)}{2\pi(\upsilon_i - f)T} \end{aligned} \tag{7.49}$$

Then, assuming that the magnitude of each multipath component, $|\alpha_i|$, is constant, we may express the Doppler spectrum as

$$S_r(f) = \lim_{T\to\infty}\frac{1}{2T}\mathrm{E}\left[E_0^2T^2\sum_{i=0}^{L-1}\sum_{j=0}^{L-1}\alpha_i\alpha_j^*\frac{\sin(2\pi(\upsilon_i-f)T)}{2\pi(\upsilon_i-f)T}\cdot\frac{\sin(2\pi(\upsilon_j-f)T)}{2\pi(\upsilon_j-f)T}\right] \tag{7.50}$$

$$= E_0^2\sum_{i=0}^{L-1}|\alpha_i|^2\mathrm{E}\left[\lim_{T\to\infty}\frac{T}{2}\left|\frac{\sin(2\pi(\upsilon_i-f)T)}{2\pi(\upsilon_i-f)T}\right|^2\right]$$

$$= E_0^2\sum_{i=0}^{L-1}|\alpha_i|^2\mathrm{E}\left[\frac{1}{4}\delta(\upsilon_i-f)\right] = \frac{E_0^2}{4}\sum_{i=0}^{L-1}|\alpha_i|^2\int_{-f_m}^{f_m}\delta(f-\upsilon)f_\upsilon(\upsilon)d\upsilon$$

$$= f_\upsilon(f)\cdot\frac{E_0^2}{4}\sum_{i=0}^{L-1}|\alpha_i|^2 = A_0^2f_\upsilon(f)$$

where $f_\upsilon(\upsilon)$ is the probability density function describing the distribution of the Doppler frequency, υ_i. It is assumed that all of the Doppler shifts are identically distributed. We may use standard techniques for determining the distribution of a function of a random variable to find the pdf of υ_i from the pdf of ϕ_i using (7.46) [Sta86]. For the moment, assume that omni-directional antennas are used at the receiver. Thus

$$f_\upsilon(\upsilon) = \frac{f_\phi(\phi_v+|\cos^{-1}(\upsilon/f_m)|)}{f_m\sqrt{1-(\upsilon/f_m)^2}}+\frac{f_\phi(\phi_v-|\cos^{-1}(\upsilon/f_m)|)}{f_m\sqrt{1-(\upsilon/f_m)^2}} \qquad |\upsilon|<f_m \tag{7.51}$$

Using (7.50), the power spectral density of the received signal is [Jak74]

$$S_r(f) = \frac{A_0^2f_\phi(\phi_v+|\cos^{-1}(f/f_m)|)}{f_m\sqrt{1-(f/f_m)^2}}+\frac{A_0^2f_\phi(\phi_v-|\cos^{-1}(f/f_m)|)}{f_m\sqrt{1-(f/f_m)^2}}$$

$$= \frac{A_0^2}{f_m\sqrt{1-(f/f_m)^2}}[f_\phi(\phi_v+|\cos^{-1}(f/f_m)|)+f_\phi(\phi_v-|\cos^{-1}(f/f_m)|)] \qquad |f|<f_m \tag{7.52}$$

Using the traditional model for DOA where the DOA is uniformly distributed on $[-\pi, \pi]$, the Doppler spectrum takes on the expression presented by Clarke [Jak74]:

$$S_r(f) = \frac{A_0^2}{\pi f_m\sqrt{1-(f/f_m)^2}} \qquad |f|<f_m \tag{7.53}$$

Alternatively, using the DOA model presented in this chapter, we may use (7.43) as the pdf for the DOA. Using (7.43) to model the distribution of the DOA, we obtain the plots shown in Figure 7–11 for $\phi_v = 0°$ and $\phi_v = 90°$ for several values of r_m. Unlike Clarke's model, for the case of $\phi_v = 0°$, for small values of r_m, the GBSB power spectral density indicates that more of the frequency content of the received signal appears at positive Doppler frequencies than at negative Doppler frequencies. This is because the mobile unit is moving toward the transmitter and there is a higher probability of receiving a multipath component from a DOA near the direct path. When the mobile unit is traveling perpendicularly to the direct path, the model presented here predicts higher spectral content near 0 Hz than Clarke's model.

As used in [Amo91] and [Amo93], when directional antennas are used at the mobile unit, the power spectral density is given by [Jak74] as

$$S_r(f) = \frac{A_0^2}{f_m\sqrt{1-(f/f_m)^2}}[G_r(\phi_v + \phi_f)f_\phi(\phi_v + \phi_f) + G_r(\phi_v - \phi_f)f_\phi(\phi_v - \phi_f)] \tag{7.54}$$

where $\phi_f = |\cos^{-1}(f/f_m)|$. $G_r(\phi)$ is the pattern of the receive antenna in the direction ϕ, measured clockwise relative to the direct path between the transmitter and receiver.

To conclude this section, we consider the effects of the GBSB model on the received signal envelope. The magnitude of the received signal will vary with time as the relative phases of multipath components change, leading to constructive and destructive interference. A fade occurs when multipath components combine destructively, leading to a drop in received power. The temporal characteristics of fading in multipath channels are important in designing error control coding schemes, interleaving techniques, and diversity combiners.

7.4 Selection of the Maximum Path Delay, τ_m

It is useful at this point to briefly discuss the methods for determining r_m. The maximum path delay, τ_m, is related to the spatial extent of scatterers in the geographical environment. Using this interpretation of the maximum path delay, τ_m will remain fixed and $r_m = \tau_m/\tau_0$ will change as the T-R separation varies. A second method is to determine r_m based on a threshold which is set relative to the direct path component power as discussed in Section 7.2. A third technique, described in [Lib95], is to determine r_m based on a desired rms delay spread for generated channel impulse responses. Finally, a fourth technique is to generate r_m by defining a

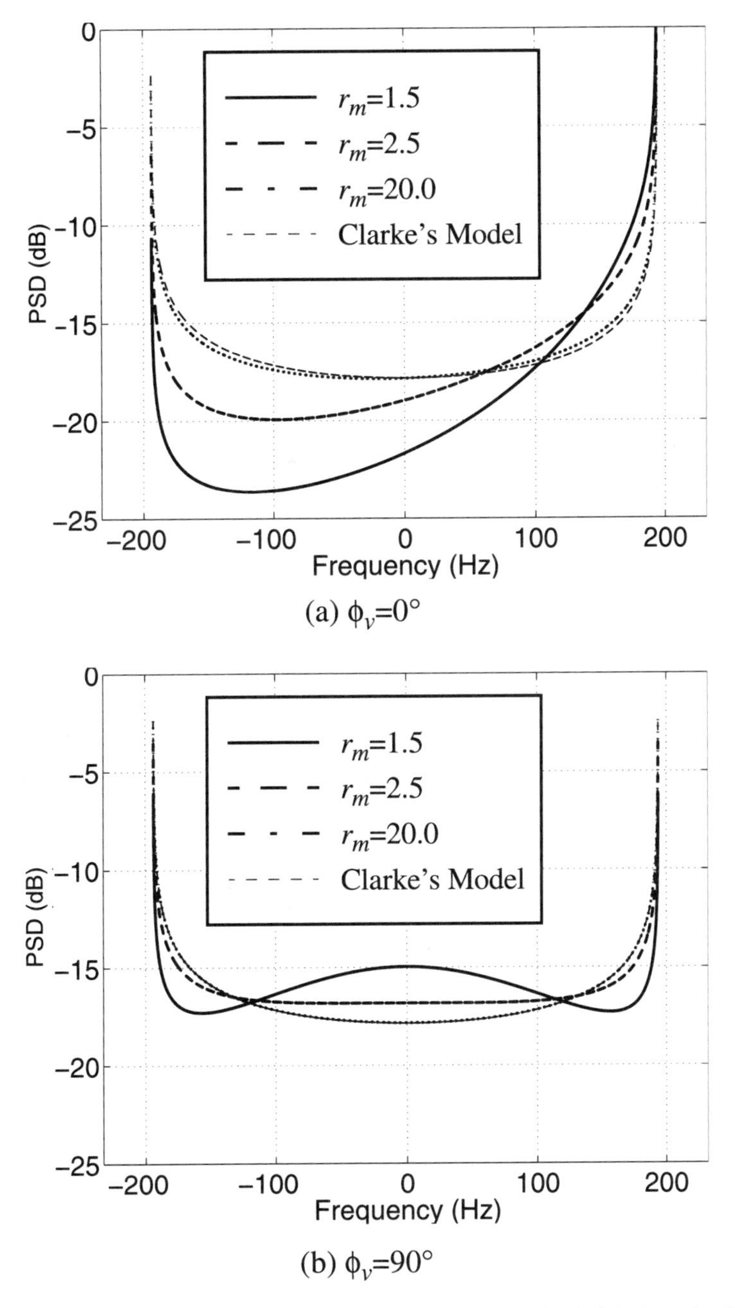

Figure 7–11 Doppler spectra produced using (7.52) with the pdf for DOA given by (7.43), for (a) $\phi_v = 0°$ such that the mobile is moving toward the transmitter, and (b) $\phi_v = 90°$ such that the mobile is moving in a direction perpendicular to the line between the transmitter and receiver. The mobile unit is travelling at 65 mph and the carrier frequency is 2 GHz.

maximum excess delay, τ_e, which is a fixed difference between τ_0 and τ_m. These techniques are summarized in Table 7–1.

Table 7–1 Four methods for selecting r_m to achieve desired channel impulse response characteristics.

Criteria	Expression
Fixed maximum delay, τ_m	$r_m = \tau_m/\tau_0$
Fixed threshold, T	$r_m = 10^{(T-L_r)/10n}$
Fixed delay spread σ_τ	$r_m = 3.244(\sigma_\tau/\tau_0)+1$
Fixed max. excess delay τ_e	$r_m = (\tau_0+\tau_e)/\tau_0$

7.5 Summary

In this chapter, a simple geometrical model was presented to develop statistics for the Time-Of-Arrival (TOA), Direction-Of-Arrival (DOA), and power of multipath components. The GBSB model is based entirely on geometry rather than measurements. Wideband channel simulation tools grounded in these models, as well as measured data are commercially available [Wir98].

The main advantage of GBSB models, including both GBSBCM and GBSMEM approaches, is that all of the assumptions underlying the model are clearly stated and well-known. In a measurement based model, it is never entirely known whether the characteristics of the environment which gave rise to the measured data match the characteristics of the environment for which the model is being applied. Even when great care is taken to describe the measurement environment, it still may not be clear which factors were important to any given aspect of the measured data. In our geometrical model, the assumptions were clearly stated and may be compared against the environment for which the model is being applied.

CHAPTER 8

Optimal Spatial Filtering and Adaptive Algorithms

In Chapter 3, adaptive spatial filtering was introduced as a way for the array to track mobile units as they move and account for time variability in the radio channel. In this chapter, optimal spatial filtering techniques and adaptive algorithms are presented. This presentation includes a variety of blind adaptive algorithms which are well-suited for CDMA.

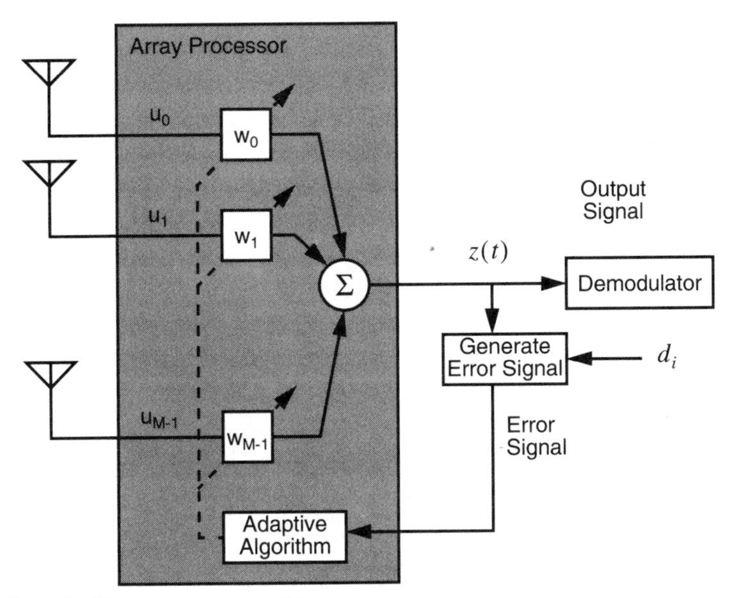

Figure 8–1 An adaptive array processor.

From Chapter 6, rewriting (6.7), the received data vector for a general multi-user system is:

$$\boldsymbol{u}(t) = \sum_{k=0}^{K-1}\sum_{i=0}^{L_k-1} \boldsymbol{a}(\phi_{k,i})\alpha_{k,i}s_k(t-\tau_{k,i}) + \boldsymbol{n}(t) \tag{8.1}$$

In Section 3.7, the optimal Minimum Mean Square Error and Least Squares approaches were derived for optimal spatial filtering, given a desired signal, $d(t)$. There are other techniques which can be used to form *statistically optimal* beam patterns based on data received by the array. In a statistically optimum beamformer, the pattern is adapted to minimize some cost function. Typically, this cost function is inversely associated with the quality of the signal at the array output so that, as the cost function is minimized, the quality of the signal is maximized at the array output. Three statistically optimal beamforming techniques are shown in Table 8–1.

Two of the techniques, the Max SNR approach which maximizes the actual signal-to-noise ratio (SNR) at the array output and the Linearly Constrained Minimum Variance (LCMV) approach, require knowledge of the Direction-Of-Arrival of the desired signal which is not typically known in mobile and portable wireless systems. The third technique, the Minimum Mean Square Error (MMSE) approach attempts to minimize the difference between the array output and some desired signal $d(t)$ [Hay91][Van88].

In each of the approaches in Table 8–1, the correlation matrix of the input data vector, $\boldsymbol{u}(t)$, plays a major role. In statistical spatial processing, the correlation matrix is

$$\boldsymbol{R} = \mathrm{E}[\boldsymbol{u}(t)\boldsymbol{u}^H(t)] \tag{8.2}$$

In the MMSE approach, we form the cross-correlation vector between the input data matrix, $\boldsymbol{u}(t)$ and $d(t)$,

$$\boldsymbol{p} = \mathrm{E}[\boldsymbol{u}(t)d_k^*(t)] \tag{8.3}$$

Then the optimal MMSE solution is

$$\boldsymbol{w}_k = \boldsymbol{R}^{-1}\boldsymbol{p} \tag{8.4}$$

Equation (8.4) has a solution provided that $\boldsymbol{R}$ is non-singular, which is the case as long as the noise component in (8.4) has a non-zero variance [Hay91].

This solution in (8.4) is called the Direct Solution of the *normal equation*. It is assumed that the correlation matrix, $\boldsymbol{R}$, is non-singular, which is the case in most real world environments. Since $\boldsymbol{u}_i$ is a zero mean random process, $\boldsymbol{R}$ is sometimes called the *spatial covariance matrix*. The MMSE approach is very useful because it allows many powerful techniques from the theory of stochastic systems and classical optimal filtering to be applied to spatial filtering. On the other hand, it is not possible to implement ensemble averaging directly, and assumptions about ergodicity and stationarity [Cou83] must typically be applied when deriving any algorithm based on MMSE approaches. An example of an adaptive algorithm which is derived based on MMSE criteria is the Least Mean Square (LMS) algorithm, presented in Chapter 3.

Table 8–1 Statistically optimum beamforming techniques [Van88][Hay91].

	MMSE [Wid67]	Max SNR [Mon80]	LCMV [Fro72]
Criteria	Minimize the difference between the output of the array and some desired response.	Maximize the ratio of the power in the desired signal component to the power in the noise component at the array output.	Minimize the variance at the output of the array subject to linear constraints. For a single constraint, this corresponds to forcing the beam pattern to be a constant in a particular direction.
Cost Function	$J(\boldsymbol{w}) = E[\|\boldsymbol{w}^H\boldsymbol{u}(t) - d(t)\|^2]$ where $y(t)$ is the array output and $d(t)$ is the desired response.	$J(\boldsymbol{w}) = \dfrac{\boldsymbol{w}^H\boldsymbol{R}_n\boldsymbol{w}}{\boldsymbol{w}^H\boldsymbol{R}_s\boldsymbol{w}}$ where $\boldsymbol{R}_n$ is the covariance matrix of the noise component of $\boldsymbol{u}(t)$ and $\boldsymbol{R}_s$ is the covariance matrix of the signal component.	$J(\boldsymbol{w}) = \boldsymbol{w}^H\boldsymbol{R}\boldsymbol{w}$ subject to the linear constraint $\boldsymbol{w}^H\boldsymbol{a}(\phi) = g$. When $g = 1$, this is called the Minimum Variance Distortionless Response (MVRD) beamformer.
Optimal Solution	$\boldsymbol{w} = \boldsymbol{R}^{-1}\boldsymbol{p}$ where $\boldsymbol{R} = \mathrm{E}[\boldsymbol{u}(t)\boldsymbol{u}^H(t)]$ and $\boldsymbol{p} = \mathrm{E}[\boldsymbol{u}(t)d^*(t)]$	$\boldsymbol{R}_n^{-1}\boldsymbol{R}_s\boldsymbol{w} = \lambda_{max}\boldsymbol{w}$ where λ_{max} is the maximum eigenvalue of $\boldsymbol{R}_s$	$\boldsymbol{w} = \boldsymbol{R}^{-1}\boldsymbol{c}[\boldsymbol{c}^H\boldsymbol{R}^{-1}\boldsymbol{c}]^{-1}g$ where $\boldsymbol{c} = \boldsymbol{a}(\phi)$ is the steering vector in the direction of the constraint.
Advantages	Knowledge of the Direction-Of-Arrival is not required	True maximization of SNR	Generalized constraint technique
Disadvantages	Generation of reference signal	Must know statistics of noise and Direction-Of-Arrival of desired signal	Must know Direction-Of-Arrival of desired component

As an example of these techniques, consider the situation in which K signals, $s_k(t)$, are incident on the array and that noise is present at each array element. Assuming that the delays of the individual multipath components are very small relative to the symbol period of $s_k(t)$, then (8.1) can be written as

$$\boldsymbol{u}(t) = \sum_{k=0}^{K-1} s_k(t)\boldsymbol{b}_k(t) + \boldsymbol{n}(t) \qquad (8.5)$$

where $\boldsymbol{b}_k$ is the *spatial signature* for subscriber k, introduced in Section 3.3.

$$\boldsymbol{b}_k(t) = \sum_{l=0}^{L_k-1} \boldsymbol{a}(\phi_{k,l})\alpha_{k,l}(t)\delta(t-\tau_{k,l}) \tag{8.6}$$

The noise term, $\boldsymbol{n}(t)$, (8.5) is a Gaussian random process with covariance matrix $\sigma_n^2\boldsymbol{I}$. It can be shown that the weight vector formed using (8.4) maximizes the SINR at the output of the array, $z(t)$. Note that while the data vector given in (8.5) is shown as a continuous time signal, adaptive arrays are typically implemented using discrete time systems so that samples are available only at time instances $t = iT_s$, where T_s is the sampling period of the discrete time system. While the continuous time notation will be used for the remainder of this chapter, the discussion is valid for discrete time systems as well.

In this example, a weight vector is formed to extract the ith signal, $s_i(t)$, at the output of the array with $d(t) = s_i(t)$. If each of the signals, $s_k(t)$ is uncorrelated so that $\mathrm{E}[s_k(t)s_k^*(t)] = 2P_k$, and $\mathrm{E}[s_k(t)s_l^*(t)] = 0$ for $k \neq l$, then

$$\boldsymbol{R} = 2\begin{bmatrix}\boldsymbol{b}_0 & \dots & \boldsymbol{b}_{K-1}\end{bmatrix}\begin{bmatrix}P_0 & & 0\\ & \ddots & \\ 0 & & P_{K-1}\end{bmatrix}\begin{bmatrix}\boldsymbol{b}_0^H\\ \vdots\\ \boldsymbol{b}_{K-1}^H\end{bmatrix} + \sigma_n^2\boldsymbol{I} = \boldsymbol{B}^H\boldsymbol{B} + \sigma_n^2\boldsymbol{I} \tag{8.7}$$

where the matrix $\boldsymbol{B}$ is defined such that the columns of $\boldsymbol{B}^H$ are the weighted spatial signatures from each user

$$\boldsymbol{B}^H = \begin{bmatrix}\sqrt{2P_0}\boldsymbol{b}_0 & \dots & \sqrt{2P_{K-1}}\boldsymbol{b}_{K-1}\end{bmatrix} \tag{8.8}$$

and $P_k = \frac{1}{2}|s_{lk}(t)|^2$.

The cross-correlation between the desired array output, $d(t) = s_i(t)$, and the data vector is

$$\boldsymbol{p}_i = 2P_i\boldsymbol{b}_i \tag{8.9}$$

then the set of weights which provides a Minimum Mean Square Error or Wiener solution which extracts the signal $s_i(t)$, is

$$\boldsymbol{w}_i = \boldsymbol{R}^{-1}\boldsymbol{p}_i = (\boldsymbol{B}^H\boldsymbol{B} + \sigma_n^2\boldsymbol{I})^{-1}2P_i\boldsymbol{b}_i \tag{8.10}$$

The Least Squares (LS) approach is based on time averages instead of ensemble averages. In LS techniques, the time average of the square of the error between the observed data and the desired response is minimized over a time window. In the LS approach, the weight vector is optimal only for the observed data over the time window that is considered; however, no assumptions about stationarity are required and algorithms based on LS can be implemented directly without additional approximations. In Chapter 3 it was shown that a Least Squares optimal solution is

$$w_k = (A^H A)^{-1} A^H d_k \tag{8.11}$$

where

$$A^H = \begin{bmatrix} u_0 \; u_1 \; \dots \; u_{P-1} \end{bmatrix} \tag{8.12}$$

and

$$d_k = \begin{bmatrix} d_{k,0} \; d_{k,1} \; \dots \; d_{k,P-1} \end{bmatrix}^T \tag{8.13}$$

It can be shown that techniques based on MMSE or LS approaches will maximize the average (either ensemble average or time average) Signal-to-Interference-and-Noise Ratio (SINR) at the array output under a broad range of conditions. There are several mechanisms through which this is accomplished. If a signal is present with no interference or multipath, then the MMSE or LS approach will result in a weight vector which directs a beam toward the DOA of the desired signal and minimizes the effects of diffuse noise.

When the noise level is very low and a small number of interfering signals (fewer than the number of array elements) are present, a weight vector will be formed which corresponds to a beam pattern with nulls in the directions of interferers. The behavior of optimal array solutions in the presence of multipath is more complicated and is described in Section 8.1. When there are more significant interferers than array elements, the array is said to be *overloaded.* When the number of significant interferers is smaller than the number of array elements, the array is said to be *underloaded.* A discussion of adaptive arrays operating under these different conditions is provided in Section 8.2.

8.1 Impact of Multipath on Optimal Spatial Filtering

Consider the case in which two multipath components from a single transmitter are incident on the antenna array. The baseband complex envelope representation of the original transmitted signal is $s_k(t)$. The signal incident on the array due to the first multipath component is weighted by $\alpha_{k,0}$ and delayed by $\tau_{k,0}$, while the second component is weighted by $\alpha_{k,1}$ and delayed by $\tau_{k,1}$. Let us assume that the channel characteristics are fixed and known so that they are not random variables.

We define the time-average of the ensemble average autocorrelation function between two multipath components as

$$\begin{aligned} \tilde{r}_m(\tau_{k,1} - \tau_{k,0}) &= \langle \mathrm{E}[\alpha_{k,0} s_k(t-\tau_{k,0}) \alpha_{k,1}^* \, s_k^* \, (t-\tau_{k,1})] \rangle \\ &= \alpha_{k,0} \alpha_{k,1}^* \, \langle r_s(t-\tau_{k,0}, t-\tau_{k,1}) \rangle \\ &= \alpha_{k,0} \alpha_{k,1}^* \, \tilde{r}_s(\tau_{k,1} - \tau_{k,0}) \end{aligned} \tag{8.14}$$

where $r_s(t, t+\tau)$ is the ensemble average autocorrelation function of the complex envelope of the transmitted signal, $s_k(t)$, as given by

$$r_s(t_1, t_2) = E[s_k(t_1) s_k^* \, (t_2)] \tag{8.15}$$

and $\tilde{r}_s(\tau) = \langle r_s(t, t+\tau)\rangle$ is the time-average of $r_s(t, t+\tau)$. The quantities, $\tilde{r}_s(\tau)$ and $r_s(t, t+\tau)$, will be equal if $s_k(t)$ is *wide sense stationary* [Cou83]. The multipath components arriving at times $\tau_{k,0}$ and $\tau_{k,1}$ are said to be *correlated* if $\tilde{r}_s(\tau_{k,1} - \tau_{k,0})$ is non-zero.

Consider a signal composed of independent binary rectangular symbols. An example of this is the DS-CDMA signal described in Chapter 1. The baseband complex envelope representation of the transmitted signal is

$$s_k(t) = A_k a_k(t) b_k(t) \tag{8.16}$$

where the power of the transmitted bandpass signal is $A_k^2/2$, $a_k(t)$ is the PN sequence given by (1.4), and $b_k(t)$ is the data signal given in (1.2). This type of signal is not wide sense stationary [Cou83]. Assume for the moment that $\tau_{k,1} \geq \tau_{k,0}$, then, computing the ensemble average autocorrelation function of $s_k(t)$, we obtain

$$r_s(t - \tau_{k,0}, t - \tau_{k,1}) = \begin{cases} A_k^2 \displaystyle\sum_{i=-\infty}^{\infty} \Psi\left(\frac{t - \tau_{k,1} - nT_c}{T_c - (\tau_{k,1} - \tau_{k,0})}\right) & 0 \leq \tau_{k,1} - \tau_{k,0} < T_c \\ 0 & \tau_{k,1} - \tau_{k,0} \geq T_c \end{cases} \tag{8.17}$$

Taking the time average of this quantity, for $0 \leq \tau_{k,1} - \tau_{k,0} < T_c$, we write

$$\tilde{r}_s(\tau_{k,1} - \tau_{k,0}) = \langle r_s(t - \tau_{k,0}, t - \tau_{k,1})\rangle = \lim_{T \to \infty} \frac{1}{T} \int_{-T/2}^{T/2} r_s(t - \tau_{k,0}, t - \tau_{k,1}) dt \tag{8.18}$$

$$= A_k^2 \lim_{T \to \infty} \sum_{i=-\infty}^{\infty} \frac{1}{T} \int_{-T/2}^{T/2} \Psi\left(\frac{t - \tau_{k,1} - nT_c}{T_c - (\tau_{k,1} - \tau_{k,0})}\right) dt$$

$$= A_k^2 \lim_{N \to \infty} \sum_{i=-N}^{N} \frac{1}{(2N+1)T_c} \int_{\tau_{k,1} + nT_c}^{\tau_{k,0} + (n+1)T_c} dt$$

$$= A_k^2 \lim_{N \to \infty} \sum_{i=-N}^{N} \frac{T_c - (\tau_{k,1} - \tau_{k,0})}{(2N+1)T_c}$$

$$= A_k^2 \frac{T_c - (\tau_{k,1} - \tau_{k,0})}{T_c}$$

We can perform a similar analysis for $-T_c < \tau_{k,1} - \tau_{k,0} \leq 0$ to obtain

$$\tilde{r}_s(\tau_{k,1}-\tau_{k,0}) = \begin{cases} A_k^2 \dfrac{T_c - |\tau_{k,1}-\tau_{k,0}|}{T_c} & |\tau_{k,1}-\tau_{k,0}| < T_c \\ 0 & |\tau_{k,1}-\tau_{k,0}| \geq T_c \end{cases} \tag{8.19}$$

Thus, the correlation between multipath components is given by

$$\tilde{r}_s(\tau_{k,1}-\tau_{k,0}) = \begin{cases} A_k^2 \alpha_{k,0}\alpha_{k,1}^* \dfrac{T_c - |\tau_{k,1}-\tau_{k,0}|}{T_c} & |\tau_{k,1}-\tau_{k,0}| < T_c \\ 0 & |\tau_{k,1}-\tau_{k,0}| \geq T_c \end{cases} \tag{8.20}$$

which has a maximum at $\tau_{k,0} = \tau_{k,1}$ and drops off linearly to zero at $\tau_{k,1}-\tau_{k,0} = \pm T_c$. Two multipath components are said to be *highly correlated* at time lag $\tau = \tau_{k,1}-\tau_{k,0}$, if $\tilde{r}_s(\tau)/\tilde{r}_s(0) \approx 1$. They are said to be *uncorrelated* if $\tilde{r}_s(\tau)/\tilde{r}_s(0) \approx 0$.

One measure of the difference in time delays of multipath components is the ensemble average *delay spread*, σ_τ, which is given by [Jak74][1]

$$\sigma_\tau^2 = E[(\tau_{k,i}-\bar{\tau})^2] \tag{8.21}$$

It is assumed that the delays of the multipath components are independent, with mean $\bar{\tau}$. We can express the expected value of the squared difference in delays between two multipath components as

$$\begin{aligned} E[(\tau_{k,1}-\tau_{k,0})^2] &= E[((\tau_{k,1}-\bar{\tau})-(\tau_{k,0}-\bar{\tau}))^2] \\ &= E[(\tau_{k,1}-\bar{\tau})^2] - 2E[(\tau_{k,1}-\bar{\tau})(\tau_{k,0}-\bar{\tau})] + E[(\tau_{k,0}-\bar{\tau})^2] \\ &= 2\sigma_\tau^2 - 2E[\tau_{k,1}\tau_{k,0} - \bar{\tau}\tau_{k,0} - \bar{\tau}\tau_{k,1} + \bar{\tau}^2] \\ &= 2\sigma_\tau^2 \end{aligned} \tag{8.22}$$

From the above discussion, we say that two multipath components are uncorrelated if they are resolvable by more than a chip duration, or $|\tau_{k,1}-\tau_{k,0}| > T_c$, which is equivalent to the condition that $(\tau_{k,1}-\tau_{k,0})^2 > T_c^2$. From (8.22), the expected value of the squared difference between two multipath component delays is $2\sigma_\tau^2$. Therefore, two multipath components are expected to be uncorrelated if $2\sigma_\tau^2 > T_c^2$ or $\sigma_\tau/T_c > 1/\sqrt{2}$.

---

1. The delay spread is defined here in terms of ensemble averages as in [Jak74]. In the literature, the delay spread is frequently defined in terms of central moments of the power delay profile as described in [Cox72] and [Rap89]. The two can be shown to be equal if the normalized averaged power delay profile is used to estimate the distribution of the multipath channel delay as shown in [Dev87a].

The *coherence bandwidth*, B_c, of the channel is defined as the difference between two frequencies such that the correlation between the envelope of the received signals at those two frequencies is 0.5 [Jak74][Ste92]. According to [Jak74], the coherence bandwidth of the channel is related to the delay spread by $B_c = 1/2\pi\sigma_\tau$. The null-to-null bandwidth of a DS-CDMA signal, using rectangular chip waveforms, is $B = 2/T_c$. Therefore we may express the ratio σ_τ/T_c as

$$\frac{\sigma_\tau}{T_c} = \frac{B}{4\pi B_c} \tag{8.23}$$

8.1.1 Flat Fading Channels

If the delay spread is small so that $\sigma_\tau < T_c/4\pi$, then the null-to-null RF signal bandwidth is less than the coherence bandwidth. This condition is referred to as *frequency-flat fading* since the signal is contained within a bandwidth over which the fading of the envelope of the signal spectrum is highly correlated. In other words, the frequency response of the multipath channel is approximately flat over the bandwidth containing the signal [Ste92]. In flat fading channels, phase coherence among the correlated multipath components allows them to interact either constructively or destructively, leading to potentially deep fades.

From Section 8.1, multipath components are expected to be correlated if the delay spread satisfies $\sigma_\tau < T_c/\sqrt{2}$. Since any channel which satisfies $\sigma_\tau < T_c/4\pi$ also satisfies $\sigma_\tau < T_c/\sqrt{2}$, a channel which exhibits flat fading will also contain correlated multipath. A channel is described as being *narrowband* if $\sigma_\tau < T_c/4\pi \approx 0.1T_c$.

It is important to note that two multipath components, arriving with small differential delay, can become decorrelated due to differing Doppler shifts. In other words, if two components arrive with small delay but distinct Doppler frequencies, an adaptive algorithm can treat both components as being uncorrelated (when a large observation time window is used relative to the reciprocal of the maximum Doppler frequency), or as correlated (when the observation time interval is short relative to the reciprocal of the maximum Doppler frequency). This decorrelation is temporal in nature, which means that the degree of decorrelation depends not only on the Doppler shift, $f_{d,i}$, but also on the time window over which signals are observed, T_w.

8.1.2 Frequency Selective/Time Dispersive Channels

If the rms delay spread is large relative to the chip period, $\sigma_\tau > T_c/4\pi$, then, from (8.23), the null-to-null signal bandwidth is larger than the coherence bandwidth of the channel. When the signal bandwidth is larger than the coherence bandwidth of the channel, *frequency selective fading* may occur in which fading in one region of the signal spectrum does not imply fading in other regions of the signal spectrum. Combining this condition with the condition that multipath components are expected to be uncorrelated if $\sigma_\tau/T_c > 1/\sqrt{2}$, we use the term *wideband* to refer to channels in which the delay spread satisfies $\sigma_\tau > T_c/\sqrt{2} \approx 0.7T_c$. In wideband channels, frequency selective fading occurs and multipath components tend to be uncorrelated.

In wideband, time dispersive channels, multipath components tend to be non-coherent with one another. Therefore, it is less likely, compared with narrowband channels, that components will be able to cancel each other. Because of this, the overall received signal strength of signals sent over wideband channels tends to vary less dramatically than power received over flat fading channels [Rap96a]. The mechanism used by the adaptive array to maximize the SINR is different, depending on the nature of the multipath.

8.1.3 Array Performance in Multipath

The MMSE optimal spatial filter, implemented using the array, performs differently in the presence of uncorrelated and correlated multipath. Figure 8–2 shows the array patterns achieved when correlated and uncorrelated multipath components are incident on a four element Linear Equally Spaced (LES) array with half wavelength spacing. In these plots, two signals are incident on the array, one Signal-of-Interest (SOI) and one Signal-Not-of-Interest (SNOI). The MMSE approach is used to find the optimal weight vector to extract the SOI.

In case (a), only a single component arrives from the SOI from $\phi = 60°$, but two uncorrelated, equal power multipath components arrive for the SNOI from $90°$ and $135°$. In this case, the array forms nulls in the directions of each SNOI component to reduce the effects of multipath at the array output. In case (b), the two multipath components for the SNOI are correlated. Note that, unlike case (a), the array does not form deep nulls in the directions of each component. Instead, the array responds by forming a weight vector $\boldsymbol{w}_0$ which is orthogonal to the spatial signature for the interfering user. In effect, the two correlated SNOI components are brought to the same amplitude by the array pattern, and a phase shift is applied by the algorithm so that the two components cancel each other at the array output.

In case (c), only a single component arrives for the SNOI, but two uncorrelated components arrive for the SOI. As discussed earlier, this can occur when the difference in arrival times between SOI multipath components is larger than a symbol or chip period. In this case, the array treats one of the SOI components as an interferer and places a null in the direction of both the SNOI and the uncorrelated SOI component. Case (d) shows the case when the two multipath components for the SOI are correlated. In this scenario, the array forms two beams and acts as an optimum combiner of the power in the two multipath components. Thus the array can be used to take advantage of path diversity in narrowband systems operating in multipath environments.

The examples in Figure 8–2 illustrate several important characteristics of adaptive antenna systems. First, when correlated multipath components are present for the desired signal, the array can combine them to take advantage of path diversity. If a conventional Rake receiver were used with an omni-directional antenna, correlated components would be considered unresolvable. An adaptive equalizer may be unable to take advantage of the two paths when the delay difference between the paths is small. Thus the adaptive antenna array can provide improvements over conventional Rake or equalized systems when correlated multipath is present.

When uncorrelated multipath is incident on the array, the MMSE array treats each component as a separate signal. If uncorrelated components are present for the SOI, then the MMSE

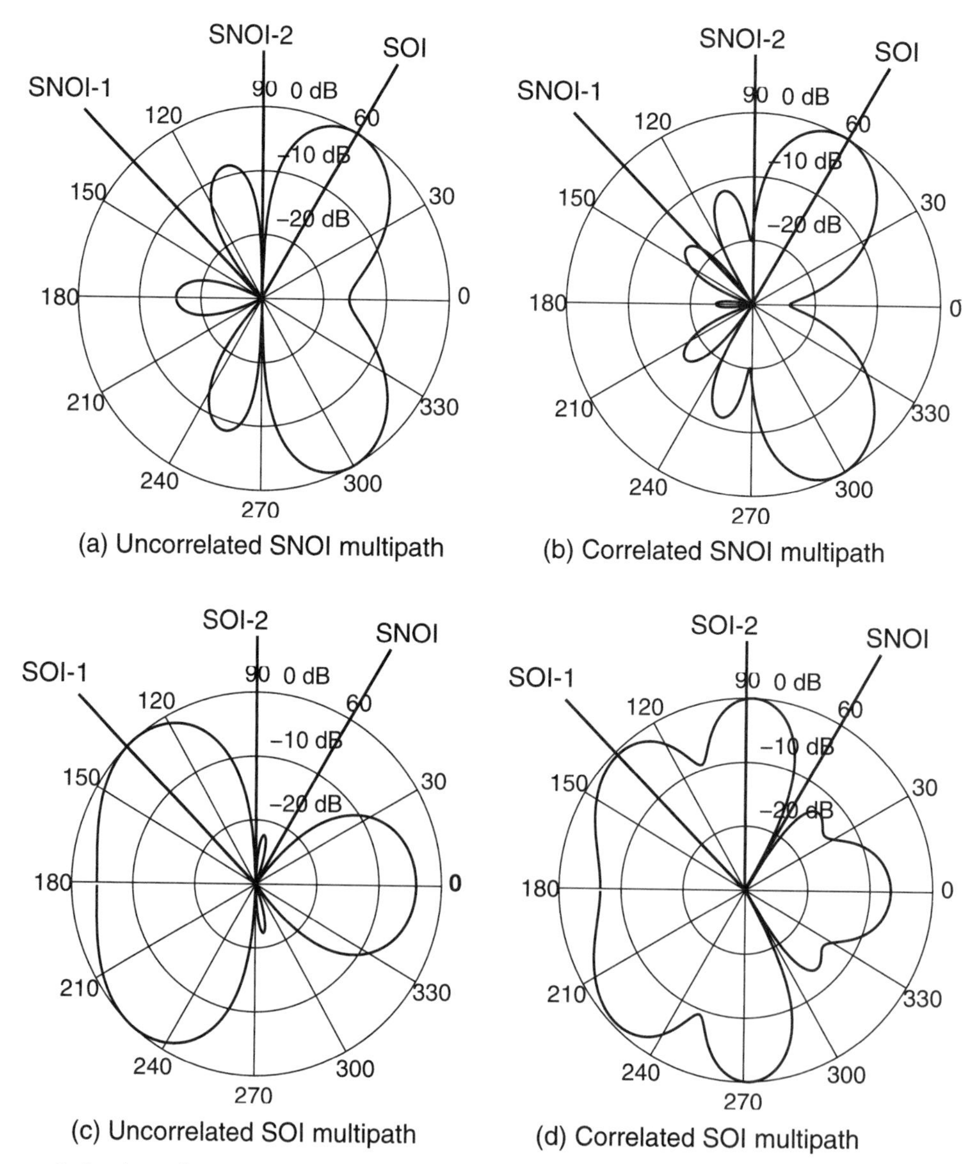

Figure 8–2 Array factor pattern resulting from the optimal MMSE array solution, given by (8.4), in the presence of (a) uncorrelated multipath for the SNOI, (b) correlated multipath for the SNOI, (c) correlated multipath for the SOI, and (d) uncorrelated multipath for the SOI [Lib95].

array will select the component with the larger power, nulling the other component. The spatial filtering Rake receiver described in Chapter 4 can be applied in this case to avoid the loss of signal power available in the multipath component.

It is interesting to note that the array can remove the effects of more multipath components if the components are correlated. The importance of this is that the adaptive array can separate many more signals from different sources in a flat-fading environment. The array can effectively cancel $M-1$ SNOI spatial signatures. If the SNOI components are all uncorrelated, then the array can completely null only $M-1$ multipath components.

8.2 Performance of Underloaded and Overloaded Adaptive Arrays

The behavior of the optimal MMSE solution, $\boldsymbol{w}_i$, used to extract a known signal component depends on whether the array is overloaded or underloaded. To illustrate this behavior, consider a four element Linear Equally Spaced (LES) array oriented along the x-axis. A Signal-Of-Interest (SOI) is incident on the array from an angle of $\phi_0 = 60°$. Initially, two interferers are added to the system at DOAs of $\phi_1 = 30°$ and $\phi_2 = 90°$. Each interfering signal, or Signal-Not-Of-Interest (SNOI) has the same incident power as the desired signal. Each SNOI may be either an interfering signal from a co-channel user or uncorrelated multipath from the same transmitter as the SOI.

The optimal weight vector formed for this configuration results in a vector $\boldsymbol{w}_0$ which has a non-zero response to $\boldsymbol{a}(\phi_0)$ and is orthogonal to $\boldsymbol{a}(\phi_1)$ and $\boldsymbol{a}(\phi_2)$. The antenna pattern corresponding to this weight vector is shown in Figure 8–3 (a). The power levels of the signals at the array output relative to the SOI are given in Table 8–2. As a third interferer is added (SNOI-3), the array is still able to form a null in the direction of all interferers, as shown in Figure 8–3 (b).

Table 8–2 Power levels of signals at the array output relative to the power level of the Signal-of-Interest. All signals incident on the array have the same power. The weight vector is the optimal MMSE solution for a four element linear array [Lib95].

	Power of SNOI Relative to SOI at Array Output			
SOI @ $\phi_0 = 60°$	**Two Interferers**	**Three Interferers**	**Four Interferers**	**Five Interferers**
SNOI-1 @ $\phi_1 = 30°$	-79.9 dB	-76.5 dB	-20.7 dB	-18.4 dB
SNOI-2 @ $\phi_2 = 90°$	-94.3 dB	-84.0 dB	-33.7 dB	-23.5 dB
SNOI-3 @ $\phi_3 = 135°$		-77.8 dB	-20.4 dB	-14.6 dB
SNOI-4 @ $\phi_4 = 150°$			-16.5 dB	-13.3 dB
SNOI-5 @ $\phi_5 = 45°$				-8.7 dB

When a fourth interferer is added (SNOI-4), the array is no longer able to form nulls in the directions of all interferers. The optimal array solution still attempts to minimize the error at the array output; however, in doing so, the array is unable to form nulls in the directions of the interfering signals SNOI-1, SNOI-2, and SNOI-3. As shown in Table 8–2, the power levels of all

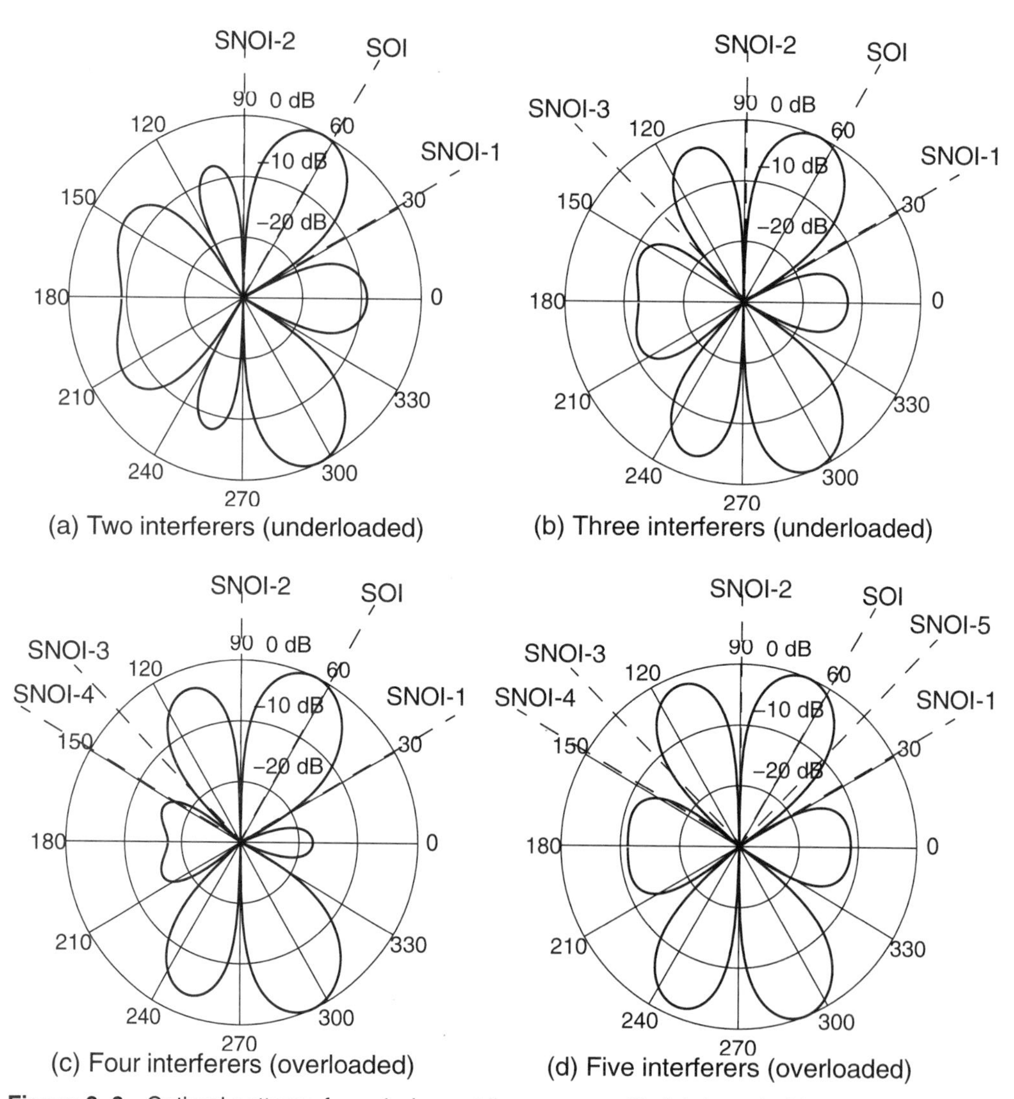

Figure 8–3 Optimal patterns for a 4 element linear array with (a) three incident signals (underloaded), (b) four incident signals (perfectly loaded), (c) five incident signals (overloaded), and (d) six incident signals.

interfering signals at the array output rise considerably as the array becomes overloaded. This trend continues as a fifth interferer is added (SNOI-5) as illustrated in Figure 8–3 (d) and Table 8–2 [Lib95].

When no multipath or interferers are present, an LES array is able to achieve a maximum gain of $10\log(M)$ with respect to the gain of each individual antenna element. As interferers are added, if the array remains underloaded, the optimal array solution will provide nulls in the directions of interferers. If the Directions-Of-Arrival of interfering signals are not very close to the DOA of the desired signal, the gain of the underloaded array with respect to noise and diffuse interference will become only slightly degraded from $10\log(M)$ as additional interferers are added. In the limiting case, as the number of interfering signals becomes very large, the array will continue to form a beam in the direction of the desired user; however, the interference nulling ability of the array will be sharply reduced. If the interfering users are uniformly distributed in DOA, then the array will act as it does when only noise and the SOI are present; namely, it will form a beam pattern using approximately equal weights on all elements which results in a pattern with an antenna gain of $10\log(M)$ dB for a linear equally spaced array. In this overloaded case, the array is not capable of significantly attenuating interfering signals arriving at DOAs near the SOI.

8.3 Adaptive Algorithms

There are several reasons why it is generally not desirable to solve the normal equations (either in an MMSE or LS sense) directly. Since the mobile environment is time-variable, the solution for the weight vector must be updated or adapted periodically. Typically, the weight vector computed at one cycle differs from the weight vector computed at the previous update by only a small amount. Also, since the data required to estimate the optimal solution is noisy, it is desirable to use an update technique which uses previous solutions for the weight vector to smooth the estimate of the optimal response, reducing the effects of noise. For these reasons, an *adaptive algorithm* is used to update the weight vector periodically. The adaptive algorithm operates either in a block mode or iterative mode. In block processing techniques, a new solution is calculated periodically using estimates of statistics obtained from the most recently available block of data. In iterative algorithms, at each iteration n, the current weight vector, $\hat{w}(n)$, is adjusted by an incremental value to form a new weight vector, $\hat{w}(n+1)$, which approximates the optimum solution w_{opt}, given by (8.4).

An adaptive array receiver is illustrated in Figure 8–1. Through adaptation, the system is able to acquire new signals and track subscribers as they move about within a cell. There are many types of adaptive algorithms, most of which are iterative and make use of past information to minimize the computations required at each update cycle.

As shown in Chapter 3, the stochastic gradient approach provides an iterative update solution for the MMSE criteria

$$w_{k,i+1} = w_{k,i} - \frac{1}{2}\mu\nabla J(w_{k,i}) \tag{8.24}$$

where $\nabla J(w)$ is the gradient of a function $J(w)$ which is given in (3.24)

$$J(w) = \mathrm{E}[|w^H u - d|^2] \tag{8.25}$$

The function $J(w)$ is the cost function which defines an *error performance surface* [Hay91]. The error performance surface is parabolic, which means that it is a second order function of the weight vector, and it has a unique minimum value. These features mean that (8.24) will always converge monotonically to a unique optimal solution for the weight vector $\boldsymbol{w}_{opt}$.

Exact calculation of the gradient, $\nabla J(\boldsymbol{w})$, required in (8.24) is cumbersome and involves matrix inversions. Therefore, other, more efficient, adaptive algorithms have been developed. Examples of two common adaptive algorithms are the Least Mean Square (LMS) algorithm, discussed in Chapter 3, and the Recursive Least Squares (RLS) algorithms. Both algorithms are illustrated in Table 8–3.

Table 8–3 The LMS, RLS, and Bussgang algorithms [Hay91].

	Least Mean Square (LMS)	**Recursive Least Squares (RLS)**	**Bussgang Algorithm**
Initialization	$\hat{w}_0 = \mathbf{0}$	$\hat{w}_0 = \mathbf{0}$ $\boldsymbol{P}_0 = \delta^{-1}\boldsymbol{I}$ δ is some small constant	$\hat{w}_0 = \begin{bmatrix} 1 & 0 & \ldots & 0 \end{bmatrix}^T$
Weight Update	$z_n = \hat{w}^H{}_n \boldsymbol{u}_n$ $e_n = d_n - z_n$ $\hat{w}_{n+1} = \hat{w}_n + \mu \boldsymbol{u}_n e^*{}_n$	$\boldsymbol{v}_n = \boldsymbol{P}_{n-1}\boldsymbol{u}_n$ $\boldsymbol{k}_n = \dfrac{\lambda^{-1}\boldsymbol{v}_n}{1 + \lambda^{-1}\boldsymbol{u}_n^H \boldsymbol{v}_n}$ $\alpha_n = d_n - \hat{w}_{n-1}^H \boldsymbol{u}_n$ $\hat{w}_n = \hat{w}_{n-1} + \boldsymbol{k}_n \alpha^*{}_n$ $\boldsymbol{P}_n = \lambda^{-1}(\boldsymbol{I} - \boldsymbol{k}_n \boldsymbol{u}_n^H)\boldsymbol{P}_{n-1}$	$z_n = \hat{w}^H{}_n \boldsymbol{u}_n$ $e_n = g(z_n) - z_n$ $\hat{w}_{n+1} = \hat{w}_n + \mu \boldsymbol{u}_n e^*{}_n$
Convergence Coefficient	Step size parameter μ, $0 < \mu < \mathrm{Trace}(\boldsymbol{R})$	Forgetting factor, λ, $0 < \lambda < 1$	Step size parameter μ

In both the LMS and RLS approaches, the desired signal must be supplied using either a training sequence or decision direction. In the training sequence approach, a brief data sequence is transmitted which is known by the receiver. The receiver uses an adaptive algorithm to estimate the weight vector during the training period, then holds the weights constant while information is being transmitted. This technique requires that the environment be stationary from one training period to the next, and it reduces channel throughput by requiring the use of channel symbols for training. In the decision directed approach, the receiver uses recreated modulated symbols based on symbol decisions, which are used as the desired signal to adapt the weight vector. The drawback to this approach is that errors in the decision directed process can lead to poorly directed weight updates, which in turn can lead to increased decision errors. This approach does not work well when extracting signals in high levels of noise, interference, or multipath.

8.3.1 Blind Adaptive Algorithms

Other techniques have been developed which do not require training sequences. These techniques, generally referred to as *blind adaptive algorithms*, adapt by attempting to restore some known property to the received signal [Hay91]. One approach to blind adaptation is the *Bussgang technique*. In the Bussgang technique, a non-linear, zero-memory estimator, $g(\bullet)$, is used to operate on the signal y_n at the output of the array combiner. The difference between $d_n = g(y_n)$ and y_n is then used to form an error function e_n, which is used to update the array weight vector as shown in Table 8–3.

$$z_n = \hat{\boldsymbol{w}}^H{}_n \boldsymbol{u}_n \tag{8.26}$$

$$e_n = g(z_n) - z_n \tag{8.27}$$

$$\hat{\boldsymbol{w}}_{n+1} = \hat{\boldsymbol{w}}_n + \mu \boldsymbol{u}_n e^*{}_n \tag{8.28}$$

Decision direction is a simple form of the Bussgang approach. However, other interesting blind techniques, which may do a better job of initially "opening the eye" of the received symbol, can also be derived from the Bussgang approach. An example of this, introduced by Godard, is the Constant Modulus Algorithm (CMA) in which $g(\bullet)$ is used to extract the phase of the input signal [Lar83][Lar86][Age89].

In the CM approach, which may be applied to signals transmitted with constant envelopes, the cost function is

$$J(\boldsymbol{w}_k) = \mathrm{E}[||\boldsymbol{w}_k^H \boldsymbol{u}_i|^p - |\alpha|^p|^q] \tag{8.29}$$

where α is the desired signal amplitude at the array output. An adaptive array which uses the CM cost-function will attempt to drive the signal at the array output to have a constant envelope with the specified amplitude, α. The exponents p and q are each equal to either 1 or 2. Using different values of p and q, it is possible to develop several different steepest-descent algorithms which have different convergence characteristics and complexity [Lar83]. There are several drawbacks to this approach, including the fact that this type of algorithm simply captures the strongest constant envelope signal present at the input, which may be an interfering signal. Also, the convergence of this algorithm is not as well-characterized as that of MMSE and LS approaches, although it has been shown that CM approaches converge under a wide range of conditions. On the other hand, the CM approach is nicely suited to reducing narrowband fading and may be easily applied to analog frequency-modulated signals. A multi-target CM algorithm was developed by Agee which circumvents the problem of the CM system capturing interference [Lar83].

For the $p = 1$, $q = 2$ form, also called the 1-2 form, we obtain the following algorithm with $\alpha = 1$

$$y(k) = \boldsymbol{w}^H(k)\boldsymbol{u}(k) \tag{8.30}$$

$$e(k) = 2\left(y(k) - \frac{y(k)}{|y(k)|}\right) \tag{8.31}$$

$$\boldsymbol{w}(k+1) = \boldsymbol{w}(k) - \mu\boldsymbol{u}(k)e^*(k) \tag{8.32}$$

Clearly, no estimate of the desired signal is required, since the new weight vector, $\boldsymbol{w}_{k+1}$, depends only on the array output, the data vector, $\boldsymbol{u}(k)$, and the previous weight vector, $\boldsymbol{w}_k$. For the other basic CMA forms, the following error functions apply [Lar83]:

"1-1"
$$e(k) = \frac{y(k)}{|y(k)|}\operatorname{sgn}(|y(k)| - 1) \tag{8.33}$$

"2-1"
$$e(k) = 2y(k)\operatorname{sgn}(|y(k)|^2 - 1) \tag{8.34}$$

"2-2"
$$e(k) = 4y(k)(|y(k)|^2 - 1) \tag{8.35}$$

8.3.2 The Least Squares Constant Modulus Algorithm

In the previous section, a number of steepest descent algorithms were presented based on the Constant Modulus Algorithm (CMA). The Least Squares Constant Modulus Algorithm (LS-CMA) was proposed by Agee using an extension of the method of nonlinear least squares, also known as Gauss's Method [Ron96a]. The extension of Gauss's method states that if a cost function can be expressed in the form

$$F(\boldsymbol{w}) = \sum_{k=1}^{K} |g_k(\boldsymbol{w})|^2 = \|\boldsymbol{g}(\boldsymbol{w})\|_2^2 \tag{8.36}$$

where

$$\boldsymbol{g}(\boldsymbol{w}) = \left[g_1(\boldsymbol{w})\ g_2(\boldsymbol{w})\ \dots\ g_K(\boldsymbol{w})\right]^T \tag{8.37}$$

then the cost function has a partial Taylor-series expansion with the sum-of-squares form:

$$F(\boldsymbol{w}+\boldsymbol{d}) \approx \|\boldsymbol{g}(\boldsymbol{w}) + \boldsymbol{D}^H(\boldsymbol{w})\boldsymbol{d}\|_2^2 \tag{8.38}$$

where $\boldsymbol{d}$ is an offset vector, and

$$\boldsymbol{D}(\boldsymbol{w}) = \left[\nabla(g_1(\boldsymbol{w}))\ \nabla(g_2(\boldsymbol{w}))\ \dots\ \nabla(g_K(\boldsymbol{w}))\right] \tag{8.39}$$

The gradient vector of $F(\boldsymbol{w}+\boldsymbol{d})$ with respect to $\boldsymbol{d}$ is

$$\begin{aligned}
\nabla_d(F(\boldsymbol{w}+\boldsymbol{d})) &= 2\frac{\partial F(\boldsymbol{w}+\boldsymbol{d})}{\partial \boldsymbol{d}^*} \\
&= 2\frac{\partial\{(\boldsymbol{g}(\boldsymbol{w}) + \boldsymbol{D}^H(\boldsymbol{w})\boldsymbol{d})^H(\boldsymbol{g}(\boldsymbol{w}) + \boldsymbol{D}^H(\boldsymbol{w})\boldsymbol{d})\}}{\partial \boldsymbol{d}^*} \\
&= 2\frac{\partial\{\|\boldsymbol{g}(\boldsymbol{w})\|_2^2 + \boldsymbol{g}^H(\boldsymbol{w})\boldsymbol{D}^H(\boldsymbol{w})\boldsymbol{d} + \boldsymbol{d}^H\boldsymbol{D}(\boldsymbol{w})\boldsymbol{g}(\boldsymbol{w}) + \boldsymbol{d}^H\boldsymbol{D}(\boldsymbol{w})\boldsymbol{D}^H(\boldsymbol{w})\boldsymbol{d}\}}{\partial \boldsymbol{d}^*} \\
&= 2\{\boldsymbol{D}(\boldsymbol{w})\boldsymbol{g}(\boldsymbol{w}) + \boldsymbol{D}(\boldsymbol{w})\boldsymbol{D}^H(\boldsymbol{w})\boldsymbol{d}\}
\end{aligned} \tag{8.40}$$

Setting $\nabla_d(F(\boldsymbol{w}+\boldsymbol{d}))$ equal to zero, the offset that minimizes the cost function $F(\boldsymbol{w}+\boldsymbol{d})$ is

$$\boldsymbol{d} = -[\boldsymbol{D}(\boldsymbol{w})\boldsymbol{D}^H(\boldsymbol{w})]^{-1}\boldsymbol{D}(\boldsymbol{w})\boldsymbol{g}(\boldsymbol{w}) \quad (8.41)$$

Adding $\boldsymbol{d}$ to $\boldsymbol{w}$ results in a new weight vector that minimizes the cost function. Therefore the weight vector can be updated by

$$\boldsymbol{w}(l+1) = \boldsymbol{w}(l) - [\boldsymbol{D}(\boldsymbol{w}(l))\boldsymbol{D}^H(\boldsymbol{w}(l))]^{-1}\boldsymbol{D}(\boldsymbol{w}(l))\boldsymbol{g}(\boldsymbol{w}(l)) \quad (8.42)$$

where l denotes the iteration number. LS-CMA is derived by applying equation (8.42) to the constant modulus function,

$$F(\boldsymbol{w}) = \sum_{k=1}^{K} ||y(k)| - 1|^2 \quad (8.43)$$

$$= \sum_{k=1}^{K} ||\boldsymbol{w}^H\boldsymbol{x}(k)| - 1|^2$$

Comparing (8.43) with (8.36), we see that

$$g_k(\boldsymbol{w}) = |y(k)| - 1 = |\boldsymbol{w}^H\boldsymbol{x}(k)| - 1 \quad (8.44)$$

Substituting (8.44) into (8.37),

$$\boldsymbol{g}(\boldsymbol{w}) = \begin{bmatrix} |y(1)| - 1 \\ |y(2)| - 1 \\ \vdots \\ |y(K)| - 1 \end{bmatrix} \quad (8.45)$$

The gradient of $g_k(\boldsymbol{w})$ is

$$\nabla(g_k(\boldsymbol{w})) = 2\frac{\partial g_k(\boldsymbol{w})}{\partial \boldsymbol{w}^*} = \boldsymbol{x}(k)\frac{y^*(k)}{|y(k)|} \quad (8.46)$$

Substituting (8.46) into (8.39), $\boldsymbol{D}(\boldsymbol{w})$ can be written as

$$\boldsymbol{D}(\boldsymbol{w}) = \left[\nabla(g_1(\boldsymbol{w}))\ \nabla(g_2(\boldsymbol{w}))\ \ldots\ \nabla(g_K(\boldsymbol{w}))\right] \quad (8.47)$$

$$= \left[\boldsymbol{x}(1)\frac{y^*(1)}{|y(1)|}\ \boldsymbol{x}(2)\frac{y^*(2)}{|y(2)|}\ \ldots\ \boldsymbol{x}(K)\frac{y^*(K)}{|y(K)|}\right]$$

$$= \boldsymbol{X}\boldsymbol{Y}_{cm}$$

where

$$\boldsymbol{X} = [\boldsymbol{x}(1)\boldsymbol{x}(2)\ldots\boldsymbol{x}(K)] \quad (8.48)$$

and

$$\boldsymbol{Y}_{cm} = \begin{bmatrix} \frac{y^*(1)}{|y(1)|} & 0 & \dots & 0 \\ 0 & \frac{y^*(2)}{|y(2)|} & & \vdots \\ \vdots & & \ddots & 0 \\ 0 & \dots & 0 & \frac{y^*(K)}{|y(K)|} \end{bmatrix} \tag{8.49}$$

Using (8.47) and (8.45), we have

$$\boldsymbol{D}(\boldsymbol{w})\boldsymbol{D}^H(\boldsymbol{w}) = \boldsymbol{X}\boldsymbol{Y}_{cm}\boldsymbol{Y}_{cm}^H\boldsymbol{X}^H = \boldsymbol{X}\boldsymbol{X}^H \tag{8.50}$$

and

$$\boldsymbol{D}(\boldsymbol{w})\boldsymbol{g}(\boldsymbol{w}) = \boldsymbol{X}\boldsymbol{Y}_{cm}\begin{bmatrix} |y(1)|-1 \\ |y(2)|-1 \\ \vdots \\ |y(K)|-1 \end{bmatrix} = \boldsymbol{X}\begin{bmatrix} y^*(1) - \frac{y^*(1)}{|y(1)|} \\ y^*(2) - \frac{y^*(2)}{|y(2)|} \\ \vdots \\ y^*(K) - \frac{y^*(K)}{|y(K)|} \end{bmatrix} = \boldsymbol{X}(\boldsymbol{y}-\boldsymbol{r})^* \tag{8.51}$$

where

$$\boldsymbol{y} = \left[y(1)\ y(2)\ \dots\ y(K)\right]^T \tag{8.52}$$

and

$$\boldsymbol{r} = \left[\frac{y(1)}{|y(1)|}\ \frac{y(2)}{|y(2)|}\ \dots\ \frac{y(K)}{|y(K)|}\right]^T = \boldsymbol{L}(\boldsymbol{y}) \tag{8.53}$$

where $\boldsymbol{L}(\boldsymbol{y})$ places a hard-limit on $\boldsymbol{y}$.

The vectors $\boldsymbol{y}$ and $\boldsymbol{r}$ are called the *output data vector* and the *complex-limited output data vector*, respectively. Substituting (8.50) and (8.51) into (8.42), we obtain

$$\begin{aligned} \boldsymbol{w}(l+1) &= \boldsymbol{w}(l) - [\boldsymbol{X}\boldsymbol{X}^H]^{-1}\boldsymbol{X}(\boldsymbol{y}(l)-\boldsymbol{r}(l))^* \\ &= \boldsymbol{w}(l) - [\boldsymbol{X}\boldsymbol{X}^H]^{-1}\boldsymbol{X}\boldsymbol{X}^H\boldsymbol{w}(l) - [\boldsymbol{X}\boldsymbol{X}^H]^{-1}\boldsymbol{X}\boldsymbol{r}^*(l) \\ &= [\boldsymbol{X}\boldsymbol{X}^H]^{-1}\boldsymbol{X}\boldsymbol{r}^*(l) \end{aligned} \tag{8.54}$$

where

$$\boldsymbol{y}(l) = [\boldsymbol{w}^H(l)\boldsymbol{X}]^T \tag{8.55}$$

then

$$\boldsymbol{r}(l) = \boldsymbol{L}(\boldsymbol{y}(l)) \tag{8.56}$$

This is called the *static Least Squares CMA algorithm*, since the algorithm iterates using a single block of K data vectors, $\{x(k)\}$. Once the weight vector, $w(l+1)$, is computed, a new estimate for the filtered output is obtained, $y(l+1)$, and a new value of $r(l+1)$ is produced. The algorithm is repeated until it converges.

In *dynamic LS-CMA*, the algorithm is not iterated over a static block of data. Instead, a window of the last K data vectors is used for each weight update, which occurs once over K samples. We denote the data block at index l as

$$X(l) = \left[x(1+lK)\ x(2+lK)\ \ldots\ x(K+lK)\right] \tag{8.57}$$

The dynamic LS-CMA algorithm is then given by

$$y(l) = [w^H(l)X(l)]^T = \left[y(1+lK)\ y(2+lK)\ \ldots\ y(K+lK)\right]^T \tag{8.58}$$

$$r(l) = L(y(l)) \tag{8.59}$$

$$w(l+1) = [X(l)X^H(l)]^{-1}X(l)r^*(l) \tag{8.60}$$

Unlike the steepest descent CMA algorithms discussed in the previous section, which update on a sample-by-sample basis, the dynamic LS-CMA approach updates on a block-by-block basis.

Finally, we can construct the sample mean estimate of the correlation matrix of the input data and the cross-correlation between the input data and the hardlimited output for the block of data available at the lth iteration:

$$\hat{R}_{xx}(l) = \frac{1}{K}X(l)X^H(l) \tag{8.61}$$

$$\hat{p}_{xr}(l) = \frac{1}{K}X(l)r^*(l) \tag{8.62}$$

Then (8.60) can be written as

$$w(l+1) = \hat{R}_{xx}^{-1}\hat{p}_{xr}(l) \tag{8.63}$$

LS-CMA and the approach used in its derivation will play an essential role in the remainder of the chapter, as we discuss adaptive spatial processing techniques that are specific to CDMA.

In addition to the CM approach, many other blind adaptive algorithms exist. Some of these techniques act by restoring spectral coherence which is present in most communication signals [Gar92b][Sch93a][Sch93b]. Many communication signals exhibit *cyclostationarity;* i.e., they tend to be correlated with frequency-shifted versions of themselves. The Self COherence REstoral (SCORE) technique, which exploits cyclostationarity, has been used to develop several new array adaptation and DOA estimation methods which have interesting properties. Other approaches based on higher order statistics have also been studied [Hay91].

8.4 Adaptive Algorithms for CDMA

In the remainder of this chapter, a number of adaptive algorithms are presented which were developed explicitly for CDMA systems. Since, in a CDMA system, multiple users share a radio channel, adaptive beamforming algorithms should have the ability to separate and extract each user's signal blindly and simultaneously; in other words, the algorithms should be multitarget-type blind algorithms. In this chapter, four algorithms are presented:

- Multitarget Least Squares Constant Modulus Algorithm (MT-LSCMA) [Age86]
- Multitarget Decision-Directed (MT-DD) Algorithm [Ron96a]
- Least Squares De-spread Respired Multitarget Array (LS-DRMTA) [Ron96a]
- Least Squares De-spread Re-spread Multitarget Constant Modulus Algorithm (LS-DRMTCMA) [Ron96a]

In a CDMA mobile communication system, multiple users occupy the same frequency band. The beamformer in the base station attempts to form a beam directed to each user, so that for one desired user, the interference from other directions is reduced. For a system with p users, the beamformer will generate p complex weight vectors.

Figure 8–4 shows the structure of a multitarget adaptive beamformer with M antenna elements and Q output ports. In Figure 8–4, $y_1(k),...,y_Q(k)$ are the outputs of ports $1,...,Q$, respectively, and $\mathbf{w}_1,...,\mathbf{w}_Q$ are the weight vectors for port $1,...,Q$. For some algorithms, the number of output ports, Q, must be less than or equal to the number of antenna elements, M, but for other algorithms such as the algorithms developed in [Ron96a], it can be greater than the number of antenna elements and equal to the number of users. From Figure 8–4, the multitarget beamformer can be viewed as a multi-input multi-output system.

There are two problems that must be addressed by the multitarget blind adaptive beamformer. The first problem is to generate different weight vectors for each port. For a multitarget non-blind adaptive beamformer, this problem can be easily solved by using different training signals in different ports. However, for a multitarget blind adaptive beamformer, if all the signals have the same properties (for example, the constant modulus property or the same cyclic feature), and a property-restoral algorithm is used, then all Q weight vectors may converge to the same spatial response. Therefore, some procedures must be implemented to ensure that each spatial processor works to extract a different signal.

The second challenge is to extract each user's signal from the outputs of each port. If the number of users is fewer than the number of antenna elements, for some algorithms, a sorting procedure must be performed to relate each user's signal to the correct port output. If the number of users is greater than M, for some algorithms which can generate, at most, only M weight vectors, in addition to the sorting procedure, another procedure may be needed to extract several users' signals from the same port output.

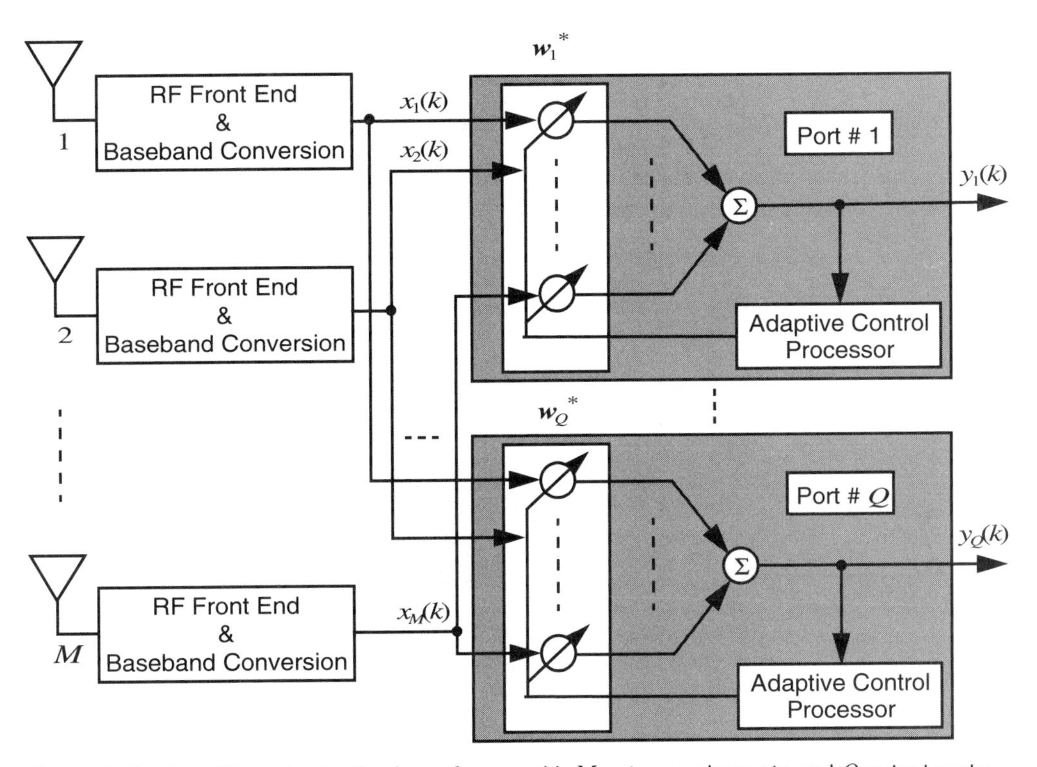

Figure 8–4 A multitarget adaptive beamformer with M antenna elements and Q output ports.

8.4.1 Multitarget Least Squares Constant Modulus Algorithm

The *Multitarget Least Squares Constant Modulus Algorithm* (MT-LSCMA) was first proposed by Agee in [Age89]. This algorithm contains three principle components: a soft-orthogonalized dynamic LS-CMA, a set of sorting and classification algorithms, and a fast acquisition algorithm. However, the MT-LSCMA [Age89] is computationally complex. Figure 8–5 shows the structure of a MT-LSCMA adaptive array.

In Figure 8–5, the number of output ports is equal to the number of antenna elements, and the weight vectors $w_1,...,w_M$ are initialized with a set of different vectors (for example, the column vectors of a $M \times M$ identity matrix). These vectors are then adapted independently by the dynamic LS-CMA which is described by equations (8.58)–(8.60). However, since the LS-CMA utilizes only the *a priori* information that the original signals have a constant envelope, for a CDMA system with all the transmitted signals having the constant modulus property, if steps are not taken, all the weight vectors in different ports may converge to the same beam pattern. To avoid this, *Gram-Schmidt Orthogonalization (GSO)* is performed.

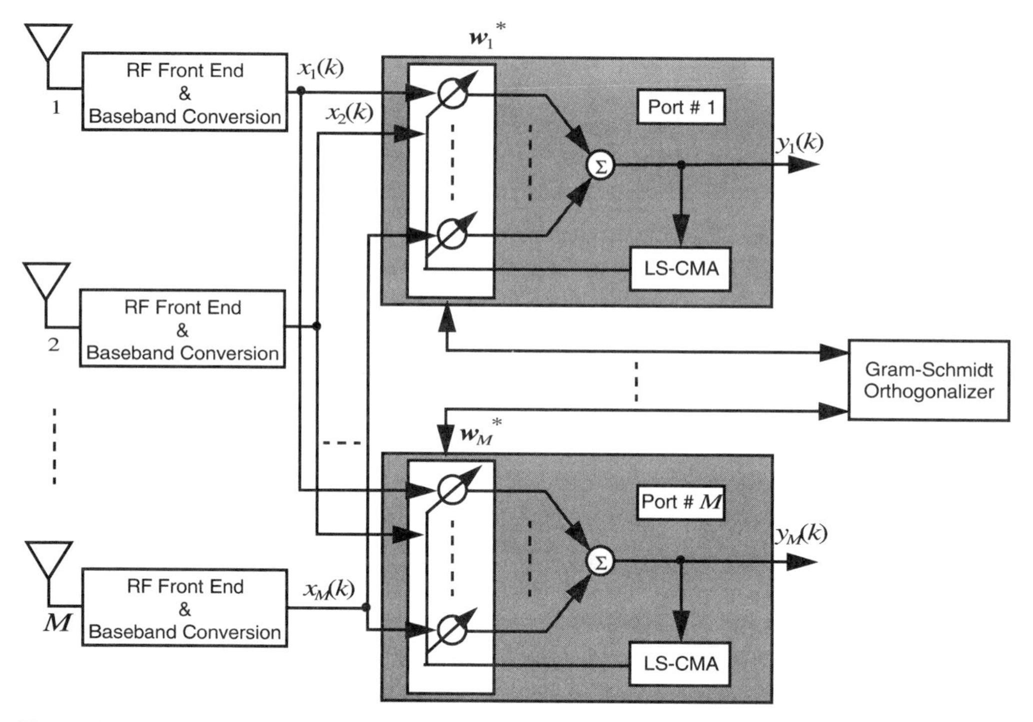

Figure 8–5 Illustration of a multitarget LS-CMA adaptive array.

8.4.2 Gram-Schmidt Orthogonalization

The GSO procedure prevents different weight vectors from converging to the same spatial response. If two weight vectors intend to converge to one with the same beam pattern, the absolute value of their *correlation coefficient* will become larger (greater than 0.5 and close to 1). The correlation coefficient between two weight vectors w_i and w_j is defined by

$$\rho_{ij} \overset{\Delta}{=} \frac{w_i^H w_j}{\|w_i\|\|w_j\|} \tag{8.64}$$

In MT-LSCMA, a threshold ρ_{tr} is set for all the correlation coefficients. After several iterations using LS-CMA, the correlation coefficients ρ_{ij}, $i = 2, \ldots, M, j = 1, \ldots, i-1$ are calculated using equation (8.64). The orthonormal basis $\hat{\mathbf{W}}$ for the range of $\mathbf{W}$ is also computed by using Gram-Schmidt orthogonalization procedure, where

$$\mathbf{W} = [w_1, w_2, \ldots, w_M] \tag{8.65}$$

$$\hat{\mathbf{W}} = [\hat{w}_1, \hat{w}_2, \ldots, \hat{w}_M] \tag{8.66}$$

The absolute values of the correlation coefficients are then compared to the threshold ρ_{tr}. If for one index $i, i = 2, 3, \ldots, M$, there exists an index $j < i$ such that $|\rho_{ij}| > \rho_{tr}$, w_i will be replaced by a scaled version of $\hat{w}_i$, namely $\|w_i\|\hat{w}_i$. After the GSO, the weight vectors are again adapted by using the LS-CMA and the above procedure is repeated until the algorithm converges.

The orthonormal basis $\hat{\mathbf{W}}$ for the range of $\mathbf{W}$ can be obtained by using the Gram-Schmidt orthogonalization procedure which is described as follows [Pro95]:

1. Given $\mathbf{W} = [w_1, w_2, \ldots, w_M]$, $i = 1$
2. $\hat{w}_i = \dfrac{w_1}{\|w_1\|}$, i=1
3. $i = i + 1$

$$\tilde{\rho}_{ij} = \frac{w_i^H \hat{w}_j}{\|w_i\|\|\hat{w}_j\|}, j = 1, \ldots, i-1 \tag{8.67}$$

$$\hat{w}_i = w_i - \sum_{j=1}^{i-1} \tilde{\rho}_{ij}\hat{w}_j \tag{8.68}$$

$$\hat{w}_i = \frac{\hat{w}_i}{\|\hat{w}_i\|} \tag{8.69}$$

4. Repeat step 3 until $i = M$.

Typically, ρ_{tr} is set close to 0.7 [Ron96a]. The GSO algorithm needs to be performed only once for several iterations of weight update and should be performed only after each weight vector has been allowed to converge to some extent.

8.4.3 Phase Ambiguity

Since MT-LS CMA uses the LS-CMA to adapt the weight vector for each port, and as shown in [Age86], the CMA-type algorithms suffer the *phase ambiguity* problem, the phase of the signal at each output port is indeterminate. This problem can be solved by three methods. The first method is to use *Ddifferential Phase-Shift Keying* (DPSK) modulation. Since, in the DPSK modulation, it is the phase difference between the current symbol and the previous symbol that determines the output data, a phase rotation does not affect the demodulated data. The second method is to send a pilot signal from the mobile to the base station and use the received pilot signal to obtain the phase rotation information. This information can then be used to compensate the phase ambiguity in the output port. The last method is to use the phase-constraint technique [Fuj92]. That is, add a phase constraint to each weight vector such that the first element of each weight vector is a real number. For a weight vector w_i after convergence, the new weight vector $\tilde{w}_i$ generated by using phase-constraint is given by

$$\tilde{w}_i = \mathbf{w}_i \exp\{-j\arg(w_{1i})\} \tag{8.70}$$

where arg( ) denotes the phase function and w_{1i} is the first element of the weight vector $\mathbf{w}_i$.

8.4.4 Sorting Procedure

In MT-LSCMA, after the algorithm converges, a sorting procedure must be performed to relate the port outputs to each user's signal. In a CDMA system, the *pseudo-noise (PN) sequence* assigned to each user can be utilized in the sorting procedure. For a CDMA system with p users, the complex envelope of the signal transmitted by the ith user can be expressed as

$$s_i(t) = \sqrt{2P_i}b_i(t)c_i(t)\exp\{-j\psi_i\}, i = 1,2,\ldots,p, \tag{8.71}$$

where P_i, $b_i(t)$, $c_i(t)$, and ψ_i are the power, the data signal, the spreading signal (PN sequence), and the random phase of the ith user signal, respectively. The data signal $b_i(t)$ is given by

$$b_i(t) = \sum_{n=-\infty}^{\infty} b_{in}\Psi\left(\frac{t-nT_b}{T_b}\right), \tag{8.72}$$

where $b_{in} \in [-1, +1]$ is the nth data bit of ith user. T_b is the bit period of the CDMA signal. The spreading signal $c_i(t)$ is given by

$$c_i(t) = \sum_{m=-\infty}^{\infty} c_{i,m}\Psi\left(\frac{t-mT_c}{T_c}\right), \tag{8.73}$$

where $c_{i,m} \in [-1, +1]$ is the mth chip of ith user. T_c is the chip period of the CDMA signal. The ratio of the bit period to the chip period is the spreading factor

$$N_c = \frac{T_b}{T_c}. \tag{8.74}$$

Usually, systems are designed to have a high spreading factor where

$$N_c \gg 1 \tag{8.75}$$

or equivalently,

$$T_b \gg T_c. \tag{8.76}$$

The PN codes for multiple users are also designed to have low cross-correlation and narrow autocorrelation function.

Assume the number of users in the system is less than or equal to the number of the output ports in the beamformer. Also, the multipath and the interference of one desired user are rejected due to the spatial filtering. Perfect power control is applied in the system. Therefore, the output of port i in the beamformer is a delayed, scaled, and phase-rotated version of $s_j(t)$ which is corrupted by noise and is given by

$$y_i(t) = \alpha_i\sqrt{2P_j}b_j(t-\tau_j)c_j(t-\tau_j)\exp\{-j(\psi_j+\gamma_i)\} + n_i(t) \tag{8.77}$$

where α_i is the scaled factor for user i such that $\alpha_i^2P_i = \alpha_j^2P_j$ for $i \neq j$, τ_j is the time delay of the jth user, γ_i is the phase shift in port i due to the phase rotation in LS-CMA, and $n_i(t)$ is the AWGN noise in port i. In equation (8.77), i may or may not be equal to j, and the sorting procedure is used to relate the user index i to the port index j. When the time delay τ_j is estimated perfectly, and the phase $\psi_j + \gamma_i$ is also estimated correctly by using the method discussed in Section 8.4.3, then the sorting procedure can be performed as shown in Figure 8–6. In Figure 8–6, a vector containing the port outputs of the beamformer is given by

$$\boldsymbol{y}(t) = [y_1(t), y_2(t), \ldots, y_M(t)]^T \tag{8.78}$$

The M output signals in $\boldsymbol{y}(t)$ are first multiplied by the delayed versions of p users' PN codes. So for each arm in Figure 8–6, there are M multiplying outputs corresponding to one user's PN code.

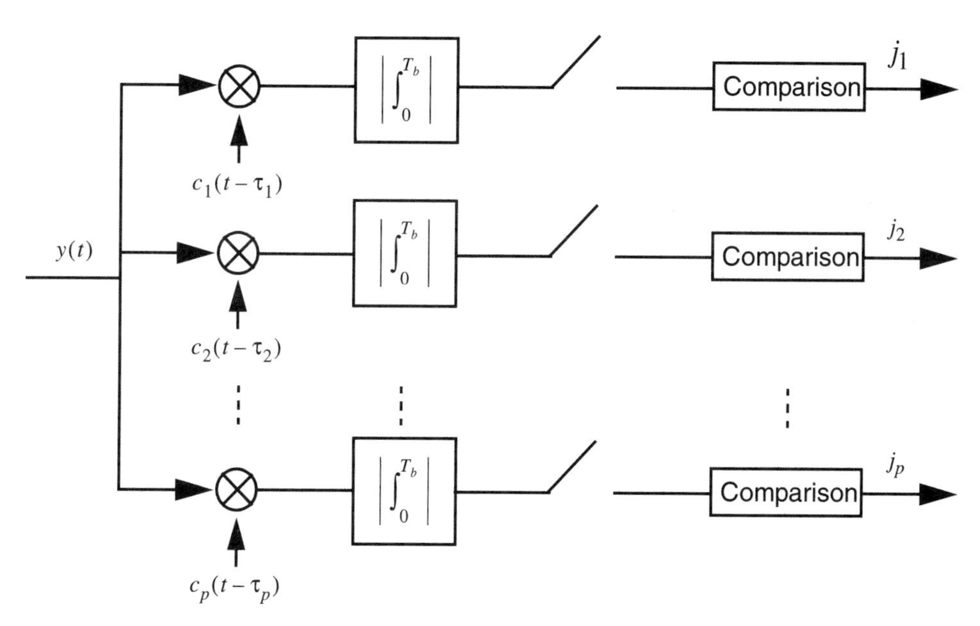

Figure 8–6 Illustration of sorting procedure in MTLSCMA for a CDMA system.

These outputs are then integrated over one bit period, the integration outputs are sampled, and their absolute values are compared with each other. The output of the jth integrator in arm i is the correlation value between $y_j(t)$ and the delayed PN code of user i, $c_i(t - \tau_i)$. Since the PN codes are designed to have low cross correlation, only the output port containing the ith user's signal will have a peak at the integration output. In arm i, the output port with the highest absolute value of the integrator output will be identified as one containing the ith user's signal, and the index of this output port, j_i, will be stored as the output of the sorting procedure.

If the number of users in the system is greater than the number of output ports, i.e., the number of antenna elements for this algorithm, then one output port may contain the signals of several users. Using the sorting procedure shown in Figure 8–6, we may relate one output port with several users. In other words, there exist i_1 and i_2, where $i_1 \neq i_2$, such that $j_{i_1} = j_{i_2}$. In this case, the output of port j_{i_1} (or j_{i_2}) will be used by the receivers for both user i_1 and user i_2 to extract their signals.

If the above procedure is performed by using a digital system, the analog signals will be replaced by their discrete time samples and the integration will be substituted by the accumulated sum.

The MT-LSCMA can be summarized as follows:

1. Initialize the M weight vectors $\boldsymbol{w}_1,\ldots,\boldsymbol{w}_M$ as the column vectors of a $M \times M$ identity matrix.
2. Adapt each weight vector independently using the LS-CMA described in section 8.3.2.
3. After several iterations in LS-CMA, perform the GSO on the resulting weight vectors as described in Section 8.4.2.
4. Repeat steps 1 and 3 until the algorithm converges.
5. Add the phase constraint or perform phase rotation compensation on the resulting weight vectors as described in Section 8.4.3.
6. Perform the sorting procedure described in Section 8.4.4 to relate the output ports to each user's signal.

8.5 Multitarget Decision-Directed Algorithm (MT-DD)

By replacing LS-CMA in the MT-LSCMA with the decision-directed (DD) algorithm, we obtain the MT-DD algorithm. The DD algorithm can be performed with either the *Steepest-Descent (SD)* method or the *Least Squares (LS)* method. If the DD algorithm in the MT-DD is performed with a steepest-decent method, we will call this multitarget-type algorithm the *Multitarget Steepest-Descent Decision-Directed (MT-SDDD) algorithm*, and if the DD algorithm in the MT-DD is performed with Least Squares method, we will call it the *Multitarget Least Squares Decision-Directed (MT-LSDD) algorithm.*

The Steepest-Descent Decision-Directed (SD-DD) algorithm is described by equation (8.83), (8.84), and (8.85). The Least Squares Decision-Directed (LS-DD) algorithm is

$$\begin{aligned} \boldsymbol{y}(l) &= [\boldsymbol{w}^H(l)\boldsymbol{X}(l)]^T \\ &= [y(1+lK), y(2+lK), \ldots y(K+lK)]^T \end{aligned} \tag{8.79}$$

$$\boldsymbol{r}(l) = [\operatorname{sgn}\{Re(y(1+lK))\}, \ldots, \operatorname{sgn}\{Re(y(K+lK))\}]^T \tag{8.80}$$

$$\boldsymbol{w}(l+1) = [\boldsymbol{X}(l)\boldsymbol{X}^H(l)]^{-1}\boldsymbol{X}(l)\boldsymbol{r}^*(l) \tag{8.81}$$

where l is the iteration number, $\boldsymbol{X}(l)$ is

$$\boldsymbol{X}(l) = \left[\boldsymbol{x}(1+lK)\ \boldsymbol{x}(2+lK)\ \ldots\ \boldsymbol{x}(K+lK)\right] \tag{8.82}$$

and K is the number of samples in each data block.

Unlike the MT-LSCMA which constrains the signal constellations on the unit circle, the MT-DD algorithm constrains the signal constellations at either +1 or -1. Since the signal of each user has a random phase, this constraint will also cause a phase ambiguity. The methods introduced in the Section 8.4.3 can be used to compensate this effect.

The MT-SDDD algorithm can be described as follows:

1. Initialize the M weight vectors $\boldsymbol{w}_1,...,\boldsymbol{w}_M$ as the column vectors of a $M \times M$ identity matrix.
2. Adapt each weight vector independently using the Steepest-Descent Decision-Directed (SD-DD) algorithm [Ron96a][Ron96b].

$$y(k) = \boldsymbol{w}^H(k)\boldsymbol{x}(k) \tag{8.83}$$

$$e(k) = y(k) - \text{sgn}(Re(y(k))) \tag{8.84}$$

$$\boldsymbol{w}(k+1) = w(k) - \mu \boldsymbol{u}(k)e^*(k) \tag{8.85}$$

3. After a number of iterations in SD-DD, perform the GSO on the resulting weight vectors as described in Section 8.4.2.
4. Repeat step 2 and 3 until the algorithm converges.
5. Add the phase constraint or perform phase rotation compensation on the resulting weight vectors as described in Section 8.4.3.
6. Perform the sorting procedure described in Section 8.4.4 to relate the output ports to each user's signal.

In step 2, the number of iterations needed before the GSO depends on the step size μ in the SD-DD algorithm. In [Ron96b], μ is set to 0.001 and the number of iterations before the GSO is set to 1000. Each iteration of the MT-SDDD algorithm is very easily computed, however, many iterations are required. In MT-LSDD and MT-LSCMA, the number of iterations needed will be around 5, but for both of these later algorithms, the matrix inverse must be calculated once (See (8.81)). Thus there is a trade-off between the convergence speed and the computational complexity.

8.6 Least Squares De-spread Re-spread Multitarget Array (LS-DRMTA)

The multitarget adaptive algorithms discussed earlier in this chapter do not utilize any information of the spreading signal of each user in the CDMA system. However, in a CDMA system, it is these spreading signals that distinguish different users occupying the same frequency band. Therefore, it is useful if the information of these spreading signals can be utilized in the multitarget adaptive algorithm. In this section and in the following section we introduce two algorithms developed by Rong and Rappaport in [Ron96a], that utilize the information of the spreading signals. The algorithm discussed in this section is called *Least Squares De-spread Re-spread Multitarget Array (LS-DRMTA)*.

8.6.1 Derivation of LS-DRMTA

In the base station of a CDMA system, the spreading signals of all the users are previously known. In the conventional receiver, to detect the ith user's data bits, the received signal is correlated with the time delayed spreading signal of the ith user, $c_i(t-\tau_i)$, and the correlation output is sent to the detector, which makes a decision based on the correlation output. There exist many techniques to estimate the time delay, τ_i, for the ith user, and we do not cover the topic here. It is assumed that the time delay for each user is detected perfectly unless specified otherwise.

If the nth data bit of the ith user is detected correctly by the detector, i.e., $\hat{b}_{in} = b_{in}$, where $\hat{b}_{in}$ is the detector output, then the waveform of the ith user's transmitted signal during time period $[(n-1)T_b, nT_b)$ can be obtained by re-spreading the detected data bit $\hat{b}_{in}$ with the PN sequence of the ith user, $c_i(t)$. This re-spread signal can then be used in the beamformer to adapt the weight vector for user i. The adaptive algorithm that uses this de-spread-and-re-spread technique is referred to as *Least Squares De-spread Re-spread Multitarget Array* (LS-DRMTA) [Ron96b]. Figure 8–7 shows the structure of a beamformer using the LS-DRMTA and Figure 8–8 shows the block diagram of the LS-DRMTA for user i. Note that in Figure 8–7, the number of beamformer output ports is now equal to the number of users in the system.

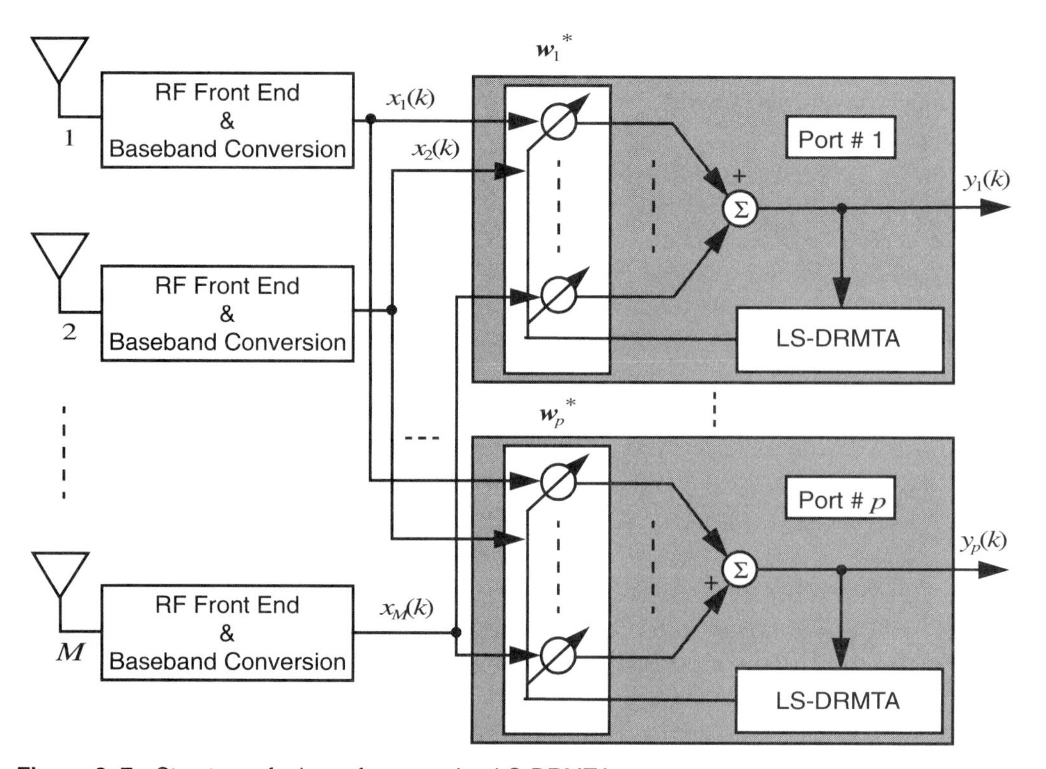

Figure 8–7 Structure of a beamformer using LS-DRMTA.

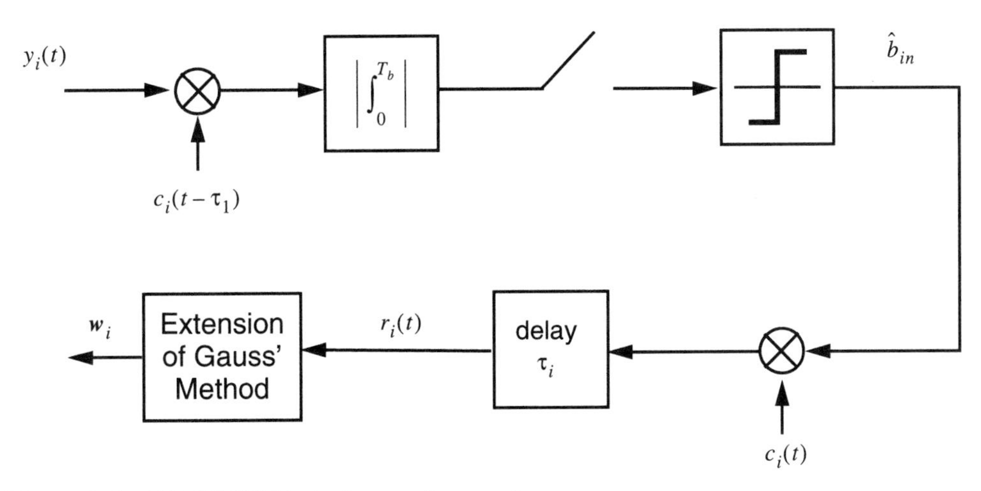

Figure 8–8 LS-DRMTA block diagram for user *i*.

In Figure 8–8, $r_i(t)$ is a time-delayed version of the re-spread signal for user *i* and is given by

$$r_i(t) = \hat{b}_{in} c_i(t-\tau_i), \quad (n-1)T_b \le t < nT_b \tag{8.86}$$

In a code-on-pulse CDMA system, the PN sequence is repeated every bit period; therefore, both $c_i(t)$ and $r_i(t)$ have a time period T_b. Let $y_i(k)$ and $r_i(k)$ denote the kth sample of $y_i(t)$ and $r_i(t)$, respectively, in a digital system; the LS-DRMTA adapts the weight vector $\boldsymbol{w}_i$ to minimize the cost function, in (8.87).

$$\begin{aligned} F(\boldsymbol{w}_i) &= \sum_{k=1}^{K} |y_i(k) - r_i(k)|^2 \\ &= \sum_{k=1}^{K} \left|\boldsymbol{w}_i^H \boldsymbol{x}(k) - r_i(k)\right|^2 \end{aligned} \tag{8.87}$$

where *K* is the data block size and is set to be equal to the number of samples in one bit period in LS-DRMTA. So if the signal is sampled with a sampling rate $R_s = N_s R_c$, where R_c is the chip rate of the CDMA signal, and N_s is an integer greater than two, the block size *K* will be equal to $N_c N_s$, where N_c is the spreading factor.

Using the extension of Gauss' method described in [Ron96a][Ron96b], we express the cost function as

$$F(\boldsymbol{w}) = \sum_{k=1}^{K} |g_k(\boldsymbol{w})|^2 \tag{8.88}$$

where

$$\begin{aligned} g_k(\boldsymbol{w}_i) &= |y_i(k) - r_i(k)| \\ &= |\boldsymbol{w}_i^H \boldsymbol{x}(k) - r_i(k)| \end{aligned} \tag{8.89}$$

which can be written as

$$g_k(\boldsymbol{w}_i) = \begin{bmatrix} |y_i(1) - r_i(1)| \\ |y_i(2) - r_i(2)| \\ \cdots \\ |y_i(k) - r_i(k)| \end{bmatrix} \tag{8.90}$$

The gradient vector of $g_k(\mathbf{w}_i)$ is given by

$$\begin{aligned} \nabla(g_k(\boldsymbol{w}_i)) &= 2\frac{\partial g_k(\boldsymbol{w}_i)}{\partial \boldsymbol{w}_i{}^*} \\ &= x(k)\frac{[y_i(k) - r_i(k)]^*}{|y_i(k) - r_i(k)|} \end{aligned} \tag{8.91}$$

Let

$$v_i(k) = y_i(k) - r_i(k) \tag{8.92}$$

Then, (8.91) can be expressed as

$$\nabla(g_k(\boldsymbol{w}_i)) = \boldsymbol{x}(k)\frac{v_i{}^*(k)}{|v_i(k)|} \tag{8.93}$$

Gauss' method states that the weight update can be written as

$$\boldsymbol{w}(l+1) = \boldsymbol{w}(l) - [\boldsymbol{D}(\boldsymbol{w}(l))\boldsymbol{D}^H(\boldsymbol{w}(l))]^{-1}\boldsymbol{D}(\boldsymbol{w}(l))\boldsymbol{g}(\boldsymbol{w}(l)) \tag{8.94}$$

where

$$\boldsymbol{D}(\boldsymbol{w}) = \left[\nabla(g_1(\boldsymbol{w})) \ \ldots \ \nabla(g_K(\boldsymbol{w}))\right] \tag{8.95}$$

Substituting equation (8.93) into (8.95), $\boldsymbol{D}(\boldsymbol{w}_i)$ can be expressed as

$$\begin{aligned} \boldsymbol{D}(\boldsymbol{w}_i) &= [\nabla(g_1(\boldsymbol{w}_i)), \nabla(g_2(\boldsymbol{w}_i)), \ldots, \nabla(g_K(\boldsymbol{w}_i))] \\ &= \left[\boldsymbol{x}(1)\frac{v_i{}^*(1)}{|v_i(1)|}, \boldsymbol{x}(2)\frac{v_i{}^*(2)}{|v_i(2)|}, \ldots, \boldsymbol{x}(K)\frac{v_i{}^*(K)}{|v_i(K)|}\right] \\ &= \boldsymbol{X}\boldsymbol{V}_{iCM} \end{aligned} \tag{8.96}$$

where

$$\boldsymbol{X} = [\boldsymbol{x}(1), \boldsymbol{x}(2), \ldots, \boldsymbol{x}(K)] \tag{8.97}$$

and

$$V_{iCM} = \begin{bmatrix} \frac{v_i^*(1)}{|v_i(1)|} & 0 & \dots & 0 \\ 0 & \frac{v_i^*(2)}{|v_i(2)|} & & \vdots \\ \vdots & & \ddots & 0 \\ 0 & \dots & 0 & \frac{v_i^*(K)}{|v_i(K)|} \end{bmatrix} \tag{8.98}$$

Using equation (8.96) and (8.90), we have

$$\begin{aligned} D(w_i)D^H(w_i) &= XV_{iCM}V_{iCM}^H X^H \\ &= XX^H \end{aligned} \tag{8.99}$$

and

$$\begin{aligned} D(w_i)g(w_i) &= XV_{iCM}\begin{bmatrix} |v_i(1)| \\ |v_i(2)| \\ \dots \\ |v_i(K)| \end{bmatrix} \\ &= X\begin{bmatrix} v_i^*(1) \\ v_i^*(2) \\ \dots \\ v_i^*(K) \end{bmatrix} \\ &= Xv_i^* \end{aligned}$$

$$D(w_i)g(w_i) = X(y_i - r_i)^* \tag{8.101}$$

where

$$v_i = [v_i(1), v_i(2), \dots, v_i(K)]^T \tag{8.102}$$

$$y_i = [y_i(1), y_i(2), \dots, y_i(K)]^T \tag{8.103}$$

$$r_i = [r_i(1), r_i(2), \dots, r_i(K)]^T \tag{8.104}$$

The vector y_i is the output data vector for user i and r_i is the estimate of the signal waveform of user i over one bit period. Substituting equation (8.99) and (8.101) into equation (8.94),

$$\begin{aligned} w_i(l+1) &= w_i(l) - [XX^H]^{-1}X(y_i(l) - r_i(l))^* \\ &= w_i(l) - [XX^H]^{-1}XX^H w_i(l) + [XX^H]^{-1}Xr_i(l)^* \\ &= [XX^H]^{-1}Xr_i^*(l) \end{aligned} \tag{8.105}$$

where $\boldsymbol{y}_i(l)$ and $\boldsymbol{r}_i(l)$ are the output data vector and estimate of signal waveform of user i over one bit period corresponding to the weight vector $\boldsymbol{w}_i$ in the lth iteration, respectively. Similar to the dynamic LS-CMA, LS-DRMTA can adapt the weight vectors using different input data blocks in each iteration. Let

$$\boldsymbol{X}(l) = [\boldsymbol{x}(1+lK), \boldsymbol{x}(2+lK), \ldots, \boldsymbol{x}((l+1)K)], \quad l = 0, 1, \ldots, L, \tag{8.106}$$

where L is the number of iterations required for the algorithm to converge, and K is the number of data samples per bit $(N_c N_s)$ if the data samples over one bit period are all used for the adaptation. In Figure 8–7, the LS-DRMTA for the ith user can be described by the following equations:

$$\begin{aligned}\boldsymbol{y}_i(l) &= [\boldsymbol{w}_i^H(l)\boldsymbol{X}(l)]^T \\ &= [y_i(1+lK), y_i(2+lK), \ldots, y_i((l+1)K)]^T\end{aligned} \tag{8.107}$$

$$\hat{b}_{il} = \operatorname{sgn}\left\{\operatorname{Re}\left(\sum_{k=1+lK}^{(1+l)K} y_i(k)c_i(k-k_{\tau_i})\right)\right\} \tag{8.108}$$

$$\boldsymbol{r}_i(l) = \hat{b}_{il}[c_i(1+lK-k_{\tau_i}), c_i(2+lK-k_{\tau_i}), \ldots, c_i((1+l)K-k_{\tau_i})]^T \tag{8.109}$$

$$\boldsymbol{w}_i(l+1) = [\boldsymbol{X}(l)\boldsymbol{X}^H(l)]^{-1}\boldsymbol{X}(l)\boldsymbol{r}_i^*(l), \tag{8.110}$$

where $c_i(k)$ is the kth sample of the spreading signal of user i, k_{τ_i} is the number of samples corresponding to τ_i, the delay of user i, and $\hat{b}_{il}$ is the estimate of lth bit for user i. The accumulated sum in equation (8.108) is equivalent to integration in the continuous time domain.

The LS-DRMTA algorithm is summarized as follows:

1. Initialize the p weight vectors $\boldsymbol{w}_1, \ldots, \boldsymbol{w}_p$ as p identical $M \times 1$ column vectors with the first element equal to 1 and the other elements equal to 0.
2. Calculate the array output vector using (8.107).
3. De-spread the ith user's signal and estimate the nth data bit using (8.108).
4. Re-spread the estimate data bit with the PN code of user i to get an estimate of the signal waveform of user i over time period $[(n-1)T_b, nT_b]$ using (8.109).
5. Adapt the weight vector $\boldsymbol{w}_i$ of user i using (8.110).
6. Repeat steps 2 through 5 to until the algorithm converges.

In equation (8.109), when $l = 0$, there may be some time indices less than zero, but since the PN sequence has a time period of T_b, or equivalently, a period of K in the discrete time domain, the index k, where $k < 0$, can be replaced by the index $k + K$.

In LS-DRMTA, if the data bit is not estimated correctly at the beginning of the algorithm, from equation (8.109) and (8.110) we see that the resulting weight vector may have a phase shift of π, but this does not affect the resulting beam pattern. The beam pattern will still have a higher gain in the DOA of the desired signal and the interference from other directions will be rejected. Therefore, if DPSK modulation is used in the system, the data bit estimation error at the beginning of the algorithm does not affect the demodulated data bit.

8.6.2 Advantages of LS-DRMTA

Since LS-DRMTA utilizes the information of each user's spreading signal to adapt the weight vectors, it has several advantages over MT-LSCMA and MT-SDDD algorithms discussed in previous sections.

The first advantage of LS-DRMTA is that there is no need to perform the GSO procedure. This can be seen by comparing Figure 8–7 with Figure 8–5. In MT-LSCMA and MT-SDDD, the weight vectors are adapted by using the same property of all the users' signals, and the GSO procedure is used to prevent different weight vectors from converging to the same value. In LS-DRMTA, however, different users' weight vectors are adapted to minimize a different cost function, which is a sum of the squared error between the port output and the re-spread signal of the desired user over one bit period. Since each user has a different spreading signal, the weight vector of each different user is updated with a different tendency and thus will not converge to one having the same beam pattern.

The second advantage of LS-DRMTA is that there is no need to perform the sorting procedure. In MT-LSCMA and MT-SDDD, the sorting procedure is used to relate each output port to each user. In LS-DRMTA, since different weight vectors are adapted with different PN sequences, the weight vector adapted by using the ith user's spreading signal will correspond to the ith user. Therefore, there is no need to perform the sorting procedure.

The third advantage of LS-DRMTA is that the number of output ports of the beamformer is not limited by the number of antenna elements of the array. In MT-LSCMA and MT-SDDD, due to the GSO procedure, the number of output ports should be fewer than or equal to the number of antenna elements. If the number of users in the system is greater than the number of antenna elements, several users must share one output port, and the interference cannot be reduced to a very low level. In LS-DRMTA, since no GSO is needed, the number of output ports can equal the number of users even if the number of users is greater than the number of antenna elements. Since each user has its own output port, the interference can be reduced to a lower level. Furthermore, if the system is expanded, more output ports can easily be added to the beamformer by adding more weight vectors adapted with the PN sequences of the new users. As shown in Figure 8–7, the new output ports use the same array input data vector, so no new RF front end and baseband conversion are needed, which will reduce the cost of the adaptive antenna system.

The fourth advantage of LS-DRMTA is that the computational complexity of the LS-DRMTA is lower than that of MT-LSCMA and MT-SDDD. In MT-LSCMA and MT-SDDD, both the GSO procedure and the sorting procedure must be performed to extract each different users' signals. From the previous discussion we see that these two procedures are computationally intensive. LS-DRMTA, on the other hand, eliminates these two procedures and therefore requires less computational complexity.

Finally, LS-DRMTA can perform under the condition where the SINR is low and both the MT-LSCMA and MT-SDDD cannot work. In LS-DRMTA, the PN sequence is used in the adaptation of the weight vector, and these PN sequences may spread the interference over a large frequency band and de-spread the desired signal. Also, the noise is averaged over one bit period in LS-DRMTA and its effect on the desired signal is reduced. So the LS-DRMTA can work in a low SINR situation.

8.7 Least Squares De-spread Re-spread Multitarget Constant Modulus Algorithm

In LS-DRMTA, we utilize the spreading signal of each user in a CDMA system to adapt the weight vectors of the beamformer. In MT-LSCMA, on the other hand, we utilize the constant modulus property of the transmitted signal to update the weight vectors. One approach is to combine the spreading signal and the constant modulus property of the transmitted signal to adapt the weight vector. The algorithm using this kind of combination in the adaptation of the weight vector is referred to as *Least Squares De-spread Re-spread Multitarget Constant Modulus Algorithm* (LS-DRMTCMA) [Ron96a][Ron96b]. Figure 8–9 shows the structure of a beamformer using the LS-DRMTCMA and Figure 8–10 shows the block diagram of the LS-DRMTCMA for user i.

8.7.1 Derivation of LS-DRMTCMA

The derivation of the LS-DRMTCMA is similar to that of the LS-DRMTA. In LS-DRMTA, the cost function that the algorithm wants to minimize is given in equation (8.111) and is repeated here for convenience.

$$F(\boldsymbol{w}_i) = \sum_{k=1}^{K} |y_i(k) - \boldsymbol{r}_i(k)|^2 \tag{8.111}$$

$$= \sum_{1=k}^{K} \left|\boldsymbol{w}_i^H \boldsymbol{x}(k) - r_i(k)\right|^2$$

where

$$r_i(k) = \hat{b}_{in} c_i(k - k_{\tau_i}),\ (n-1)K \le k < nK \tag{8.112}$$

In LS-DRMTCMA, the cost function that the algorithm minimizes has the same form as that shown in equation (8.111). However, as illustrated in Figure 8–10, $r_i(k)$ now becomes the sum of the weighted re-spread signal and the weighted complex-limited output.

$$r_i(k) = a_{PN} r_{iPN}(k) + a_{CM} r_{iCM}(k), \tag{8.113}$$

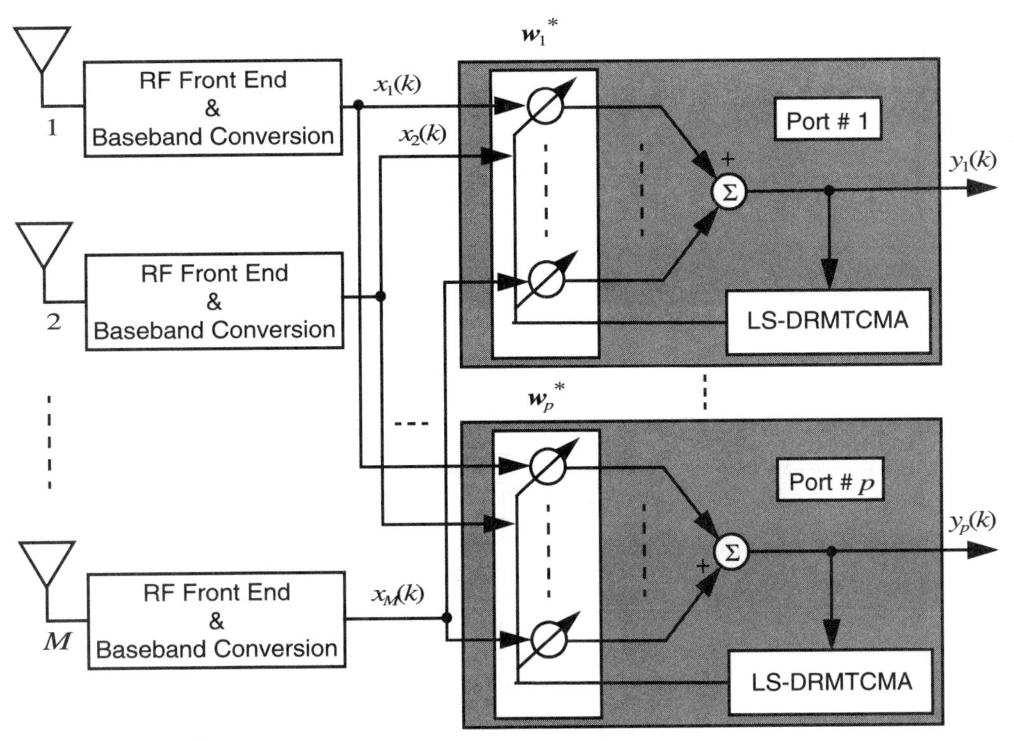

Figure 8–9 Structure of a beamformer using LS-DRMTCMA.

where $r_{iPN}(k)$ is the re-spread signal of user i and is given in equation (8.112), r_{iCM} is the complex-limited output of user i and can be expressed as

$$r_{iCM}(k) = \frac{y_i(k)}{|y_i(k)|} \tag{8.114}$$

The terms, a_{PN} and a_{CM}, are the real positive weight coefficients for the re-spread signal and the complex-limited output of user i, respectively. The coefficients a_{PN} and a_{CM} should satisfy the condition

$$a_{PN} + a_{CM} = 1, \qquad a_{PN}, a_{CM} > 0 \tag{8.115}$$

Following the derivation in 8.6.1, we obtain the following equations for LS-DRMTCMA:

$$\begin{aligned} \boldsymbol{y}_i(l) &= [\boldsymbol{w}_i^H(l)\boldsymbol{X}(l)]^T \\ &= [y_i(1 + lK), y_i(2 + lK), \ldots, y_i((l+1)K)]^T \end{aligned} \tag{8.116}$$

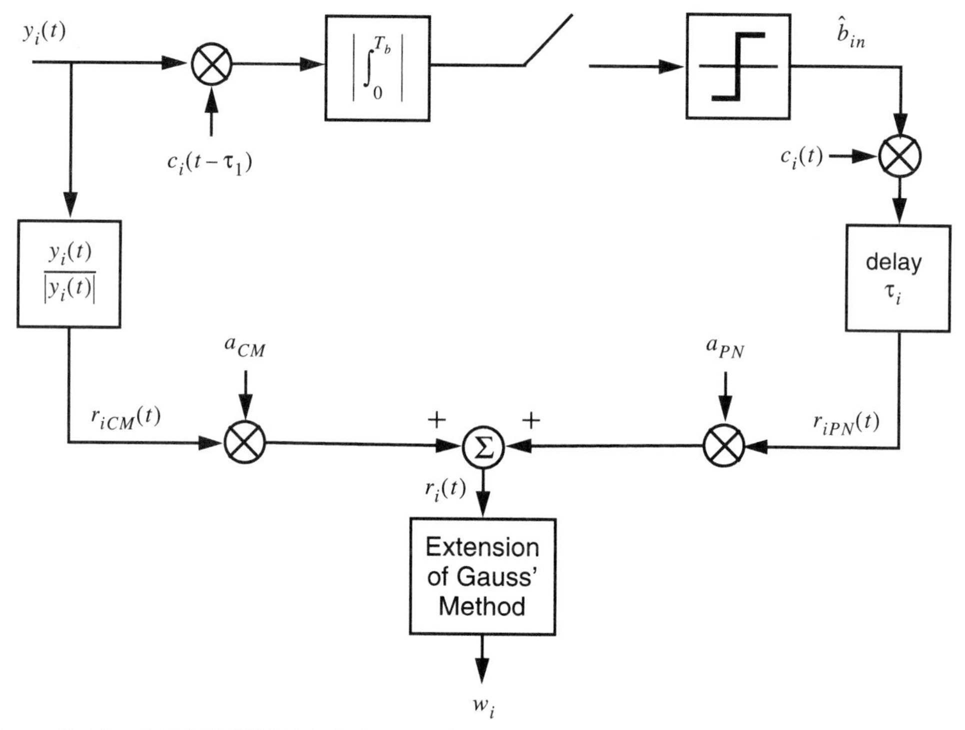

Figure 8–10 LS-DRMTCMA block diagram for user *i*.

$$\hat{b}_{il} = \mathrm{sgn}\left\{\mathrm{Re}\left(\sum_{k=1+lK}^{(1+l)K} y_i(k)c_i(k-k_{\tau_i})\right)\right\} \tag{8.117}$$

$$\boldsymbol{r}_{iPN}(l) = \hat{b}_{il}[c_i(1+lK-k_{\tau_i}), c_i(2+lK-k_{\tau_i}), \ldots, c_i((1+l)K-k_{\tau_i})]^T \tag{8.118}$$

$$\boldsymbol{r}_{iCM}(l) = \left[\frac{y(1+lK)}{|y(1+lK)|}, \frac{y(2+lK)}{|y(2+lK)|}, \ldots, \frac{y((1+l)K)}{|y((1+l)K)|}\right]^T \tag{8.119}$$

$$\boldsymbol{r}_i(l) = a_{PN}\mathbf{r}_{iPN}(l) + a_{CM}\mathbf{r}_{iCM}(l) \tag{8.120}$$

$$\boldsymbol{w}_i(l+1) = [\boldsymbol{X}(l)\boldsymbol{X}^H(l)]^{-1}\boldsymbol{X}(l)\boldsymbol{r}_i^*(l) \tag{8.121}$$

From the above equations we see that if a_{CM} is set to zero, the LS-DRMTCMA becomes the LS-DRMTA; therefore, LS-DRMTA can be viewed as a special case of LS-DRMTCMA. Also we see that if a_{PN} is set to zero and the GSO procedure is performed during the adaptation, the algorithm becomes MT-LSCMA. The choice of a_{PN} and a_{CM} can affect the resulting beam pattern and thus the performance of the system.

We can summarize the LS-DRMTCMA as follows:

1. Initialize the p weight vectors $\boldsymbol{w}_1,...,\boldsymbol{w}_p$ as p identical $M \times 1$ column vectors with the first element equal to 1 and the other elements equal to 0.
2. Calculate the array output vector using (8.116).
3. De-spread the ith user's signal and estimate the nth data bit using (8.117).
4. Re-spread the estimate data bit with the PN code of user i to get an estimate of the signal waveform of user i over time period $[(n-1)T_b, nT_b]$ using (8.118).
5. Calculate the complex-limited output vector of user i using (8.119).
6. Calculate the reference signal vector for user i by summing up the weighted re-spread signal vector and the weighted complex-limited output vector using (8.120).
7. Adapt the weight vector $\boldsymbol{w}_i$ of user i using (8.121).
8. Repeat step 2 and 7 until the algorithm converges.

8.7.2 Advantages of LS-DRMTCMA

Since the LS-DRMTCMA utilizes both the PN sequence and the constant modulus property of the transmitted signal, it possesses all the advantages of LS-DRMTA and has additional advantages. Most importantly, it can achieve a much lower BER than LS-DRMTA [Ron96b]. However, since LS-DRMTCMA utilizes the complex-limited output vector of each user in the adaptation of the weight vectors, this advantage is achieved at the expense of additional complexity.

8.8 Summary

In this chapter, the essential features of optimal and adaptive spatial processing systems were presented. The performance of optimal spatial processors under different multipath conditions was illustrated using a number of examples for overloaded and underloaded arrays. These results show the degree to which smart antenna performance is dependent on the characteristics of the radio channel.

In the second half of the chapter a series of adaptive algorithms were presented, several of which were specifically developed to exploit the underlying signal characteristics of CDMA signals, including the PN sequence and the constant modulus nature of phase modulated signals. Other CDMA-specific algorithms have also been developed in the literature which take advantage of different features of the CDMA signal. For example, the Seemingly-Positive (SP) approach exploits the fact that the difference between the best output of the Walsh function matched filter in the CDMA uplink, shown in Figure 4–2, and other matched filter outputs will be enhanced as the spatial filter nears its optimal solution [Lu98].

The adaptive spatial processing algorithms presented here allow the smart antenna to react to the time-dynamic radio channel, and track subscribers and interferers as they move about. In the next chapter, techniques for esimating the Direction-Of-Arrival of multipath components are discussed. DOA estimation can be an essential procedure for both position location systems and downlink beamforming strategies.

CHAPTER 9

Direction-Of-Arrival Estimation Algorithms

This chapter provides a detailed overview of the various methods available for estimation of Direction-Of-Arrival (also called Angle-Of-Arrival) of a radio signal using an antenna array. Promising methods for determining the position of a mobile user are also given. The use of adaptive arrays is critical in position location applications [Ken96], and, given the recent mandate by the FCC to require 125m accuracy on wireless emergency calls by the year 2001 [Ree98][Rap96b], there is intense interest in determining the Direction-Of-Arrival of RF signals in wireless systems.

The array-based Direction-Of-Arrival (DOA) estimation techniques considered here are broadly divided into four different types: *conventional techniques*, *subspace based techniques*, *maximum likelihood techniques* and the *integrated techniques* which combine property restoral techniques with subspace based approaches. Conventional methods are based on classical beamforming techniques discussed in Chapter 3 and require a large number of elements to achieve high resolution. Subspace based methods are high resolution sub-optimal techniques which exploit the eigen structure of the input data matrix. Maximum likelihood techniques are optimal techniques which perform well even under low signal-to-noise ratio conditions, but are often computationally very intensive. A promising method for CDMA is the integrated approach which uses property-restoral based techniques to separate multiple signals and estimate their spatial signatures from which their Directions-Of-Arrival can be determined using subspace techniques [Rap98].

9.1 Conventional Methods for DOA Estimation

Conventional methods for Direction-Of-Arrival estimation are based on the concepts of beamforming and null-steering and do not exploit the nature of the model of the received signal vector $\boldsymbol{u}(k)$ or the statistical model of the signals and noise. Given the knowledge of the array manifold, an array can be steered electronically as described in Chapter 3. Conventional DOA estimation techniques electronically steer beams in all possible directions and look for peaks in the output power [Sch93d]. The conventional methods discussed here are the delay-and-sum method (classical beamformer) and Capon's minimum variance method.

9.1.1 Delay-and-Sum Method

The delay-and-sum method, also referred to as the classical beamformer method or Fourier method, is one of the simplest techniques for DOA estimation. Figure 9–1 shows the classical narrowband beamformer structure, where the output signal $y(k)$ is given by a linearly weighted sum of the sensor element outputs. That is,

$$y(k) = \boldsymbol{w}^H \boldsymbol{u}(k) \tag{9.1}$$

The total output power of the conventional beamformer can be expressed as

$$P_{cbf} = E[|y(k)|^2] = E[|\boldsymbol{w}^H \boldsymbol{u}(k)|^2] = \boldsymbol{w}^H E[\boldsymbol{u}(k)\boldsymbol{u}^H(k)]\boldsymbol{w} = \boldsymbol{w}^H \boldsymbol{R}_{uu} \boldsymbol{w} \tag{9.2}$$

where $\boldsymbol{R}_{uu}$ is the autocorrelation matrix of the array input data as defined in (8.2). Equation (9.2) plays a central role in all of the conventional DOA estimation algorithms. The autocorrelation matrix $\boldsymbol{R}_{uu}$ contains useful information about both the array response vectors and the signals themselves, and it is possible to estimate signal parameters by careful interpretation of $\boldsymbol{R}_{uu}$.

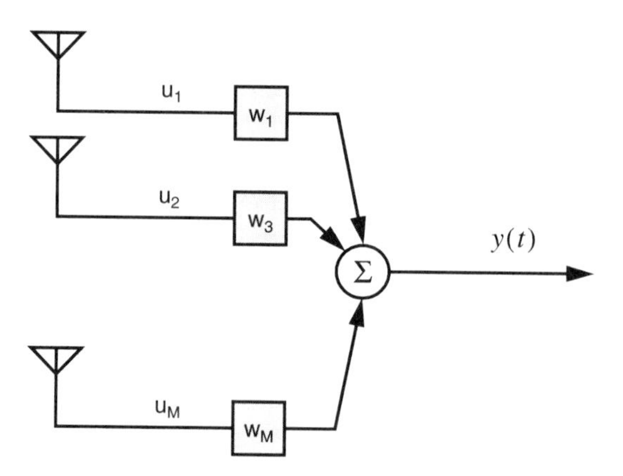

Figure 9–1 Illustration of the classical beamforming structure.

Consider a signal $s(k)$ impinging on the array at an angle ϕ_0. Following the narrowband input data model expressed in (8.5), the power at the beamformer output can be expressed as

$$\begin{aligned} P_{cbf}(\phi_0) &= E[|\boldsymbol{w}^H \boldsymbol{u}(k)|^2] = E[|\boldsymbol{w}^H(\boldsymbol{a}(\phi_0)s(k) + \boldsymbol{n}(k))|^2] \\ &= (|\boldsymbol{w}^H \boldsymbol{a}(\phi_o)|^2(\sigma_s^2 + \sigma_n^2)) \end{aligned} \tag{9.3}$$

where $\boldsymbol{a}(\phi_0)$ is the steering vector associated with the DOA angle ϕ_0, $\boldsymbol{n}(k)$ is the noise vector at the array input, and $\sigma_s = E[s(k)^2]$ and $\sigma_n = E[n(k)^2]$ are the signal power and noise power, respectively. It is clearly seen from (9.3) that the output power is maximized when $\boldsymbol{w} = \boldsymbol{a}(\phi_0)$. Therefore, of all the possible weight vectors, the receiver antenna has the highest gain in the direction ϕ_0, when $\boldsymbol{w} = \boldsymbol{a}(\phi_0)$. This is because $\boldsymbol{w} = \boldsymbol{a}(\phi_0)$ aligns the phases of the signal components arriving from ϕ_0 at the sensors, causing them to add constructively.

In the classical beamforming approach to DOA estimation, the beam is scanned over the angular region of interest in discrete steps by forming weights $\boldsymbol{w} = \boldsymbol{a}(\phi)$ for different ϕ, and the output power is measured. Using equation (9.3), the output power at the classical beamformer as a function of the Angle-Of-Arrival is given by

$$P_{cbf}(\phi) = \boldsymbol{w}^H \boldsymbol{R}_{uu} \boldsymbol{w} = \boldsymbol{a}^H(\phi) \boldsymbol{R}_{uu} \boldsymbol{a}(\phi) \tag{9.4}$$

Therefore, if we have an estimate of the input autocorrelation matrix and know the steering vectors $\boldsymbol{a}(\phi)$ for all ϕ's of interest (either through calibration or analytical computation), it is possible to estimate the output power as a function of the Angle-Of-Arrival ϕ. The output power as a function of Angle-Of-Arrival is often termed the *spatial spectrum*. Clearly, the Directions-Of-Arrival can be estimated by locating peaks in the spatial spectrum defined in (9.4).

The delay-and-sum method has many disadvantages. The width of the beam and the height of the sidelobes limit the effectiveness when signals arriving from multiple directions and/or sources are present, because the signals over a wide angular region contribute to the measured average power at each look direction. Hence, this technique has poor resolution. Although it is possible to increase the resolution by adding more sensor elements, increasing the number of sensors increases the number of receivers and the amount of storage required for the calibration data, i.e., $\boldsymbol{a}(\phi)$.

9.1.2 Capon's Minimum Variance Method

The delay-and-sum method works on the premise that pointing the strongest beam in a particular direction yields the best estimate of power arriving in that direction. In other words, all the degrees of freedom available to the array were used in forming a beam in the required look direction. This works well when there is only one signal present. But when there is more than one signal present, the array output power contains contribution from the desired signal as well as the undesired ones from other directions.

Capon's minimum variance technique [Cap69] attempts to overcome the poor resolution problems associated with the delay-and-sum method. The technique uses some (not all) of the degrees of freedom to form a beam in the desired look direction while simultaneously using the remaining degrees of freedom to form nulls in the direction of interfering signals. This technique minimizes the contribution of the undesired interferences by minimizing the output power while maintaining the gain along the look direction to be constant, usually unity. That is,

$$\min_{\boldsymbol{w}} E[|y(k)|^2] = \min_{\boldsymbol{w}} \boldsymbol{w}^H \boldsymbol{R}_{uu} \boldsymbol{w} \qquad \text{subject to} \quad \boldsymbol{w}^H \boldsymbol{a}(\phi_0) = 1 \tag{9.5}$$

The weight vector obtained by solving (9.5) is often called the *minimum variance distortionless response* (MVDR) beamformer weights since, for a particular look direction, it minimizes the variance (average power) of the output signal while passing the signal arriving in the look direction without distortion (unity gain and zero phase shift). Equation (9.5) represents a constraint optimization problem which can be solved using the method of Lagrange multipliers. This approach converts the constraint optimization problem into an unconstrained one, thereby allowing the use of Least Squares techniques to determine the solution. Using a Lagrange multiplier, the weight vector that solves (9.5) can be shown to be [Hay91]

$$\boldsymbol{w} = \frac{\boldsymbol{R}_{uu}^{-1}\boldsymbol{a}(\phi)}{\boldsymbol{a}^{H}(\phi)\boldsymbol{R}_{uu}^{-1}\boldsymbol{a}(\phi)} \tag{9.6}$$

Now the output power of the array as a function of the Angle-Of-Arrival, using Capon's beamforming method, is given by Capon's spatial spectrum,

$$P_{Capon}(\phi) = \frac{1}{\boldsymbol{a}^{H}(\phi)\boldsymbol{R}_{uu}^{-1}\boldsymbol{a}(\phi)} \tag{9.7}$$

By computing and plotting Capon's spectrum over the whole range of ϕ, the DOA's can be estimated by locating the peaks in the spectrum.

Although it is not a maximum likelihood (ML) estimator, Capon's method is sometimes referred to as an ML estimator, since for any choice of ϕ, $P_{Capon}(\phi)$ is the maximum likelihood estimate of the power of a signal arriving from the direction ϕ in the presence of white Gaussian noise having arbitrary spatial characteristics [Cap79].

Figure 9–2 illustrates the performance improvement obtained by Capon's method over the delay-and-sum method. Computer simulations show that using a six element uniformly spaced linear array with half wavelength interelement spacing, Capon's method is able to distinguish between the two signals arriving at 90 and 100 degrees, respectively, while the delay-and-sum method fails to differentiate between the two signals.

Though it provides a better resolution when compared to the delay-and-sum method, Capon's method suffers from many disadvantages. Capon's method fails if other signals that are correlated with the Signal-of-Interest are present because it inadvertently uses that correlation to reduce the processor output power without spatially nulling it [Sch93d]. In other words, the correlated components may be combined destructively in the process of minimizing the output power. Also, Capon's method requires the computation of a matrix inverse, which can be expensive for large arrays.

9.2 Subspace Methods for DOA Estimation

Though many of the classical beamforming based methods such as Capon's minimum variance method are often successful and are widely used, these methods have some fundamental limitations in resolution. Most of these limitations arise because they do not exploit the structure of the input data model given in (8.5). Schmidt [Sch79] and Bienvenu and Kopp [Bie79]

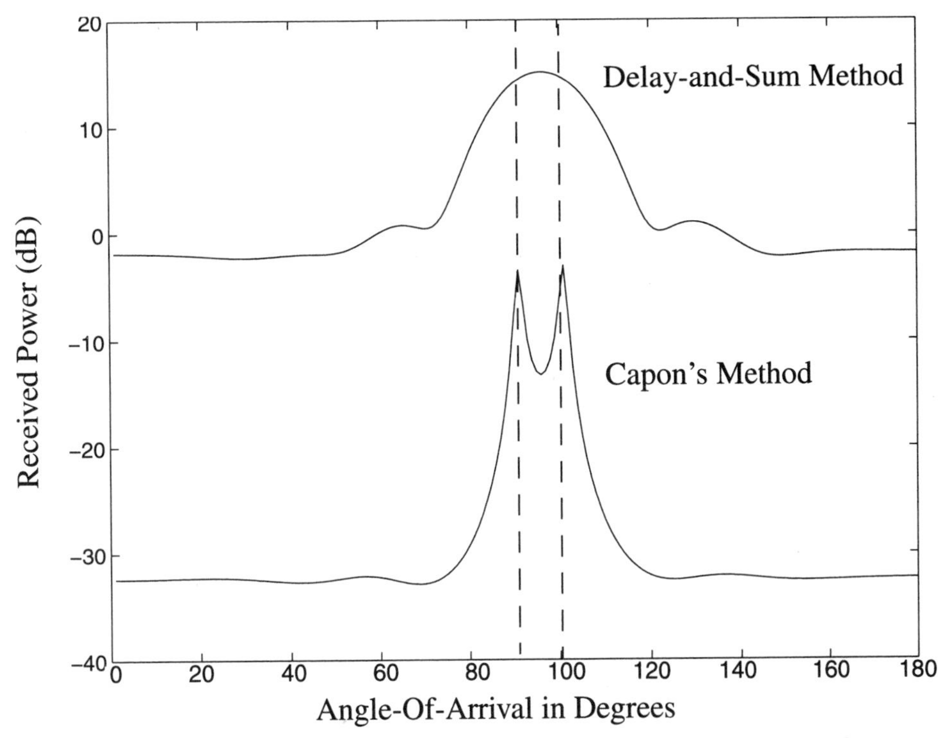

Figure 9–2 Comparison of resolution performance of delay-and-sum method and Capon's minimum variance method. Two signals of equal power at an SNR of 20 dB arrive at a 6-element uniformly spaced array with an interelement spacing equal to half a wavelength at angles 90 and 100 degrees, respectively.

were the first to exploit the structure of a more accurate data model for the case of sensor arrays of arbitrary form. Schmidt derived a complete geometric solution to the DOA estimation problem in the absence of noise, and extended the geometric concepts to obtain a reasonable approximation to the solution in the presence of noise. The technique proposed by Schmidt is called the MUltiple SIgnal Classification (MUSIC) algorithm, and it has been thoroughly investigated since its inception [Bar84][Sto89][Pil89a]. The geometric concepts upon which MUSIC is founded form the basis for a much broader class of *subspace-based* algorithms [Pau93][Joh86]. Apart from MUSIC, the primary contributions to the subspace-based algorithms include the Estimation of Signal Parameters via Rotational Invariance Technique (ESPRIT) proposed by Roy *et al.* [Pau86b][Roy89][Roy90] and the minimum-norm method proposed by Kumaresan and Tufts [Kum83].

9.2.1 The MUSIC Algorithm

The MUSIC algorithm proposed by Schmidt in 1979 [Sch79][Sch86a] is a high resolution multiple signal classification technique based on exploiting the eigenstructure of the input covariance matrix. MUSIC is a signal parameter estimation algorithm which provides information about the number of incident signals, Direction-Of-Arrival (DOA) of each signal, strengths and cross correlations between incident signals, noise power, etc. While the MUSIC algorithm provides very high resolution, it requires very precise and accurate array calibration. The MUSIC algorithm has been implemented and its performance has been experimentally verified [Sch86b].

The development of the MUSIC algorithm is based on a geometric view of the signal parameter estimation problem. Following the narrowband data model discussed in Chapter 8, if there are D signals incident on the array, the received input data vector at an M-element array can be expressed as a linear combination of the D incident waveforms and noise. That is,

$$\boldsymbol{u}(t) = \sum_{l=0}^{D-1} \boldsymbol{a}(\phi_l)s_l(t) + \boldsymbol{n}(t) \tag{9.8}$$

$$\boldsymbol{u}(t) = \left[\boldsymbol{a}(\phi_0)\ \boldsymbol{a}(\phi_1)\ \dots\ \boldsymbol{a}(\phi_{D-1})\right]\begin{bmatrix} s_0(t) \\ \\ s_{D-1}(t) \end{bmatrix} + \boldsymbol{n}(t) = \boldsymbol{A}\boldsymbol{s}(t) + \boldsymbol{n}(t) \tag{9.9}$$

where $\boldsymbol{s}^T(t) = [s_0(t)\ s_1(t)\ \dots\ s_{D-1}(t)]$ is the vector of incident signals, $\boldsymbol{n}(t) = [n_0(t)\ n_1(t)\ \dots\ n_{D-1}(t)]$ is the noise vector, and $\boldsymbol{a}(\phi_j)$ is the array steering vector corresponding to the Direction-Of-Arrival of the jth signal. For simplicity, we will drop the time argument from $\boldsymbol{u}$, $\boldsymbol{s}$, and $\boldsymbol{n}$ from this point onward.

In geometric terms, the received vector $\boldsymbol{u}$ and the steering vectors $\boldsymbol{a}(\phi_j)$ can be visualized as vectors in M dimensional space. From (9.9), it is seen that the received vector $\boldsymbol{u}$ is a particular linear combination of the array steering vectors, with $s_0, s_1, \dots, s_{D-1}$ being the coefficients of the combination. In terms of the above data model, the input covariance matrix $\boldsymbol{R}_{uu}$ can be expressed as

$$\boldsymbol{R}_{uu} = E[\boldsymbol{u}\boldsymbol{u}^H] = \boldsymbol{A}E[\boldsymbol{s}\boldsymbol{s}^H]\boldsymbol{A}^H + E[\boldsymbol{n}\boldsymbol{n}^H] \tag{9.10}$$

$$\boldsymbol{R}_{uu} = \boldsymbol{A}\boldsymbol{R}_{ss}\boldsymbol{A}^H + \sigma_n^2\boldsymbol{I} \tag{9.11}$$

where $\boldsymbol{R}_{ss}$ is the *signal correlation matrix* $E[\boldsymbol{s}\boldsymbol{s}^H]$.

The eigenvalues of $\boldsymbol{R}_{uu}$ are the values, $\{\lambda_0, \dots, \lambda_{M-1}\}$ such that

$$|\boldsymbol{R}_{uu} - \lambda_i\boldsymbol{I}| = 0 \tag{9.12}$$

Using (9.11), we can rewrite this as

$$|\boldsymbol{A}\boldsymbol{R}_{ss}\boldsymbol{A}^H + \sigma_n^2\boldsymbol{I} - \lambda_i\boldsymbol{I}| = |\boldsymbol{A}\boldsymbol{R}_{ss}\boldsymbol{A}^H - (\lambda_i - \sigma_n^2)\boldsymbol{I}| = 0 \tag{9.13}$$

Therefore the *eigenvalues*, v_i, of $\boldsymbol{A}\boldsymbol{R}_{ss}\boldsymbol{A}^H$ are

$$v_i = \lambda_i - \sigma_n^2 \tag{9.14}$$

Since $\boldsymbol{A}$ is comprised of steering vectors which are linearly independent, it has full column rank, and the signal correlation matrix $\boldsymbol{R}_{ss}$ is non-singular as long as the incident signals are not highly correlated.

A full column rank $\boldsymbol{A}$ and nonsingular $\boldsymbol{R}_{ss}$ guarantees that, when the number of incident signals D is less than the number of array elements M, the $M \times M$ matrix $\boldsymbol{A}\boldsymbol{R}_{ss}\boldsymbol{A}^H$ is positive semidefinite with rank D.

From elementary linear algebra, this implies that $M - D$ of the eigenvalues, v_i, of $\boldsymbol{A}\boldsymbol{R}_{ss}\boldsymbol{A}^H$ are zero. From (9.14), this means that $M - D$ of the eigenvalues of $\boldsymbol{R}_{uu}$ are equal to the noise variance, σ_n^2. We then sort the eigenvalues of $\boldsymbol{R}_{uu}$ such that λ_0 is the largest eigenvalue, and λ_{M-1} is the smallest eigenvalue. Therefore,

$$\lambda_D, \ldots, \lambda_{M-1} = \sigma_n^2 \tag{9.15}$$

In practice, however, when the autocorrelation matrix $\boldsymbol{R}_{uu}$ is estimated from a finite data sample, all the eigenvalues corresponding to the noise power will not be identical. Instead they will appear as a closely spaced cluster, with the variance of their spread decreasing as the number of samples used to obtain an estimate of $\boldsymbol{R}_{uu}$ is increased. Once the multiplicity, K, of the smallest eigenvalue is determined, an estimate of the number of signals, $\hat{D}$, can be obtained from the relation $M = D + K$. Therefore, the estimated number of signals is given by

$$\hat{D} = M - K \tag{9.16}$$

The eigenvector associated with a particular eigenvalue, λ_i, is the vector $\boldsymbol{q}_i$ such that

$$(\boldsymbol{R}_{uu} - \lambda_i\boldsymbol{I})\boldsymbol{q}_i = \boldsymbol{0} \tag{9.17}$$

For eigenvectors associated with the $M - D$ smallest eigenvalues, we have

$$(\boldsymbol{R}_{uu} - \sigma_n^2\boldsymbol{I})\boldsymbol{q}_i = \boldsymbol{A}\boldsymbol{R}_{ss}\boldsymbol{A}^H\boldsymbol{q}_i + \sigma_n^2\boldsymbol{I} - \sigma_n^2\boldsymbol{I} = \boldsymbol{0} \tag{9.18}$$

$$\boldsymbol{A}\boldsymbol{R}_{ss}\boldsymbol{A}^H\boldsymbol{q}_i = \boldsymbol{0} \tag{9.19}$$

Since $\boldsymbol{A}$ has full rank and $\boldsymbol{R}_{ss}$ is nonsingular, this implies that

$$\boldsymbol{A}^H\boldsymbol{q}_i = \boldsymbol{0} \tag{9.20}$$

or

$$\begin{bmatrix} \boldsymbol{a}^H(\phi_0)\boldsymbol{q}_i \\ \boldsymbol{a}^H(\phi_1)\boldsymbol{q}_i \\ \vdots \\ \boldsymbol{a}^H(\phi_{D-1})\boldsymbol{q}_i \end{bmatrix} = \begin{bmatrix} 0 \\ 0 \\ \vdots \\ 0 \end{bmatrix} \tag{9.21}$$

This means that the eigenvectors associated with the $M-D$ smallest eigenvalues are orthogonal to the D steering vectors that make up $\boldsymbol{A}$.

$$\{\boldsymbol{a}(\phi_0), \ldots, \boldsymbol{a}(\phi_{D-1})\} \perp \{\boldsymbol{q}_D, \ldots, \boldsymbol{q}_{M-1}\} \tag{9.22}$$

This is the essential observation of the MUSIC approach. It means that one can estimate the steering vectors associated with the received signals by finding the steering vectors which are most nearly orthogonal to the eigenvectors associated with the eigenvalues of $\boldsymbol{R}_{uu}$ that are approximately equal to σ_n^2.

This analysis shows that the eigenvectors of the covariance matrix $\boldsymbol{R}_{uu}$ belong to either of the two orthogonal subspaces, called the principal eigen subspace (*signal subspace*) and the non-principal eigen subspace (*noise subspace*). The steering vectors corresponding to the Directions-Of-Arrival lie in the signal subspace and are hence orthogonal to the noise subspace. By searching through all possible array steering vectors to find those which are perpendicular to the space spanned by the non-principal eigenvectors, the DOAs, ϕ_ι's can be determined.

To search the noise subspace, we form a matrix containing the noise eigenvectors:

$$\boldsymbol{V}_n = \left[\boldsymbol{q}_D\ \boldsymbol{q}_{D+1}\ \cdots\ \boldsymbol{q}_{M-1}\right] \tag{9.23}$$

Since the steering vectors corresponding to signal components are orthogonal to the noise subspace eigenvectors, $\boldsymbol{a}^H(\phi)\boldsymbol{V}_n\boldsymbol{V}_n^H\boldsymbol{a}(\phi) = 0$ for ϕ corresponding to the DOA of a multipath component. Then the DOAs of the multiple incident signals can be estimated by locating the peaks of a MUSIC spatial spectrum given by

$$P_{MUSIC}(\phi) = \frac{1}{\boldsymbol{a}^H(\phi)\boldsymbol{V}_n\boldsymbol{V}_n^H\boldsymbol{a}(\phi)} \tag{9.24}$$

or, alternatively,

$$P_{MUSIC}(\phi) = \frac{\boldsymbol{a}^H(\phi)\boldsymbol{a}(\phi)}{\boldsymbol{a}^H(\phi)\boldsymbol{V}_n\boldsymbol{V}_n^H\boldsymbol{a}(\phi)} \tag{9.25}$$

Orthogonality between $\boldsymbol{a}(\phi)$ and $\boldsymbol{V}_n$ will minimize the denominator and hence will give rise to peaks in the MUSIC spectrum defined in (9.24) and (9.25). The $\hat{D}$ largest peaks in the MUSIC spectrum correspond to the directions of arrival of the signals impinging on the array.

Once the Directions-Of-Arrival, ϕ_i, are determined from the MUSIC spectrum, the signal covariance matrix $\boldsymbol{R}_{ss}$ can be determined from the following relation [Sch86a].

$$\boldsymbol{R}_{ss} = (\boldsymbol{A}^H\boldsymbol{A})^{-1}\boldsymbol{A}^H(\boldsymbol{R}_{uu} - \lambda_{min}\boldsymbol{I})\boldsymbol{A}(\boldsymbol{A}^H\boldsymbol{A})^{-1} \tag{9.26}$$

From (9.26), the powers and cross correlations between the various input signals can be readily obtained.

The MUSIC algorithm may be summarized as follows:

1. Collect input samples $\boldsymbol{u}_k$, $k = 0, \ldots, K-1$, and estimate the input covariance matrix

$$\hat{\boldsymbol{R}}_{uu} = \frac{1}{K}\sum_{k=0}^{K-1}\boldsymbol{u}_k\boldsymbol{u}_k^H \tag{9.27}$$

2. Perform eigen decomposition on $\hat{\boldsymbol{R}}_{uu}$

$$\hat{\boldsymbol{R}}_{uu}\boldsymbol{V} = \boldsymbol{V}\Lambda \tag{9.28}$$

where $\Lambda = diag\{\lambda_0, \lambda_1, \ldots, \lambda_{M-1}\}$, $\lambda_0 \geq \lambda_1 \geq \ldots \geq \lambda_{M-1}$ are the eigenvalues and $\boldsymbol{V} = \begin{bmatrix}\boldsymbol{q}_0 \; \boldsymbol{q}_1 \; \cdots \; \boldsymbol{q}_{M-1}\end{bmatrix}$ are the corresponding eigenvectors of $\hat{\boldsymbol{R}}_{uu}$.

3. Estimate the number of signals $\hat{D}$, from the multiplicity K, of the smallest eigenvalue λ_{min} as

$$\hat{D} = M - K \tag{9.29}$$

4. Compute the MUSIC spectrum

$$\hat{P}_{MUSIC}(\phi) = \frac{\boldsymbol{a}^H(\phi)\boldsymbol{a}(\phi)}{\boldsymbol{a}^H(\phi)\boldsymbol{V}_n\boldsymbol{V}_n^H\boldsymbol{a}(\phi)} \tag{9.30}$$

where $\boldsymbol{V}_n = \begin{bmatrix}\boldsymbol{q}_D \; \boldsymbol{q}_{D+1} \; \cdots \; \boldsymbol{q}_{M-1}\end{bmatrix}$.

5. Find the $\hat{D}$ largest peaks of $\hat{P}_{MUSIC}(\phi)$ to obtain estimates of the Direction-Of-Arrival.

Figure 9–3 shows a comparison between the resolution performance of MUSIC and the Capon's minimum variance method. As seen clearly from the plot, MUSIC can resolve closely spaced signals which cannot be detected by Capon's method. Simulation results show that two signals arriving at angles 90 and 95 degrees, respectively, at the input of a 6-element uniformly spaced linear array can be detected by MUSIC, whereas Capon's minimum variance method fails to differentiate between the two signals [Muh96b].

It should be noted that, unlike the conventional methods, the MUSIC spatial spectrum does not estimate the signal power associated with each arrival angle. Instead, when the ensemble average of the array input covariance matrix is known exactly, under uncorrelated and identical noise conditions, the peaks of $P_{MUSIC}(\phi)$ are guaranteed to correspond to the true Directions-Of-Arrival. Since these peaks are distinct irrespective of the actual separation between arrival angles, in principle, with perfect array calibration, these estimators can distinguish and resolve arbitrarily closely spaced signals. When impinging signals $s_l(t)$ from (9.8) are highly correlated, MUSIC fails because R_{ss} becomes singular. In Section 9.4.2, techniques are presented to handle highly correlated signals.

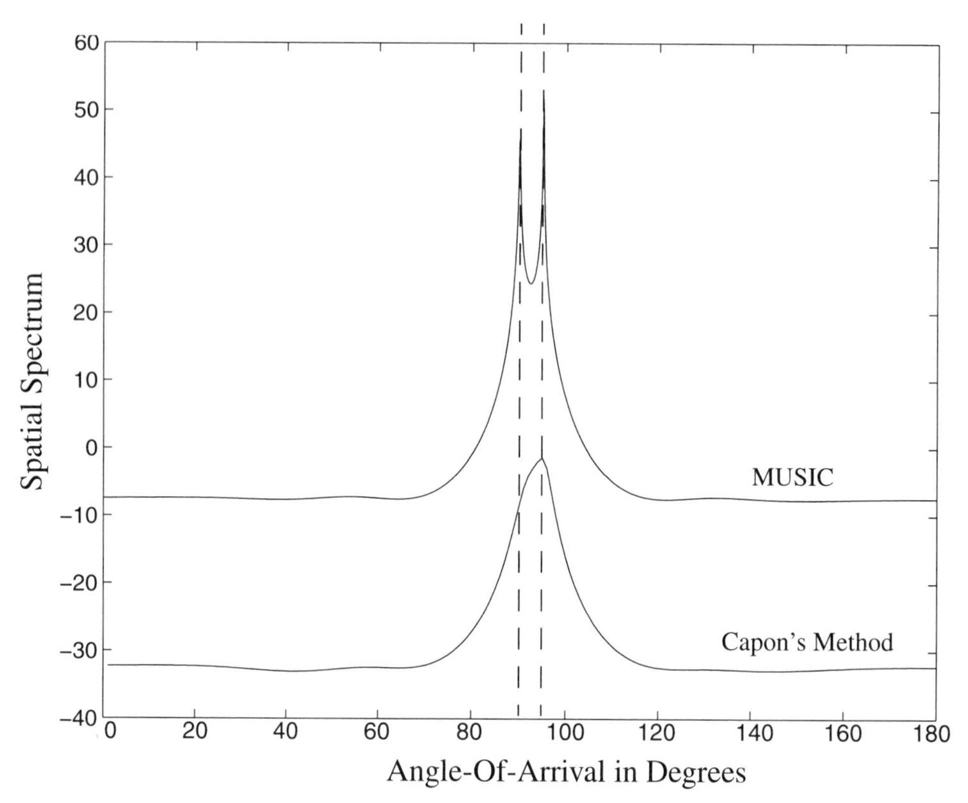

Figure 9–3 Comparison of MUSIC and Capon's minimum variance method. Two signals of equal power at an SNR of 20 dB arrive at a 6-element uniformly spaced array with an interelement spacing equal to half a wavelength at angles 90 and 95 degrees, respectively [Muh96a].

9.2.2 Improvements to the MUSIC Algorithm

Various modifications to the MUSIC algorithm have been proposed to increase its resolution performance and decrease the computational complexity. One such improvement is the Root-MUSIC algorithm developed by Barabell [Bar83], which is based on polynomial rooting and provides higher resolution, but is applicable only to a uniform spaced linear array. Another improvement proposed by Barabell uses the properties of signal space eigenvectors (principal eigenvectors) to define a rational spectrum function with improved resolution capability [Bar83].

Cyclic MUSIC, which exploits the spectral coherence properties of the signal to improve the performance of the conventional MUSIC algorithms, has been proposed in [Sch89]. Fast Subspace Decomposition techniques have also been studied to decrease the computational complexity of MUSIC [Xu94b].

9.2.3 Root-MUSIC Algorithm

For the case of a uniformly spaced linear array with interelement spacing d, the mth element of the steering vector $\boldsymbol{a}(\phi)$ may be expressed as (see Chapter 3):

$$a_m(\phi) = \exp\left(j2\pi m\left(\frac{d}{\lambda}\right)\cos\phi\right); \qquad m = 1 \ldots M \tag{9.31}$$

The MUSIC spectrum given by (9.24) is an all-pole function of the form

$$\begin{aligned} P_{MUSIC}(\phi) &= \frac{1}{\boldsymbol{a}^H(\phi)\boldsymbol{V}_n\boldsymbol{V}_n^H\boldsymbol{a}(\phi)} \\ &= \frac{1}{\boldsymbol{a}^H(\phi)\boldsymbol{C}\boldsymbol{a}(\phi)} \end{aligned} \tag{9.32}$$

where $\boldsymbol{C} = \boldsymbol{V}_n\boldsymbol{V}_n^H$. Using equation (9.31), the denominator of (9.32) may be written as

$$P_{MUSIC}^{-1}(\phi) = \sum_{m=1}^{M}\sum_{n=1}^{M}\exp\left(-j\frac{2\pi md}{\lambda}\cos\phi\right)C_{mn}\exp\left(j\frac{2\pi nd}{\lambda}\cos\phi\right) \tag{9.33}$$

where C_{mn} is the entry in the mth row and n^{th} column of $\boldsymbol{C}$. Combining the two summations into one, (9.33) can be simplified as

$$P_{MUSIC}^{-1}(\phi) = \sum_{l=-M+1}^{M-1} C_l \exp\left(-j\frac{2\pi d}{\lambda}l\cos\phi\right) \tag{9.34}$$

where $C_l = \sum_{m-n=l} C_{mn}$ is the sum of the entries of $\boldsymbol{C}$ along the lth diagonal.

By defining a polynomial $D(z)$ as follows,

$$D(z) = \sum_{l=-M+1}^{M+1} C_l z^{-l} \tag{9.35}$$

evaluating the MUSIC spectrum $P_{MUSIC}(\phi)$ becomes equivalent to evaluating the polynomial $D(z)$ on the unit circle, and the peaks in the MUSIC spectrum exist because the roots of $D(z)$ lie close to the unit circle. Ideally, with no noise, the poles will lie exactly on the unit circle at locations determined by the Direction-Of-Arrival. In other words, a pole of $D(z)$ at $z = z_1 = |z_1|\exp(j\arg(z_1))$ will result in a peak in the MUSIC spectrum at

$$\cos\phi = \left(\frac{\lambda}{2\pi d}\right)\arg(z_1) \tag{9.36}$$

Barabell [Bar83] showed through simulations that the ROOT-MUSIC algorithm has better resolution than the spectral MUSIC algorithm, especially at low SNR conditions.

9.2.4 Cyclic MUSIC Algorithm

Cyclic MUSIC is a signal selective direction finding algorithm which exploits the *spectral coherence* of the received signal as well as the *spatial coherence*. By exploiting spectral correlation along with MUSIC, it is possible to resolve signals spaced more closely than the resolution threshold of the array when only one of them is a Signal-of-Interest (SOI) [Sch89][Sch94]. Cyclic MUSIC also circumvents the requirement that the total number of signals impinging on the array (including both SOI and interference) must be less than the number of sensor elements [Gar88].

Consider an array of M sensors which receives D_α signals which exhibit spectral correlation at a cycle frequency α, and an arbitrary number of interferers that do not exhibit spectral correlation at that particular frequency. For example, this could be the case where a desired user, with a particular spectral correlation and number of multipath components, is to be detected in a heavy co-channel interference environment. Let $s_i(t)$, $i = 0,\ldots, D_\alpha - 1$ be the desired signals, and $\boldsymbol{n}(t)$ the noise and interference vector incident on the array. The received signal vector $\boldsymbol{u}(t)$ can then be expressed as

$$\boldsymbol{u}(t) = \sum_{i=0}^{D_\alpha - 1} \boldsymbol{a}(\phi_i)s_i(t) + \boldsymbol{n}(t) \tag{9.37}$$

$$= \boldsymbol{A}\boldsymbol{s}(t) + \boldsymbol{n}(t)$$

Since only the desired signals exhibit spectral correlation at α, the cyclic autocorrelation matrix $\boldsymbol{R}_{uu}^{\alpha}(\tau)$ of the received signal $\boldsymbol{u}(t)$ defined as

$$\boldsymbol{R}_{uu}^{\alpha} = \left\langle \boldsymbol{u}\left(t+\frac{\tau}{2}\right)\left\{\boldsymbol{u}\left(t-\frac{\tau}{2}\right)\exp(j2\pi\alpha t)\right\}^H \right\rangle_\infty \tag{9.38}$$

can be expressed as

$$\boldsymbol{R}_{uu}^{\alpha}(\tau) = \boldsymbol{A}\boldsymbol{R}_{ss}^{\alpha}(\tau)\boldsymbol{A}^H \tag{9.39}$$

where $\boldsymbol{R}_{ss}^{\alpha}(\tau)$ is the cyclic autocorrelation matrix of the desired signals, defined as

$$\boldsymbol{R}_{ss}^{\alpha} = \left\langle \boldsymbol{s}\left(t+\frac{\tau}{2}\right)\left\{\boldsymbol{s}\left(t-\frac{\tau}{2}\right)\exp(j2\pi\alpha t)\right\}^H \right\rangle_\infty \tag{9.40}$$

where

$$\langle\cdot\rangle_\infty = \lim_{T\to\infty}\int_{-\frac{T}{2}}^{\frac{T}{2}}(\cdot)dt \tag{9.41}$$

Clearly, the matrix $R_{uu}^{\alpha}(\tau)$ has rank D_{α}. For $D_{\alpha} < M$, the null space of $R_{uu}^{\alpha}(\tau)$ is spanned by the eigenvectors $V_{n,\alpha}$ corresponding to its zero eigenvalues,

$$R_{uu}^{\alpha}(\tau)V_{n,\alpha} = 0 \tag{9.42}$$

If the signals are not fully correlated, $R_{ss}^{\alpha}(\tau)$ has full rank equal to D_{α}. Since A is also full rank, it follows from (9.39) and (9.42) that the null space of $R_{uu}^{\alpha}(\tau)$ is orthogonal to the direction vectors of the desired signals. That is,

$$V_{n,\alpha}^{H}a(\phi_i) = 0, \qquad i = 0, \ldots, D_{\alpha} - 1 \tag{9.43}$$

Using (9.43) as a measure of orthogonality, a cyclic MUSIC spectrum similar to (9.25) can be defined as follows:

$$P_{CYCLIC\text{-}MUSIC}(\phi) = \frac{a^H(\phi)(a(\phi))}{a^H(\phi)V_{n,\alpha}V_{n,\alpha}^{H}a(\phi)} \tag{9.44}$$

The Direction-Of-Arrival of the desired signals can be computed by searching through all ϕ for the D_{α} highest peaks of $P_{CYCLIC\text{-}MUSIC}(\phi)$.

9.2.5 The ESPRIT Algorithm

The Estimation of Signal Parameters via Rotational Invariance Techniques (ESPRIT) algorithm is another subspace-based DOA estimation technique developed by Roy *et. al.* [Pau86b][Roy89][Roy90]. ESPRIT dramatically reduces the computational and storage requirements of MUSIC and does not involve an exhaustive search through all possible steering vectors to estimate the Direction-Of-Arrival. Unlike MUSIC, ESPRIT does not require that the array manifold vectors be precisely known; hence, the array calibration requirements are not stringent. ESPRIT derives its advantages by requiring that the sensor array have a structure that can be decomposed into two equal-sized identical subarrays with the corresponding elements of the two subarrays displaced from each other by a fixed translational (not rotational) distance. That is, the array should possess a displacement (translational) invariance, and the sensors should occur in matched pairs with identical displacement. Fortunately, there are many practical situations where these conditions are satisfied, such as in the case of a uniform linear array.

Consider a planar array of arbitrary geometry composed of $m = M/2$ sensor pairs or doublets, as shown in Figure 9–4. (It should be noted that, although the array is assumed to be composed of $M/2$ sensor doublets in this example, it is possible to have M-element arrays such as the uniformly spaced linear array to be composed of M-1 overlapping doublets). To describe mathematically the effect of the translational invariance of the sensor array, it is convenient to describe the array as being composed of two identical subarrays, X_0 and X_1, physically displaced (not rotated) from each other by a known displacement (translational) Δx. The signals received at the ith doublet can then be expressed as

$$u_{0,i}(t) = \sum_{k=0}^{D-1} s_k(t)a_i(\phi_k) + n_{0i}(t) \tag{9.45}$$

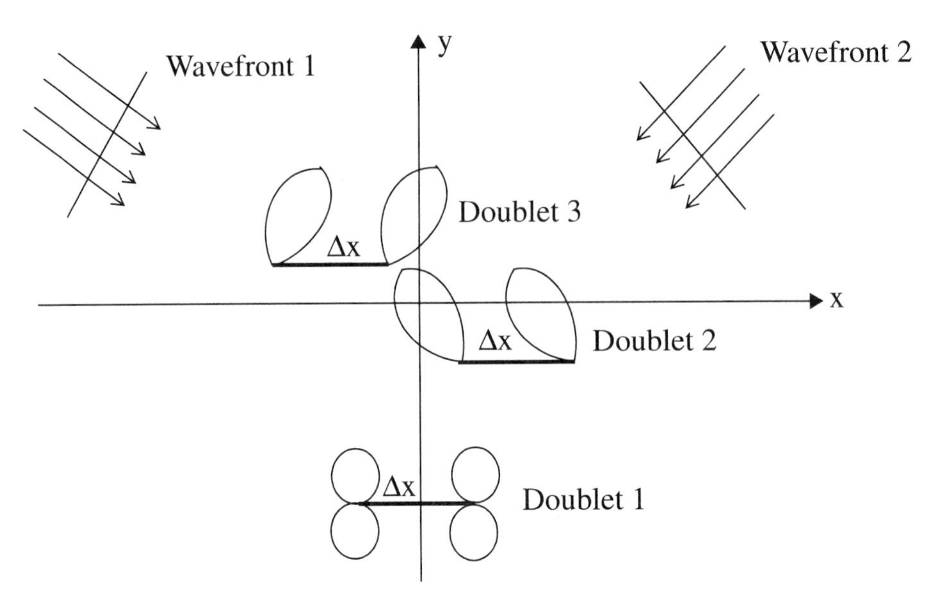

Figure 9–4 Illustration of ESPRIT array geometry [Roy90].

$$u_{1,i}(t) = \sum_{k=0}^{D-1} s_k(t)\exp[j\beta\Delta x\cos(\phi_k)]a_i(\phi_k) + n_{1,i}(t) \tag{9.46}$$

where ϕ_k is the Direction-Of-Arrival of the kth source relative to the direction of the translational Δx, and D is the number of signals incident on the array. Now, using matrix and vector notation, the received signal vector at the two subarrays can be written as follows:

$$\boldsymbol{u}_0(t) = \boldsymbol{A}\boldsymbol{s}(t) + \boldsymbol{n}_0(t) \tag{9.47}$$

$$\boldsymbol{u}_1(t) = \boldsymbol{A}\boldsymbol{\Phi}\boldsymbol{s}(t) + \boldsymbol{n}_1(t) \tag{9.48}$$

where $\boldsymbol{\Phi}$ is a $D \times D$ diagonal unitary matrix whose diagonal elements represent the phase delays between the doublet sensors for the D signals. The matrix $\boldsymbol{\Phi}$ relates the measurements from subarray $\boldsymbol{u}_0$ to those from subarray $\boldsymbol{u}_1$, and is given by

$$\boldsymbol{\Phi} = diag\{\exp(j\gamma_0), \exp(j\gamma_1), \ldots, \exp(j\gamma_{D-1})\}, \text{ where } \gamma_k = \beta\Delta x\cos(\phi_k) \tag{9.49}$$

Though in the complex field, the matrix $\boldsymbol{\Phi}$ is a simple scaling operator, it is similar to the real two-dimensional rotation operator. The total array output vector $\boldsymbol{u}(t)$ can be written as

$$\boldsymbol{u}(t) = \begin{bmatrix} \boldsymbol{u}_0(t) \\ \boldsymbol{u}_1(t) \end{bmatrix} = \bar{\boldsymbol{A}}\boldsymbol{s}(t) + \boldsymbol{n}(t) \tag{9.50}$$

where

$$\bar{A} = \begin{bmatrix} A \\ A\Phi \end{bmatrix} \quad \text{and} \quad n(t) = \begin{bmatrix} n_0(t) \\ n_1(t) \end{bmatrix} \tag{9.51}$$

The basic idea behind ESPRIT is to exploit the rotational invariance of the underlying signal subspace induced by the translational invariance of the sensor array [Roy89]. The relevant signal subspace is the one that contains the outputs from the two subarrays u_0 and u_1. Simultaneous sampling of the output of the arrays leads to two sets of vectors, V_0 and V_1, that span the same signal subspace.

The signal subspace can be obtained from the knowledge of the input covariance matrix $R_{uu} = \bar{A}R_{ss}\bar{A} + \sigma_n^2 I$. If $D \le M$, the M-D smallest eigenvalues of R_{uu} are equal to σ_n^2. The D eigenvectors V_s corresponding to the D largest eigenvalues satisfy the relation,

$$Range\{V_s\} = Range\{\bar{A}\} \tag{9.52}$$

Now, since $Range\{V_s\} = Range\{\bar{A}\}$, there must exist a unique nonsingular T such that $V_s = \bar{A}T$. Further, the invariance structure of the array allows the decomposition of V_s into $V_0 \in C^{M \times D}$ and $V_1 \in C^{M \times D}$ such that $V_0 = AT$ and $V_1 = A\Phi T$. This implies that

$$Range\{V_0\} = Range\{V_1\} = Range\{A\} \tag{9.53}$$

Since V_0 and V_1 share a common column space, the rank of $V_{01} = [V_0 \mid V_1]$ is D. This implies that there exists a unique rank-D matrix $F \in C^{2D \times D}$, such that

$$0 = [V_0 \mid V_1]F = V_0F_0 + V_1F_1 = ATF_0 + A\Phi TF_1 \tag{9.54}$$

F spans the null space of V_{01}. By defining $\Psi = -F_0F_1^{-1}$, (9.54) can be rearranged to obtain

$$AT\Psi = A\Phi T \tag{9.55}$$

which implies

$$AT\Psi T^{-1} = A\Phi \tag{9.56}$$

Now, assuming A to be full rank, which is true as long as the Directions-Of-Arrival of each signal is distinct, (9.56) implies that

$$T\Psi = \Phi T \tag{9.57}$$

From (9.57), it is evident that the eigenvalues of Ψ must be equal to the diagonal elements of Φ, and the columns of T are the eigenvectors of Ψ. This is the *key* relationship in the development of ESPRIT. The signal parameters are obtained as nonlinear functions of the eigenvalues of the operator Ψ that maps (rotates) one set of vectors V_0 that span an m-dimensional signal subspace into another set of vectors V_1.

In practice, with only a finite number of noisy measurements available, the conditions in equations (9.52) and (9.53) are not satisfied. Hence finding a Ψ such that $\hat{V}_0\Psi = \hat{V}_1$ is not possible. Therefore, one must resort to a Least Squares solution, which minimizes the residual error. Assuming that the set of equations is overdetermined, the Least Squares solution is given by

$$\Psi = (\hat{V}_0^H \hat{V}_0)^{-1} \hat{V}_0^H \hat{V}_1 \tag{9.58}$$

Once Ψ is obtained, its eigenvalues which correspond to the diagonal elements of Φ can be easily computed. Since the diagonal elements of Φ are related to the Angle-Of-Arrival via equation (9.49), they can then be directly computed.

Since both $\hat{V}_0$ and $\hat{V}_1$ are equally noisy, the problem is better solved using the *Total Least Squares criterion* (TLS). This amounts to replacing the zero matrix in (9.54) by a matrix of errors whose Frobenius norm (i.e., total Least Squared error) is to be minimized. The TLS ESPRIT algorithm does this and may be summarized as follows:

1. Obtain an estimate $\hat{R}_{uu}$ of R_{uu} from the measurements u.
2. Perform eigen decomposition on $\hat{R}_{uu}$, i.e.,

$$\hat{R}_{uu} = V \Lambda V, \tag{9.59}$$

   where $\Lambda = diag\{\lambda_0, \ldots, \lambda_{M-1}\}$ and $V = [q_0, \ldots, q_{M-1}]$ are the eigenvalues and eigenvectors, respectively.
3. Using the multiplicity, K, of the smallest eigenvalue λ_{min}, estimate the number of signals $\hat{D}$, as $\hat{D} = M - K$
4. Obtain the signal subspace estimate $\hat{V}_s = [\hat{V}_0, \ldots, \hat{V}_{\hat{D}-1}]$ and decompose it into subarray matrices,

$$\hat{V}_s = \begin{bmatrix} \hat{V}_0 \\ \hat{V}_1 \end{bmatrix} \tag{9.60}$$

5. Compute the eigen decomposition $(\lambda_1 > \ldots > \lambda_{2\hat{D}})$

$$\hat{V}_{01}^H \hat{V}_{01} = \begin{bmatrix} \hat{V}_0^H \\ \hat{V}_1^H \end{bmatrix} \begin{bmatrix} \hat{V}_0 & \hat{V}_1 \end{bmatrix} = V \Lambda V^H \tag{9.61}$$

   and partition V into $\hat{D} \times \hat{D}$ submatrices,

$$V = \begin{bmatrix} V_{11} & V_{12} \\ V_{21} & V_{22} \end{bmatrix} \tag{9.62}$$

6. Calculate the eigenvalues of $\Psi = -V_{12} V_{22}^{-1}$,

$$\hat{\Phi}_k = \text{eigenvalues of}(-V_{12} V_{22}^{-1}), \qquad \forall k = 0, \ldots, \hat{D} - 1 \tag{9.63}$$

7. Estimate the Angle-Of-Arrival as

$$\hat{\phi}_k = \cos^{-1}\left[c \frac{(\arg(\hat{\Phi}_k))}{\beta \Delta x}\right] \tag{9.64}$$

As seen from the above discussion, ESPRIT eliminates the search procedure inherent in most DOA estimation methods. ESPRIT produces the DOA estimates directly in terms of the eigenvalues.

9.3 Maximum Likelihood Techniques

Maximum Likelihood (ML) techniques were some of the first techniques to be investigated for DOA estimation. Since ML techniques were computationally intensive, they were less popular than suboptimal subspace techniques. However, in terms of performance, the ML techniques are superior to the subspace based techniques, especially in low signal-to-noise ratio conditions or when the number of signal samples is small [Zis88]. Moreover, unlike subspace based techniques, ML based techniques can also perform well in correlated signal conditions.

To derive the ML estimator, data collected over a block of N snapshots is formulated as

$$\boldsymbol{U} = \boldsymbol{A}(\Phi)\boldsymbol{S} + \boldsymbol{N} \tag{9.65}$$

where $\boldsymbol{U} = [\boldsymbol{u}(0),\ldots., \boldsymbol{u}(N-1)]$ is the array data input vector matrix of dimension $M \times N$, $\boldsymbol{A}(\Phi) = [\boldsymbol{a}(\phi_0),\ldots., \boldsymbol{a}(\phi_{D-1})]$ is the spatial signature matrix of dimension $M \times D$, $\boldsymbol{S} = [\boldsymbol{s}(0),\ldots., \boldsymbol{s}(N-1)]$ is the signal waveform matrix of dimension $D \times N$, and $\boldsymbol{N} = [\boldsymbol{n}(0),\ldots., \boldsymbol{n}(N-1)]$ is the noise matrix of dimension $M \times N$. In order to derive the ML estimate of the Angles-Of-Arrival ϕ_0, ..., ϕ_{D-1} of the D sources, some assumptions are made about the signals and noise. First, it is assumed that the number of signals is known or estimated and is smaller than the number of sensors. Second, the set of D steering vectors is assumed to be linearly independent. The noise component is assumed to be an ergodic complex-valued Gaussian process of zero mean and covariance $\sigma^2\boldsymbol{I}$, where σ^2 is an unknown scalar and $\boldsymbol{I}$ is the identity matrix. Finally, it is assumed that the noise samples are statistically independent. It should be noted that the ML estimator is meaningful even when the assumptions made about noise do not hold, in which case it coincides with the Least Squares estimator [Zis88].

The derivation of the ML estimator described here regards the signals to be sample functions of unknown deterministic sequences rather than random processes. Based on the assumptions made about the nature of noise, the joint probability density function of the sampled data as given by equation (9.65) can be expressed as [Zis88]

$$f(\boldsymbol{U}) = \prod_{k=0}^{N-1} \frac{1}{\pi \det[\sigma^2 \boldsymbol{I}]} \exp\left(-\frac{1}{\sigma^2}|\boldsymbol{u}(k) - \boldsymbol{A}(\Phi)\boldsymbol{s}(k)|^2\right) \tag{9.66}$$

where det[ ] denotes the determinant. Ignoring the constant terms, the log likelihood function is given by

$$J = -ND\log\sigma^2 - \frac{1}{\sigma^2}\sum_{k=0}^{N-1}|\boldsymbol{u}(k) - \boldsymbol{A}(\Phi)\boldsymbol{s}(k)|^2 \tag{9.67}$$

To compute the maximum likelihood estimator, the log likelihood function of (9.67) must be maximized with respect to the unknown parameters.

Because the logarithm is a monotonic function, maximizing (9.68) is equivalent to the following minimization problem:

$$\min_{(\Phi, S)} \left\{ \sum_{k=0}^{N-1} |\boldsymbol{u}(k) - \boldsymbol{A}(\Phi)\boldsymbol{s}(k)|^2 \right\} \tag{9.68}$$

Fixing Φ and minimizing with respect to $\boldsymbol{S}$, yields the well-known Least Squares solution

$$\hat{\boldsymbol{s}}(k) = (\boldsymbol{A}^H(\Phi)\boldsymbol{A}(\Phi))^{-1}\boldsymbol{A}^H(\Phi)\boldsymbol{u}(k) \tag{9.69}$$

Substituting (9.69) into (9.68), we obtain

$$\min \sum_{k=0}^{N-1} |\boldsymbol{u}(k) - \boldsymbol{P}_{A(\Phi)}\boldsymbol{u}(k)|^2 \tag{9.70}$$

where $\boldsymbol{P}_{A(\Phi)}$ is the projection operator which projects vectors onto the space spanned by the columns of $\boldsymbol{A}(\Phi)$, and is given by

$$\boldsymbol{P}_{A(\Phi)} = \boldsymbol{A}(\Phi)(\boldsymbol{A}^H(\Phi)\boldsymbol{A}(\Phi))^{-1}\boldsymbol{A}^H(\Phi) \tag{9.71}$$

Therefore, the ML estimate of the Directions-Of-Arrival $\Phi = \{\phi_1, \ldots, \phi_{D-1}\}$ is obtained by maximizing the log-likelihood function

$$J(\Phi) = \sum_{k=0}^{N-1} |\boldsymbol{P}_{A(\Phi)}\boldsymbol{u}(k)|^2 \tag{9.72}$$

Equation (9.72) can be interpreted in a geometric way such that the ML technique appears as a variant of the subspace based method. Viberg and Otterson [Vib91] presented a generalized framework to highlight the similarities between the various subspace based DOA estimation techniques and the maximum likelihood technique. In geometric terms, (9.72) the ML estimator is obtained by searching over the array manifold for the D steering vectors that form a D-dimensional signal subspace which is closest to the vectors $\{\boldsymbol{u}(k), k = 0, \ldots, N-1\}$, where closeness is measured by the modulus of the projection of the vectors onto this subspace.

It can be shown that equation (9.72) can be equivalently written as

$$J(\Phi) = \mathrm{trace}[\boldsymbol{P}_{A(\Phi)}\hat{\boldsymbol{R}}_{uu}] \tag{9.73}$$

where $\hat{\boldsymbol{R}}_{uu}$ is the sample covariance matrix

$$\hat{\boldsymbol{R}}_{uu} = \frac{1}{N}\sum_{k=0}^{N-1} \boldsymbol{u}(k)\boldsymbol{u}^H(k) \tag{9.74}$$

The maximization of the log-likelihood function in (9.73) is a nonlinear, multidimensional maximization problem which is computationally very intensive. Many computationally efficient algorithms have been developed to simplify the solution to the maximization problem [Fed88] [Zis88] [Li93].

The *Alternating Projection Algorithm* developed by Ziskind and Wax [Zis88] is an iterative technique which reduces the maximization problem from a multi-dimensional problem to a one-dimensional problem. The idea is to perform the maximization with respect to a single parameter while holding the remaining parameters fixed. That is, the value of ϕ_i at the $(n+1)^{\text{th}}$ iteration is obtained by solving the following one-dimensional maximization problem:

$$\hat{\phi}_i^{(n+1)} = \arg\max_{\phi_i} \operatorname{trace}\left[\boldsymbol{P}_{[A(\hat{\Phi}_i^{(n)}),\, \boldsymbol{a}(\phi_i)]} \hat{\boldsymbol{R}}_{uu}\right] \tag{9.75}$$

where $\hat{\Phi}_i^{(n)}$ denotes the vector comprising of the Direction-Of-Arrival angle estimates of all signals other than the one being computed.

$$\hat{\Phi}_i^{(n)} = [\hat{\Phi}_0^{(n)}, \ldots, \hat{\Phi}_{i-1}^{(n)}, \hat{\Phi}_{i+1}^{(n)}, \ldots, \hat{\Phi}_{D-1}^{(n)}] \tag{9.76}$$

Since the log-likelihood function $J(\Phi)$ may have multiple local maximas, proper initialization is critical for global convergence. The initialization procedure suggested by Ziskind and Wax begins by solving the maximization problem for a single source:

$$\hat{\phi}_0^{(0)} = \arg\max_{\phi_0} \operatorname{trace}[\boldsymbol{P}_{\boldsymbol{a}(\phi_0)} \hat{\boldsymbol{R}}_{uu}] \tag{9.77}$$

Using the estimated ϕ_0, ϕ_1 is calculated as,

$$\hat{\phi}_1^{(0)} = \arg\max_{\phi_1} \operatorname{trace}\left[\boldsymbol{P}_{[\boldsymbol{a}(\hat{\phi}_0^{(0)}),\, \boldsymbol{a}(\phi_1)]} \hat{\boldsymbol{R}}_{uu}\right] \tag{9.78}$$

Continuing in this fashion, $\Phi^{(0)} = [\hat{\phi}_0^{(0)} \hat{\phi}_1^{(0)}, \ldots, \hat{\phi}_{D-1}^{(0)}]$ is computed. After proper initialization, the alternating projection algorithm can be used to maximize the log-likelihood function.

Further reduction in computational complexity can be achieved by taking advantage of the properties of the projection matrix [Zis88]. It can be shown that, by using the properties of the projection matrix, (9.75) can be written as

$$\hat{\Phi}_i^{(n+1)} = \arg\max_{\phi_i} \operatorname{trace}\left[\boldsymbol{P}_{\boldsymbol{a}(\phi_i)_{A(\hat{\Phi}_i^{(n)})}} \hat{\boldsymbol{R}}_{uu}\right] \tag{9.79}$$

where

$$\boldsymbol{a}(\phi_i)_{A(\hat{\Phi}_i^{(n)})} = (\boldsymbol{I} - \boldsymbol{P}_{A(\hat{\Phi}_i^{(n)})})\boldsymbol{a}(\phi_i) \tag{9.80}$$

By defining a unit vector,

$$\boldsymbol{b}(\phi_i, \hat{\Phi}_i^{(n)}) = \frac{\boldsymbol{a}(\phi_i)_{A(\hat{\Phi}_i^{(n)})}}{\left\|\boldsymbol{a}(\phi_i)_{A(\hat{\Phi}_i^{n})}\right\|} \tag{9.81}$$

equation (9.79) can be rewritten as

$$\hat{\Phi}_i^{(n+1)} = \max_{\phi_i} \boldsymbol{b}^H(\phi_i, \hat{\Phi}_i^{(n)})\boldsymbol{R}_{uu}\boldsymbol{b}(\phi_i, \hat{\Phi}_i^{(n)}) \tag{9.82}$$

The Alternating Projection based maximum likelihood estimator algorithm may be summarized as follows:

1. Initialization

$$\hat{\phi}_i^{(0)} = \max_{\phi_i} \boldsymbol{b}^H(\phi_i, \hat{\Phi}_i^{(0)})\boldsymbol{R}_{uu}\boldsymbol{b}(\phi_i, \hat{\Phi}_i^{(0)}), \quad \text{for } i = 0, \ldots, D\text{-}1 \tag{9.83}$$

2. Main Loop

$$\text{Until } \left|\hat{\phi}_i^{(n)} - \hat{\phi}_i^{(n-1)}\right| < \varepsilon, \text{ do}$$

$$\hat{\phi}_i^{(n)} = \max_{\phi_i} \boldsymbol{b}^H(\phi_i, \hat{\Phi}_i^{(n)})\boldsymbol{R}_{uu}\boldsymbol{b}(\phi_i, \hat{\Phi}_i^{(n)}), \quad \text{for } i = 0, \ldots, D\text{-}1. \tag{9.84}$$

9.4 DOA Estimation under Coherent Signal Conditions

As mentioned in the Section 9.2.1, the MUSIC algorithm works on the premise that the signals impinging on the array are not fully correlated or coherent. Only under uncorrelated conditions does the source covariance matrix $\boldsymbol{R}_{SS}$ satisfy the full rank condition which is the basis of the MUSIC eigen decomposition. The performance of MUSIC degrades severely in a coherent or highly correlated signal environment as encountered in multipath propagation where multiple versions of the same signal arrive within the resolvable chip or symbol duration. Many modifications to the MUSIC algorithm have been proposed to make it work in the presence of coherent signals. Many of these techniques involve modification of the covariance matrix through a preprocessing scheme called *spatial smoothing*. One method of spatial smoothing proposed by Evans *et al* [Eva82] and further expanded by Shan *et al* [Sha85] is based on averaging the covariance matrix of identical overlapping arrays. This method requires an array of identical elements built with some form of periodic structure, such as the uniformly spaced linear array. An adaptive spatial smoothing technique was proposed by Takao and Kikuma [Tak87], which is useful for interference cancellation in multipath environments. Another form of spatial smoothing proposed by Haber and Zoltowski [Hab86] involves moving the entire array structure during the time interval in which the covariances are estimated. A similar technique based on moving the array was proposed by Li and Compton [Li94]. Spatial smoothing techniques always impose restrictions on the type and structure of the array. For the general case, coherent signal detection involves employing a multidimensional search through all possible linear combination of steering vectors to find those orthogonal to the noise subspace [Zol86].

9.4.1 Spatial Smoothing Techniques

The idea behind the spatial smoothing scheme proposed by Evans *et al*, [Eva82] is to let a linear uniform array with M identical sensors be divided into overlapping forward subarrays of size p, such that the sensor elements $\{0, ..., p-1\}$ form the first forward subarray and sensors $\{1, ..., p\}$ form the second forward subarray, etc. Let $\boldsymbol{u}_k(t)$ denote the vector of the received signals at the kth forward subarray. Based on the notation of equation (9.9) we can model the signals received at each subarray as

$$\boldsymbol{u}_k^f(t) = \boldsymbol{A}\boldsymbol{F}^{(k-1)}\boldsymbol{s}(t) + \boldsymbol{n}_k(t) \tag{9.85}$$

where $\boldsymbol{F}^{(k)}$ denotes the kth power of the diagonal matrix

$$\boldsymbol{F} = \text{diag}\{\exp(-j\beta\cos\phi_0), \ldots, \exp(-j\beta\cos\phi_{D-1})\} \tag{9.86}$$

The covariance matrix of the kth forward subarray is therefore given by

$$\boldsymbol{R}_k^f = \boldsymbol{A}\boldsymbol{F}^{(k-1)}\boldsymbol{R}_{ss}\boldsymbol{F}^{H(k-1)}\boldsymbol{A}^H + \sigma_n^2\boldsymbol{I} \tag{9.87}$$

where $\boldsymbol{R}_{ss}$ is the covariance matrix of the sources.

Based on the above, the forward averaged spatially smoothed covariance matrix $\boldsymbol{R}^f$ is defined as the sample mean of the subarray covariance matrices:

$$\boldsymbol{R}^f = \frac{1}{L}\sum_{k=0}^{L-1}\boldsymbol{R}_k^f \tag{9.88}$$

where $L=M-p+1$ is the number of subarrays. Now, substituting (9.87) in (9.88), we obtain

$$\boldsymbol{R}^f = \boldsymbol{A}\left(\frac{1}{L}\sum_{k=0}^{L-1}\boldsymbol{F}^{(k-1)}\boldsymbol{R}_{ss}^f(\boldsymbol{F}^{(k-1)})^H\right)\boldsymbol{A}^H + \sigma_n^2\boldsymbol{I} \tag{9.89}$$

where $\boldsymbol{R}^f_{ss}$ is the modified covariance matrix of the signals, given by

$$\boldsymbol{R}_{ss}^f = \frac{1}{L}\sum_{k=0}^{L-1}\boldsymbol{F}^{(k-1)}\boldsymbol{R}_{ss}(\boldsymbol{F}^{(k-1)})^H \tag{9.90}$$

For $L \geq D$, the covariance matrix $\boldsymbol{R}^f_{ss}$ will be nonsingular regardless of the coherence of the signals [Pil89a].

The price paid for detection of coherent signals using forward averaging spatial smoothing is the reduction in the array aperture. An M element array can detect only $M/2$ coherent signals using MUSIC with forward averaging spatial smoothing as opposed to $M-1$ noncoherent signals that can be detected by conventional MUSIC.

Pillar and Kwon [Pil89a] proved that by making use of a set of forward and conjugate backward subarrays simultaneously, it is possible to detect up to $2M/3$ coherent signals. In this scheme, in addition to splitting the array into overlapping forward subarrays, it is also split into overlapping backward arrays such that the first backward subarray is formed using elements $\{M, M-1, ..., M-p+1\}$, the second subarray is formed using elements $\{M-1, M-2, ..., M-p\}$, and so on.

Similar to (9.85), the complex conjugate of the received signal vector at the kth backward subarray can be expressed as

$$\begin{aligned} \boldsymbol{u}_k^b &= [u^*_{M-k+1},\ u^*_{M-k},\ \ldots,\ u^*_{p-k+1}]^T \\ &= \boldsymbol{A}\boldsymbol{F}^{k-1}(\boldsymbol{F}^{M-1}\boldsymbol{s})^* + \boldsymbol{n}_k^*, \qquad (0 \le k \le L-1) \end{aligned} \tag{9.91}$$

where $\boldsymbol{F}$ is defined in (9.86). The covariance matrix of the kth backward subarray is therefore given by

$$\boldsymbol{R}_k^b = \boldsymbol{A}\boldsymbol{F}^{k-1}\boldsymbol{R}_{\tilde{s}s}(\boldsymbol{F}^{k-1})^H\boldsymbol{A}^H + \sigma_n^2\boldsymbol{I} \tag{9.92}$$

where

$$\begin{aligned} \boldsymbol{R}_{\tilde{s}s} &= \boldsymbol{F}^{-(M-1)}E[\boldsymbol{s}^*\boldsymbol{s}^T](\boldsymbol{F}^{-(M-1)})^H \\ &= \boldsymbol{F}^{-(M-1)}\boldsymbol{R}_{ss}^*(\boldsymbol{F}^{-(M-1)})^H \end{aligned} \tag{9.93}$$

Now the spatially smoothed backward subarray matrix $\boldsymbol{R}^b$ can be defined as

$$\boldsymbol{R}^b = \frac{1}{L}\sum_{k=0}^{L-1}\boldsymbol{R}_k^b = \boldsymbol{A}\boldsymbol{R}_{ss}^b\boldsymbol{A}^H + \sigma_n^2\boldsymbol{I} \tag{9.94}$$

It can be shown that the backward spatially smoothed covariance matrix $\boldsymbol{R}^b$ will be of full rank as long as $\boldsymbol{R}_{ss}^b$ is non-singular, and the non-singularity of $\boldsymbol{R}_{ss}^b$ is guaranteed whenever $L \ge D$ [Pil89a].

Now the forward/conjugate backward smoothed covariance matrix $\hat{R}$ is defined as the mean of $\boldsymbol{R}^f$ and $\boldsymbol{R}^b$, i.e,

$$\hat{R} = \frac{\boldsymbol{R}^f + \boldsymbol{R}^b}{2} \tag{9.95}$$

Using an M element array, applying MUSIC on $\hat{R}$, it is possible to detect up to $2M/3$ coherent signals [Pil89a].

Figure 9–5 shows a comparison between conventional MUSIC and MUSIC with forward/backward spatial smoothing in a coherent multipath signal environment. Simulations with three coherent signals impinging on a 6-element uniform linear array at 60, 90, and 120 degrees show that MUSIC fails almost completely, whereas with a spatial smoothing preprocessing scheme, all of the three multipath signals are detected clearly.

9.4.2 Multidimensional MUSIC

It was stated in Section 9.2.1 that in the presence of coherent signals, the signal correlation matrix $\boldsymbol{R}_{ss}$ becomes singular, and hence violates the premise on which the MUSIC derivation is

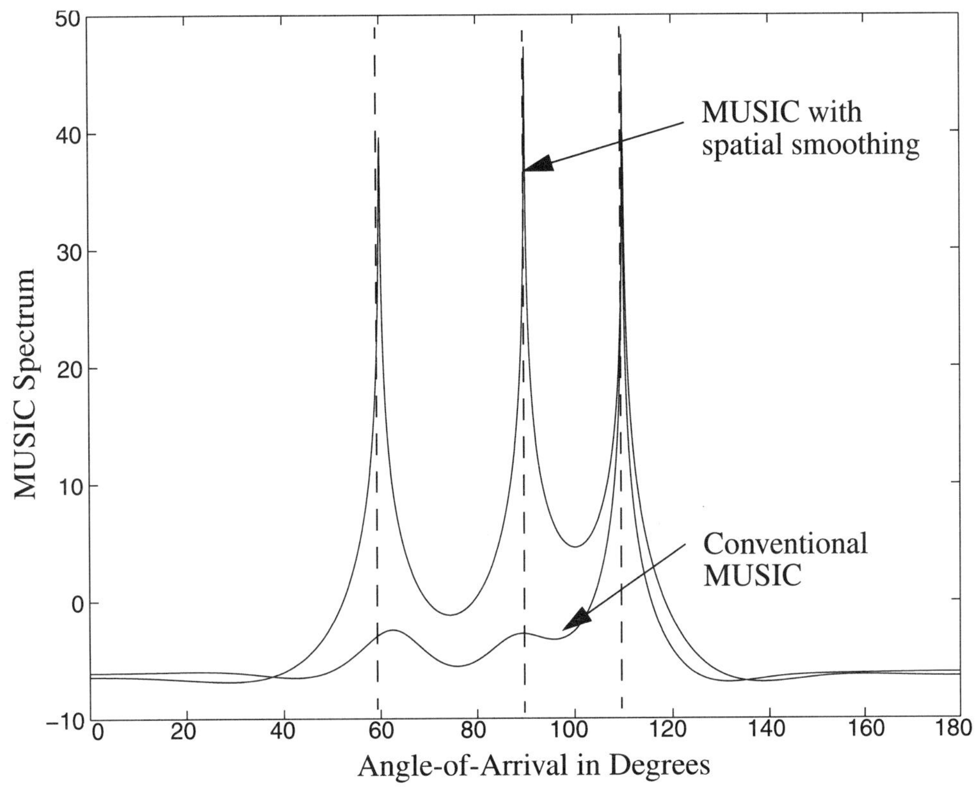

Figure 9–5 Comparison of MUSIC with and without forward/backward averaging in coherent multipath. Three coherent signals of equal power with SNRs of 20 dB arrive at a 6-element uniformly spaced array with an interelement spacing equal to half a wavelength at angles 60, 90, and 120 degrees, respectively [Muh96a].

based. However, if all of the coherent signals (typically all multipath components associated with a single source within a resolvable chip) are grouped together as a single signal, the signal correlation matrix can retain its full rank. However, now the direction vector matrix $\boldsymbol{A}$ will not consist of steering vectors corresponding to distinct Directions-Of-Arrival. Instead the columns of $\boldsymbol{A}$ will consist of spatial signatures associated with each source (group of coherent signals). Essentially, we can have $\boldsymbol{R}_{ss}$ as full rank by applying the data model of (8.5) in Chapter 8. Now the column vectors of $\boldsymbol{A}$ are linear combinations of one or more steering vectors corresponding to one or more Directions-Of-Arrival. Once the full rank status of $\boldsymbol{R}_{ss}$ is maintained, the MUSIC algorithm is valid, and the signal subspace is spanned by the spatial signature vectors

$$\boldsymbol{a}_0, \quad \boldsymbol{a}_1, \ \ldots, \ \boldsymbol{a}_{D-1} \tag{9.96}$$

which are orthogonal to the noise subspace. Computing the MUSIC spectrum now involves searching through all possible spatial signature vectors to find peaks in the spectrum. Since spatial signatures are linear combinations of steering vectors, this essentially involves a search in an

N_{mp} dimensional space, where N_{mp} is the number of components associated with a single source (group). The multi-dimensional MUSIC spectrum is given by

$$P_{MD\_MUSIC}(\phi_0, \phi_1, \ldots, \phi_{N_{mp}-1}) = \{\min_{c}[c^H a^H(\phi_1, \phi_2, \ldots, \phi_{N_{mp}}) V_n V_n^H a(\phi_1, \phi_2, \ldots, \phi_{N_{mp}}) c]\}^{-1} \tag{9.97}$$

where the vector c is defined as

$$c = \left[1, \frac{c_1}{c_0}, \frac{c_2}{c_0}, \ldots, \frac{c_{N_{mp}-1}}{c_0}\right]^T \tag{9.98}$$

where the values $\{c_i\}$ weight one steering vector relative to another.

As clearly seen from equation (9.97), as the number of multipath components increases, the complexity of the multidimensional search increases exponentially. The computational complexity of MD-MUSIC makes real-time implementation extremely difficult for more than two dimensions.

9.5 The Iterative Least Squares Projection Based CMA

The Iterative Least Squares Projection Based CMA (ILSP-CMA) is a property restoral based algorithm which can be used to jointly detect the spatial signatures and the waveforms associated with multiple sources incident on a receiver array. The ILSP-CMA is a data-efficient and cost-efficient technique which overcomes many of the problems associated with the Multi-target CMA and Multi-stage CMA algorithms [Par95].

Consider an M-element array with signals from D sources incident on it. Over a block of N snapshots, the array output can be expressed as

$$U = AS + N \tag{9.99}$$

where $U = [u(0), \ldots, u(N-1)]$ is the array input data matrix of dimension $M \times N$, $A = [a(\phi_0), \ldots, a(\phi_{D-1})]$ is the spatial signature matrix of dimension $M \times D$, $S = [s(0), \ldots, s(N-1)]$ is the signal waveform matrix of dimension $D \times N$, and $N = [n(0), \ldots, n(N-1)]$ is the noise matrix of dimension $M \times N$. Given this block formulation, the ILSP-CMA algorithm provides a means of jointly estimating the spatial signature matrix A and the signal waveforms S, given N snapshots of the array input data matrix U.

By modeling the unknown signal waveforms as deterministic quantities to be estimated, and assuming that the number of signals is known or has been estimated, the log-likelihood function of the array output data is given by [Tal94]

$$J = -\text{constant} - MN\log\sigma_n^2 - \frac{1}{\sigma_n^2}\sum_{k=0}^{N-1}\|u(k) - As(k)\|^2 \tag{9.100}$$

where σ_n^2 is the noise power. The maximum-likelihood (ML) estimator maximizes J with respect to the unknown spatial signatures A and $s(k)$, $k = 0, \ldots, N-1$ to yield the following minimization problem [Tal94]:

$$\min_{A,S} = \|U - AS\|_F^2 \tag{9.101}$$

where $\|.\|_F^2$ is the squared Frobenius norm, and the elements of $\boldsymbol{S}$ are constrained to have a constant modulus. This is a nonlinear separable optimization problem which can be solved in two steps [Gol73].

The ILSP-CMA is an efficient algorithm to solve the minimization problem of (9.101) [Par95]. Let

$$f(\boldsymbol{A}, \boldsymbol{S}) = \|\boldsymbol{U} - \boldsymbol{A}\boldsymbol{S}\|_F^2 \tag{9.102}$$

be a function of continuous matrix variables $\boldsymbol{A}$ and $\boldsymbol{S}$. Given an initial estimate $\hat{\boldsymbol{A}}$ of $\boldsymbol{A}$, the minimization of $f(\hat{\boldsymbol{A}}, \boldsymbol{S})$ with respect to continuous $\boldsymbol{S}$ is a Least Squares problem. This is a separable problem in $\boldsymbol{S}$, and the estimate of $\boldsymbol{S}$ is given by the Least Squares solution of (9.102), which is given by,

$$\hat{\boldsymbol{S}} = (\hat{\boldsymbol{A}}^H\hat{\boldsymbol{A}})^{-1}\hat{\boldsymbol{A}}^H\boldsymbol{U} \tag{9.103}$$

Each element of the solution $\hat{\boldsymbol{S}}$ is then divided by its absolute value to make each signal unit modulus. That is, each signal is projected onto the unit circle. A better estimate of $\boldsymbol{A}$ is then obtained by minimizing $f(\boldsymbol{A}, \hat{\boldsymbol{S}})$ with respect to $\boldsymbol{A}$, keeping $\hat{\boldsymbol{S}}$ fixed. This is again a Least Squares problem and the new estimate of $\boldsymbol{A}$ is given by the Least Squares solution,

$$\hat{\boldsymbol{A}} = \boldsymbol{U}\hat{\boldsymbol{S}}^H(\hat{\boldsymbol{S}}\hat{\boldsymbol{A}}^H)^{-1} \tag{9.104}$$

Since the first element in each unknown spatial signature is a real value, it can be made equal to unity by dividing the elements of the spatial signature vectors by their first element. This process is continued until $\hat{\boldsymbol{A}}$ converges.

The ILSP-CMA algorithm may be summarized as follows:

1. Given $\boldsymbol{A}_0$ (start with a random $\boldsymbol{A}_0$), $n = 0$
2. $n = n+1$
   - $\boldsymbol{S}_n = (\boldsymbol{A}_{n-1}^H\boldsymbol{A}_{n-1})^{-1}\boldsymbol{A}_{n-1}^H\boldsymbol{U}$
   - Project all elements of $[\boldsymbol{S}_n]$ to the closest values on the unit circle (hard limit)
   - $\boldsymbol{A}_n = \boldsymbol{U}\boldsymbol{S}_n^H(\boldsymbol{S}_n\boldsymbol{S}_n^H)^{-1}$
3. Divide each column of $\boldsymbol{A}_n$ by its first element value.
4. Repeat steps 2 and 3 until $\boldsymbol{A}_n$ and $\boldsymbol{A}_{n-1}$ are close enough.

9.6 The Integrated Approach to DOA Estimation

In Section 9.5 it was shown how to obtain an estimate of multiple signals and their spatial signatures using the ILSP-CMA algorithm. If a signal has only one component, its spatial signature is identical to the steering vector corresponding to its Direction-Of-Arrival. Therefore, if we have an estimate of the spatial signature of a signal with a single component, we can estimate its Direction-Of-Arrival by observing its spatial signature. The Direction-Of-Arrival can be estimated by searching through all possible steering vectors and determining the one closest in two norms to the estimated spatial signature. Mathematically, the Direction-Of-Arrival ϕ is given by

$$\phi = \arg[\min_{\phi}\|\boldsymbol{a}(\phi) - \boldsymbol{a}_{ss}\|_2] \tag{9.105}$$

where $\boldsymbol{a}(\phi)$ is the steering vector corresponding to Direction-Of-Arrival ϕ, and $\boldsymbol{a}_{ss}$ is the estimated spatial signature.

Xu and Liu proposed a novel technique to estimate the Directions-Of-Arrival of the direct and multipath components of a signal from the spatial signature [Xu95]. In subspace based algorithms, in order to determine the Directions-Of-Arrival, a covariance matrix whose signal subspace is the span of $\boldsymbol{A}$ must be constructed. If an estimate of the spatial signature matrix $\boldsymbol{A}$ is available, we can form a *spatial signature covariance matrix* $\boldsymbol{R}_{aa} = \boldsymbol{A}\boldsymbol{A}^H$ on which the eigen decomposition may equivalently be performed to obtain the Directions-Of-Arrival. In the presence of coherent signals, forward/backward averaging is required to form a spatially smoothed spatial signature covariance matrix before eigen decomposition.

When the spatial signature estimate of a source exists, Xu and Liu proposed a technique to estimate the Directions-Of-Arrival of the various components of the signal [Xu95]. By applying the standard forward/backward spatial smoothing techniques discussed in Section 9.4.1 [Pil89a] to the spatial signature vector $\boldsymbol{a}_{ss}$, a smoothed spatial signature covariance matrix can be formed as

$$\boldsymbol{R}_{fb}(K) = \boldsymbol{R}_f(K) + \boldsymbol{J}\boldsymbol{R}_f^*(K)\boldsymbol{J} \tag{9.106}$$

where $\boldsymbol{J}$ is the permutation matrix with all zeros except ones in the anti-diagonal elements, K is the smoothing factor (number of subarrays), and

$$\boldsymbol{R}_f(K) = \sum_{i=0}^{K-1}(\boldsymbol{a}_{ss}(i{:}M-K+i)\boldsymbol{a}_{ss}(i{:}M-K+1)^H) \tag{9.107}$$

Subspace based algorithms such as MUSIC and ESPRIT can now be applied to this smoothed spatial signature covariance matrix, and up to $2M^2/3$ DOAs from coherent sources can be estimated. Note that by using ILSP-CMA to estimate M spatial signatures, and applying subspace techniques such as MUSIC or ESPRIT to the smoothed spatial signature covariance matrices of each spatial signature, it is possible to estimate up to $2M^2/3$ DOA's using an M-element array. In a situation where there are multiple cochannel users with each user having multiple components, this technique can determine the Directions-Of-Arrival of multiple components of multiple users and associate each component with the correct user [Muh96b].

Computer simulations were run to study the performance of ILSP-CMA for spatial signature estimation along with MUSIC with forward and conjugate backward averaging. Figure 9–6 shows the estimated MUSIC spectrum for the case of a uniformly spaced linear array with six elements with an interelement spacing of a half wavelength. For spatial smoothing the array was divided into two overlapping 5-element subarrays. Six uncorrelated narrowband signals, each with a direct and three multipath components, are incident on the array. The multipath components are 10 dB below the direct component and the direct signal-to-noise ratio was 20 dB. The first signal had components arriving at 80, 110, 140, and 170 degrees, the second signal had components at 20, 90, 130, and 150 degrees, the third signal had components at 30, 60, 90, and

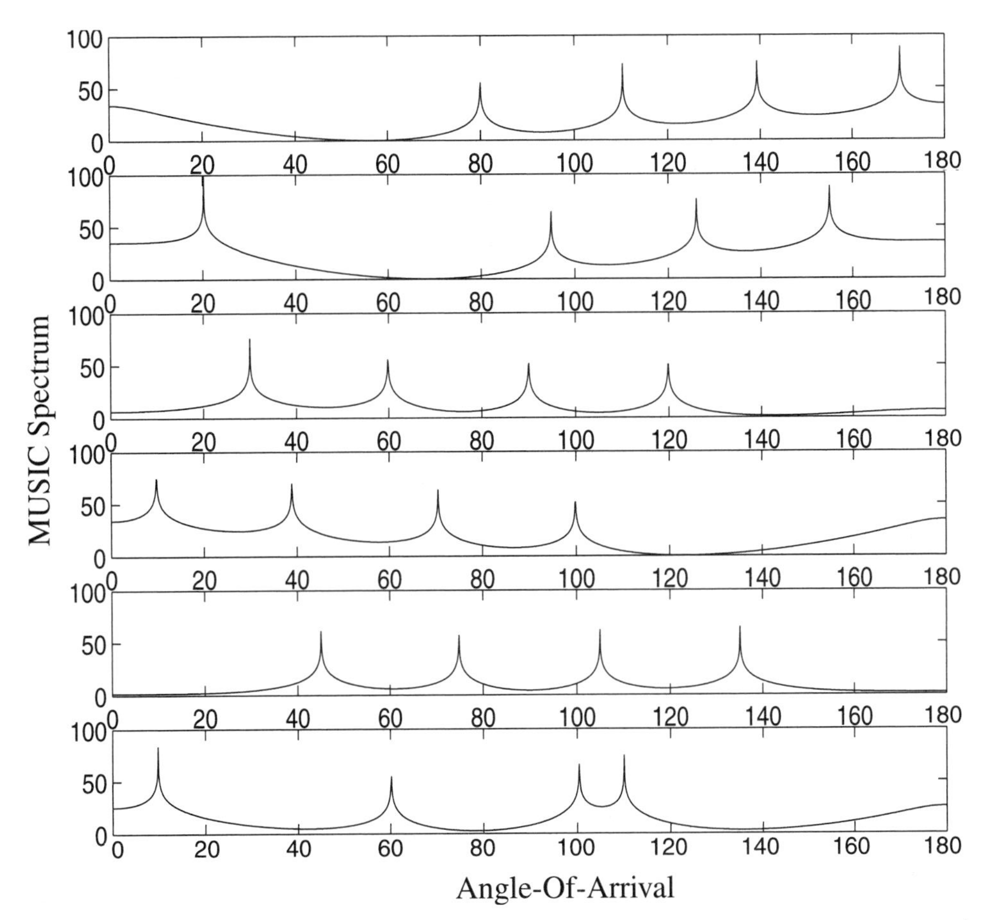

Figure 9–6 Example of spatial spectrum estimated using ILSP-CMA for spatial signature estimation, followed by MUSIC with forward/backward averaging. The six element array is able to resolve 24 direct and multipath components and associate each component to the appropriate signal (user) [Muh96a].

120 degrees, the fourth signal had components at 10, 40, 70, and 100 degrees, the fifth signal had components at 45, 75, 105, and 135 degrees, and the sixth signal had components at 10, 60, 100, and 110 degrees. Figure 9–6 shows that ILSP-CMA, along with forward and conjugate backward averaged MUSIC, is able to resolve a total of 24 signal components [Muh96b].

9.7 Detection of Number of Sources in Eigen Decomposition

Many of the DOA estimation algorithms described earlier require that the number of sources be known or estimated. Detection of the number of sources impinging on the array is a key step in most of the superresolution DOA estimation techniques. In the eigen decomposition based techniques, an estimate of the number of sources is obtained from an estimate of the number of repeated smallest eigenvalues. Since in practice the input sample covariance matrix is formed using a finite set of samples, the smallest eigenvalues are not exactly equal. Various statistical methods have been proposed to test for the equality or closeness of eigenvalues, which can be used to estimate the number of sources.

9.7.1 The SH, MDL and AIC Criteria

Anderson [And63] showed that a useful statistic for testing the closeness of eigenvalues is

$$L(d) = -N(M-d)\log\left\{\frac{\left(\prod_{i=d+1}^{M} \hat{\lambda}_i\right)^{\frac{1}{(M-d)}}}{\frac{1}{M-d}\sum_{i=d+1}^{M} \hat{\lambda}_i}\right\} \tag{9.108}$$

where $\hat{\lambda}_i$ is the estimated k^{th} eigenvalue, d is the hypothesized estimate of the number of signals, M is the number of elements in the array, and N is the size of the sample data block. In (9.108), the closeness of the eigenvalues is measured as the ratio of their geometric mean to their arithmetic mean. By setting up a subjective threshold γ_d, a sequential hypothesis (SH) test can be performed and the first d such that $L(d) < \gamma_d$ can be taken as the estimate $\hat{D}$ of the number of signals D.

While the above sequential hypothesis testing to determine the number of sources is computationally attractive, the need to set up a subjective threshold is a major disadvantage. Wax and Kailath [Sch93d] proposed two other detection schemes based on the application of the Akaike information theoretic criteria (AIC) [Sch93d] and the Rissanen Minimum Descriptive Length (MDL) criteria [Ris78]. These methods do not require a subjective threshold, and the number of sources is determined as the value for which the AIC or MDL criteria is minimized.

In the AIC-based approach, the number of signals $\hat{D}$, is determined as the value of $d \in \{0, 1, \ldots M-1\}$ which minimizes the following criterion

$$AIC(d) = -\log\left\{\frac{\prod_{i=d+1}^{M} \lambda_i^{\frac{1}{(M-d)}}}{\frac{1}{M-d}\sum_{i=d+1}^{M} \lambda_i}\right\}^{(M-d)N} + d(2M-d) \tag{9.109}$$

where λ_i are the eigenvalues of the sample covariance matrix $\hat{\boldsymbol{R}}_{uu}$, N is the number of snapshots used to compute $\hat{\boldsymbol{R}}_{uu}$, and M is the number of elements in the array. The first term in equation (9.109) is derived directly from the log-likelihood function, and the second term is the penalty factor added by the AIC criterion.

In the MDL-based approach, the number of signals is determined as the argument which minimizes the following criterion:

$$MDL(d) = -\log\left\{\frac{\prod_{i=d+1}^{M} \lambda_i^{\frac{1}{(M-d)}}}{\frac{1}{M-d}\sum_{i=d+1}^{M} \lambda_i}\right\}^{(M-d)N} + \frac{1}{2}d(2M-d)\log N \tag{9.110}$$

Here again, the first term is derived directly from the log-likelihood function, and the second term is the penalty factor added by the MDL criterion. Wax and Kailath [Wax85] showed through simulations that the MDL criterion yields a consistent estimate of the number of signals, and the AIC yields an inconsistent estimate that tends, asymptotically, to overestimate the number of signals.

Xu *et al* [Xu94a] showed that the SH, AIC and MDL procedures cannot be applied directly to situations where spatial smoothing preprocessing is involved. The spatial smoothing preprocessing operation complicates the source order detection process. Xu *et al* modified the AIC and MDL criterion so that they can be applied correctly to situations where spatial smoothing is performed. Essentially, this leads to a change in the penalty function associated with the AIC and MDL criterion. If the penalty functions (i.e., the second term in (9.109) and (9.110)) are denoted $p_{AIC}(d)$ and $p_{MDL}(d)$, respectively, they must be modified as shown in Table 9–1 for the various spatial smoothing cases.

Table 9–1 Modified Values of the Penalty function in AIC and MDL Detection Schemes

Spatial Smoothing Type	$P_{AIC^{(d)}}$	$P_{MDL^{(d)}}$
Forward only	$d(2M\text{-}2d\text{+}1)$	$0.5d(2M\text{-}2d\text{+}1)\log N$
Forward/conjugate Backward	$0.5d(2M\text{-}d\text{+}1)$	$0.25d(2M\text{-}d\text{+}1)\log N$

9.7.2 Order Estimation Using Transformed Gerschgorin Radii

Wu *et al* [Wu95] proposed a new technique for source order estimation based on the effective use of the Gerschgorin radii of a unitary transformed input covariance matrix. The Gerschgorin theorem on eigenvalues of a matrix provides a method for estimating the location of eigenvalues from the values of the matrix elements. For an $M \times M$ matrix $\mathbf{A}=\{a_{ij}\}$, Gerschgorin proved that all of the eigenvalues of the matrix are contained in the union of M disks $\boldsymbol{O}_i$, $i = 1, \ldots, M$. These disks are centered at a_{ii}, and have radii, called the Gerschgorin radii r_i, equal to the sum of the magnitudes of all elements of the ith row vector, excluding the ith element. That is,

$$r_i = \sum_{j=1, j \neq i}^{M} |a_{ij}| \tag{9.111}$$

In other words, the eigenvalues of a matrix are located within the Gerschgorin disks which represent the collection of points in the complex plane whose distance to a_{ii} is, at most, r_i. That is, $\boldsymbol{O}_i$ represents the collection of complex numbers z with the property of

$$r_i \geq |z - a_{ii}| \tag{9.112}$$

Wu *et al* observed that at low signal to noise ratio conditions, the eigenvalues of the input covariance matrix $\boldsymbol{R}_{\text{uu}}$ are spread across a large range and the Gerschgorin disks for the matrix tightly overlap. Through a proper unitary transformation which preserves the eigenvalues, the overlap in the Gerschgorin disks can be reduced and hence can be effectively used for source number detection. The idea is to rotate the covariance matrix so that its Gerschgorin disks can be formed into two distinct signal and noise constellations. The source collection with larger Gerschgorin radii will contain exactly M largest signal eigenvalues, and the noise collection with small Gerschgorin radii will contain the remaining noise eigenvalues. That is, a unitary transformation should be chosen so that the noise Gerschgorin disks of the transformed covariance matrix are small and as far away from the signal Gerschgorin disks as possible. Once this is achieved, it is relatively easy to estimate the source order by classification of disks.

In the proposed method, the covariance matrix is first partitioned as follows:

$$\boldsymbol{R}_{uu} = \begin{bmatrix} r_{11} & r_{12} & \cdots & r_{1M} \\ r_{21} & r_{22} & \cdots & r_{2M} \\ & & \cdots & \\ r_{M1} & r_{M2} & \cdots & r_{MM} \end{bmatrix} = \left[\begin{array}{c|c} \boldsymbol{R}_1 & \boldsymbol{r} \\ \hline \boldsymbol{r}^H & r_{MM} \end{array}\right] \tag{9.113}$$

where $\boldsymbol{R}_1$ is the leading principal submatrix of $\boldsymbol{R}_{\text{uu}}$. The reduced covariance matrix $\boldsymbol{R}_1$ can also be decomposed by its eigen-structure as

$$\boldsymbol{R}_1 = \boldsymbol{U}_1 \boldsymbol{D}_1 \boldsymbol{U}_1^H \tag{9.114}$$

where $\boldsymbol{U}_1$ is the $M-1 \times M-1$ unitary matrix formed by the eigenvectors of $\boldsymbol{R}_1$ as

$$\boldsymbol{U}_1 = [\boldsymbol{v}'_1, \boldsymbol{v}'_2, \ldots, \boldsymbol{v}'_D, \ldots, \boldsymbol{v}'_{M-1}] \tag{9.115}$$

and $\boldsymbol{D}_1$ is the diagonal matrix constructed from the corresponding eigenvalues as

$$\boldsymbol{D}_1 = diag(\lambda'_1 \lambda'_1 \ldots \lambda'_D \ldots \lambda'_{M-1}) \tag{9.116}$$

where $\lambda'_1 \geq \lambda'_2 \geq \ldots \geq \lambda'_D \geq \ldots \geq \lambda'_{M-1}$. If the eigenvalues of the original covariance matrix $\boldsymbol{R}_{uu}$ are denoted $\lambda_1, \lambda_2, \ldots, \lambda_D, \ldots, \lambda_M$, it can be shown that the eigenvalues of $\boldsymbol{R}_{uu}$ and $\boldsymbol{R}_1$ satisfy the *interlacing property*,

$$\lambda_1 \geq \lambda'_1 \geq \lambda_2 \geq \lambda'_2 \geq \ldots \geq \lambda_D \geq \lambda'_D \geq \ldots \geq \lambda_{M-1} \geq \lambda'_{M-1} \geq \lambda_M \tag{9.117}$$

By defining a unitary transformation $\boldsymbol{U}$

$$\boldsymbol{U} = \begin{bmatrix} \boldsymbol{U}_1 & 0 \\ 0^T & 1 \end{bmatrix} \tag{9.118}$$

the transformed input covariance matrix becomes

$$\boldsymbol{Q}_{uu} = \boldsymbol{U}^H \boldsymbol{R}_{uu} \boldsymbol{U} = \begin{bmatrix} \boldsymbol{U}_1^H \boldsymbol{R}_{uu} \boldsymbol{U}_1 & \boldsymbol{U}_1^H \boldsymbol{r} \\ \boldsymbol{r}^H \boldsymbol{U}_1 & r_{MM} \end{bmatrix} = \begin{bmatrix} \boldsymbol{D}_1 & \boldsymbol{U}_1^H \boldsymbol{r} \\ \boldsymbol{r}^H \boldsymbol{U}_1 & r_{MM} \end{bmatrix} \tag{9.119}$$

which can be shown to be equal to [Wu95]

$$\boldsymbol{Q}_{uu} = \begin{bmatrix} \lambda_1 & 0 & 0 & \dots & 0 & \rho_1 \\ 0 & \lambda_2' & 0 & \dots & 0 & \rho_2 \\ 0 & 0 & \lambda_3' & \dots & 0 & \rho_3 \\ \dots & \dots & \dots & \dots & \dots & \dots \\ 0 & 0 & 0 & \dots & \lambda_{M-1}' & \rho_{M-1} \\ \rho_1^* & \rho_2^* & \rho_3^* & \dots & \rho_{M-1} & r_{MM} \end{bmatrix} \tag{9.120}$$

where

$$\rho_i = \boldsymbol{v}_i^{'H} \boldsymbol{r}, \quad i = 1, 2, \dots, M-1 \tag{9.121}$$

The Gerschgorin disks of the transformed covariance matrix $\boldsymbol{Q}_{uu}$ possess the Gerschgorin radii

$$r_i = |\rho_i| = |\boldsymbol{v}_i^{'H} \boldsymbol{r}| \text{ for } i = 1, 2, \dots, M-1 \tag{9.122}$$

By incorporating the Gerschgorin radii information into the log-likelihood function, Wu *et al* derived a new source order estimator function which was called the Gerschgorin Likelihood Estimator (GLE). The GLE function is given by [Wu95]

$$GLE(d) \approx -N(M-1-k)\log\left[\frac{\left(\prod_{i=d+1}^{M-1} \lambda_i'\right)^{\frac{1}{M-d-1}}}{\frac{1}{M-d-1}\sum_{i=d+1}^{M-1} \lambda_i'}\right] - N\log\left(r_{MM} - \sum_{i=1}^{d} \frac{r_i^2}{\lambda_i'}\right) \tag{9.123}$$

where N is the number of data snapshots.

By applying the AIC and MDL penalty factors to equation (9.123), two source order estimation functions can be obtained [Muh96a][Rap98]. They are called the Gerschgorin AIC (GAIC) and Gerschgorin MDL (GMDL) criteria respectively, and are given by

$$GAIC(d) \approx -N(M-1-k)\log\left[\frac{\left(\prod_{i=d+1}^{M-1}\lambda_i'\right)^{\frac{1}{M-d-1}}}{\frac{1}{M-d-1}\sum_{i=d+1}^{M-1}\lambda_i'}\right] - N\log\left(r_{MM} - \sum_{i=1}^{d}\frac{r_i^2}{\lambda_i'}\right) + d^2 + d \tag{9.124}$$

$$GMDL(d) \approx -N(M-1-k)\log\left[\frac{\left(\prod_{i=d+1}^{M-1}\lambda_i'\right)^{\frac{1}{M-d-1}}}{\frac{1}{M-d-1}\sum_{i=d+1}^{M-1}\lambda_i'}\right] - N\log\left(r_{MM} - \sum_{i=1}^{d}\frac{r_i^2}{\lambda_i'}\right) + 0.5(d^2 + d)\log N \tag{9.125}$$

The number of sources is determined as the argument which minimizes these functions. It is shown in [Wu95] that the GAIC and GMDL estimators are more consistent than the simple AIC and MDL estimators. Further, since the GAIC and GMDL techniques involve the eigen decomposition of an $M-1 \times M-1$ matrix as opposed to a $M \times M$ matrix required in AIC and MDL techniques, they are computationally more efficient. However, since they are based on a submatrix of $\boldsymbol{R}_{uu}$, they can detect only up to $M-2$ sources, as opposed to $M-1$ sources that can be detected using the regular AIC and MDL criteria.

All the source estimators discussed so far are based on the assumptions of Gaussian and spatially white noise. Wu *et al* have also derived modified versions of GAIC and GMDL which they call MGAIC and MGMDL respectively, which yield consistent estimates under non-white noise conditions and when only a few snapshots of data are available [Wu95].

9.8 Summary

In this chapter, we described various techniques for estimating the Direction-Of-Arrival of radio signals impinging on an antenna array. A survey of the most promising and classic algorithms used for DOA estimation was presented. A discussion of the source order estimation algorithms was presented in Section 9.7. The final chapter of this text explores how these DOA techniques may be applied to the timely subject of position location.

CHAPTER 10

RF Position Location Systems

As discussed in [Rap96b], [Miz96b], and [Ree98] radio frequency *Position Location* (PL) systems can be classified into two broad categories: *Direction Finding* (DF) and *Range-Based* systems. An extensive list of PL system classifications is provided in [Law76]. These DF and range-based PL systems and techniques may be used individually or in combination, in a number of different configurations, to produce a PL solution.

Direction finding systems estimate the position location of a mobile source by measuring the Direction-Of-Arrival (DOA), or Angle-Of-Arrival (AOA), of the source's signal, using methods described in Chapter 9. The DOA measurement restricts the location of the source along a line in the estimated DOA. When multiple DOA measurements are made simultaneously by multiple base stations, a triangulation method may be used to form a location estimate of the source at the intersection of these *Lines-Of-Bearing* (LOB). In theory, DF systems require only two receiving sensors to locate a mobile user, but in practice, finite angular resolution, multipath, and noise often dictate the need for more than two sensors.

Range-based PL systems may be categorized as either ranging, range-sum, or range-difference systems. The type of measurement used in each of these systems identifies a geometry for the position location solution. Several base station receivers are used to simultaneously record propagation delay, or range, measurements between a mobile transmitter and the fixed receivers.

Ranging PL systems locate a mobile source by measuring the absolute or differential distance between a source and a receiver. The distance, d, traveled by a propagating wave is given by $d = c\tau$, where τ is the propagation delay and c is the speed of light, 3×10^8 m/s. Thus, range estimates are made by measuring the Time-Of-Arrival (TOA) of the signal propagating between the mobile user and each base station receiver. The TOA estimate defines a sphere of constant range around the receiver. The intersection of multiple spheres produced by multiple range measurements from multiple base station receivers provides the position location estimate of the mobile user. Consequently, ranging systems are also known as *TOA* or *spherical* PL systems. Most practical ranging systems are unable to measure the exact range between the user and a base station directly, and as a result, range measurements contain a bias term. This bias term can be calculated using an additional range measurement by an additional base station. Ranging systems of this type are often called *pseudo-range* systems [Par96].

Range-sum PL systems measure the relative sum of ranges between the source and the fixed receivers. These systems measure the *Time-Sum-Of-Arrival* (TSOA) of the propagating signal from a mobile user to two base station receivers to produce a range-sum measurement. The range-sum estimate defines an ellipsoid with foci at two receivers, and when multiple range-sum measurements are obtained, the position location estimate of the user occurs at the intersection of the ellipsoids. Consequently, range-sum PL systems are also known as *TSOA* or *elliptical* PL systems.

Range-difference PL systems measure the relative difference in ranges between a source and receiver. These systems measure the *Time Difference Of Arrival* (TDOA) of the propagating signal from a mobile to two base station receivers to produce a range-difference measurement. The range-difference measurement defines a hyperboloid of constant range-difference with the base stations at the foci. When multiple range-difference measurements are obtained, producing multiple hyperboloids, the position location estimate of the user occurs at the intersection of the hyperboloids. Consequently, range-difference PL systems are also known as *TDOA* or *hyperbolic* PL systems.

Range-based systems can also be classified as either *multilateration* or *trilateration* systems. Multilateration PL systems are those systems which utilize measurements from four or more base station receivers to estimate the three-dimensional (3-D) location of the mobile user. In a multilateration hyperbolic system, four or more base stations produce three or more range-difference measurements. Trilateration PL systems are those which utilize measurements from three base station receivers to estimate the two-dimensional (2-D) location of the mobile transmitter. In a trilateration ranging PL system, three range measurements are produced from the three base stations, whereas in a trilateration hyperbolic PL system, two range-difference measurements are produced from three base stations. The additional measurement by ranging systems is required to reduce the ambiguities due to multipath, signal degradation, and noise.

Position location systems which are able to measure the change in frequency of a transmitted signal are called *Doppler systems*. When either the source or receiver is moving, the received signal is subjected to a Doppler shift. This change in frequency is proportional to the relative direction and velocity of movement. By measuring the change in frequency, the rate of change between a mobile and base station can be determined. If the trajectory of a moving base station, such as a satellite-based PCS base station, is known, then the user's location can be uniquely determined from the Doppler frequency changes. Doppler systems generally require that the velocity of either the receiver station or user be high enough to generate an easily resolvable Doppler shift. Consequently, systems of this type are generally applicable to non-geosynchronous satellite-based PL systems or fixed terrestrial applications which estimate the position of high velocity vehicles in high tier cellular and PCS systems.

10.1 Direction Finding PL Systems

Direction finding (DF) systems utilize antenna arrays and Direction-Of-Arrival (DOA) estimation techniques to determine the direction of the signal of interest. The DOA measurement restricts the source location along a line in the estimated DOA, which is called a Line-of-Bearing (LOB). When multiple DOA measurements from multiple base stations are used in a triangulation configuration, the location estimate of the source is obtained at the intersection of these lines. Figure 10–1 illustrates the two-dimensional (2-D) PL solution of DF systems. While only two DOA estimates are required to estimate the PL of a source, multiple DOA estimates are commonly used to improve the estimation accuracy.

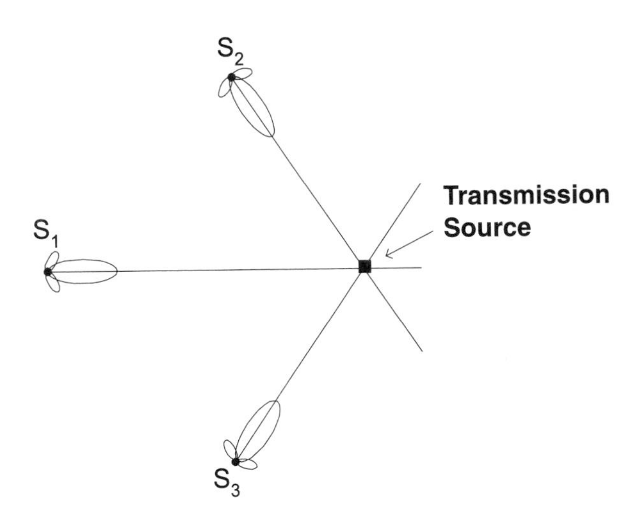

Figure 10–1 2-D Direction Finding Position Location Solution, where the Lines-of-Bearing intersect the source.

Direction-Of-Arrival estimation is performed by algorithms discussed in Chapter 9. Antenna arrays which either use superresolution techniques or arrays with large apertures and a large number of elements (i.e., long baseline) are able to provide very high angular resolution.

DOA estimation can suffer from mobile radio channel impairments. In DF systems which are not able to estimate the delay of multipath components, the direction to the source is often approximated by using the DOA of the strongest arriving signal component. In systems that are able to estimate multipath component delays, as in the case of the spatial processing Rake receiver, described in Chapter 4, the direction to the source is estimated using the first arriving resolvable multipath component.

In shadowed environments often encountered in urban areas, the true LOS signal path may be obstructed by the surrounding environment and only multipath components of the signal may be detectable. In this case, the DOA estimate will be in the direction of either the strongest or the earliest arriving resolvable multipath component, which leads to errors in the DOA estimate. Depending on the transmitter-receiver distance, these errors in the DOA estimate can lead to dramatic errors in the PL estimate. Even if the LOS signal is available, as described in Chapter 8, multipath can complicate accurate estimation of the DOA of the direct path. While angular accuracy of a few degrees is possible with these techniques, additional position location estimates based on ranging are usually needed for highly reliable position location. Furthermore, practical algorithms for DOA PL often have accuracies which depend on the geometric relationships between the receivers and the source, even when a line-of-sight exists to each receiver [McG89].

10.2 True Ranging PL Systems

Ranging PL systems measure the absolute distance between a mobile source and a set of base stations through the use of Time-Of-Arrival (TOA) measurements. The TOA measurements are related to range estimates that define a sphere around the receiver. When measurements are made from receivers with known locations, the spheres described by the range measurements intersect at a unique point indicating the position location estimate of the source. Figure 10–2 illustrates the three-dimensional (3-D) solution of the ranging PL system, where three sensors (S_1, S_2, S_3) are used to receive the signal from the source. If the spheres described by the range measurements intersect at more than one point, the result is an ambiguous solution to the position location estimate. Redundant range measurements, resulting in a multilateration ranging PL estimation, are commonly made to reduce or eliminate PL ambiguities.

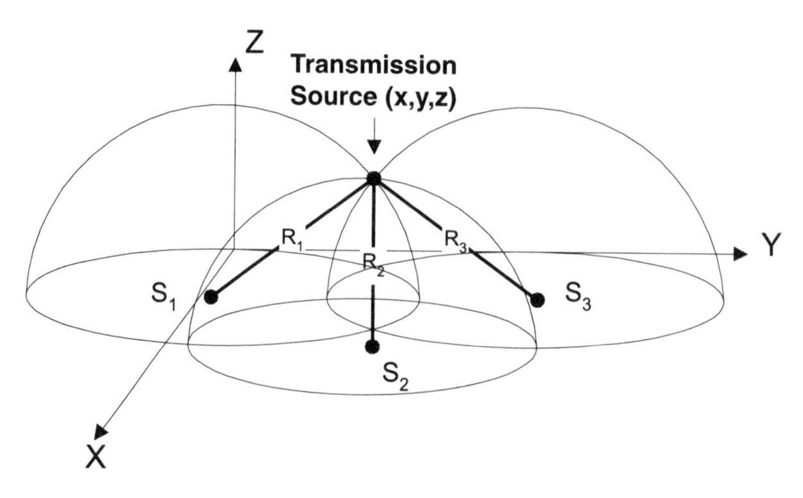

Figure 10–2 3-D Ranging Position Location Solution.

To illustrate the ranging PL concept, consider a 3-D ranging PL system using N base stations. The Time-of-Arrival of a signal at each receiver is estimated and related to the range measurement by the relationship

$$R_i = c\tau_i \tag{10.1}$$

where R_i is the range measurement, c is the signal propagation speed and τ_i is the TOA estimate at the ith receiver. The mathematical relationships between range measurements at N base stations, the coordinates of the known base station locations, and the coordinates of the source are

$$R_i = \sqrt{(X_i - x)^2 + (Y_i - y)^2 + (Z_i - z)^2} \text{ for } i=1, 2, 3, \ldots, N \tag{10.2}$$

where (X_i, Y_i, Z_i) are the geographical coordinates of the ith base station receiver, R_i is the ith range estimate to the source, and (x,y,z) is the location of the subscriber. Equation (10.2) defines a $N \times 3$ set of nonlinear equations whose solution is the location coordinates of the source. If the number of unknowns, or coordinates of the source to be solved, is equal to the number of range measurements, the set of equations is *consistent* and a unique solution exists. However, if redundant measurements produce more range measurements than the number of unknowns, then the system may be *inconsistent* and a unique solution may or may not exist. This generally requires that an error criterion be selected and iterative techniques be employed to produce a solution. A Least Squares (LS) fit is commonly used to simultaneously solve these equations for both the position location and error coefficients.

Accurate time or phase measurements in ranging PL systems require strict clock synchronization between the source and the base stations. This is accomplished through the use of stable clocks, such as rubidium or cesium standard clocks, or received signals from GPS satellites at the base station receivers. In IS-95, system time is accurately tracked at each base station via GPS. This system time is sent to each mobile via the paging channel, so it is possible to implement a highly stable clock at all mobiles and base stations, over short time spans. As shown in Figure 2–17, the mobile subscriber in an IS-95 system can measure the one-way path delay to every base station in its active set by comparing the received short code PN-sequence offset with the chip offset corresponding to zero propagation delay, which can be derived from known system time. Alternatively, as illustrated in Figure 2–17, a base station can measure the round trip propagation delay, 2τ, by comparing the chip-offset of the short code received from a subscriber with the local system time at the base station.

A disadvantage of the ranging PL technique is that accuracy is very dependent on system geometry. Highest accuracies are attained when all ranging spheres intersect at right angles with each other, which unambiguously defines the location of the source. Degradation in performance is experienced as the intersections deviate from this angle. For systems with fixed receivers and moving sources, such as cellular and PCS systems, the optimum situation is rarely attained. Another disadvantage of this PL technique is that the errors in the TOA estimate common to all receivers will degrade the PL estimate.

10.3 Elliptical PL Systems

Elliptical PL systems locate a source by the intersection of ellipsoids described by the range-sum measurements between multiple receiver sensors. Figure 10–3 illustrates the 2-D solution of an elliptical location system. The range-sum is determined from the sum of the arrival times at multiple receivers. The relationship between range-sum, $R_{i,j}$, and the TOA between base station receivers is given by

$$R_{i,j} = c\tau_{i,j} = c(\tau_i + \tau_j) = R_i + R_j \tag{10.3}$$

where c is the signal propagation speed and $\tau_{i,j}$ is the sum of the TOAs at receiver i and j. The range-sum measurement restricts the possible source locations to an ellipsoid. The ellipsoids that describe the range-sum between receivers is given by

$$R_{i,j} = \sqrt{(X_i - x)^2 + (Y_i - y)^2 + (Z_i - z)^2} + \sqrt{(X_j - x)^2 + (Y_j - y)^2 + (Z_j - z)^2} \tag{10.4}$$

where (X_i, Y_i, Z_i) and (X_j, Y_j, Z_j) define the locations of base station receivers i and j, and (x, y, z) is the position location estimate of the source. A source location can be uniquely determined by the intersection of three or more ellipsoids. Redundant range-sum measurements can be made to improve the accuracy and resolve location solution ambiguities. This method offers the advantage of not requiring high precision clocks at the mobile.

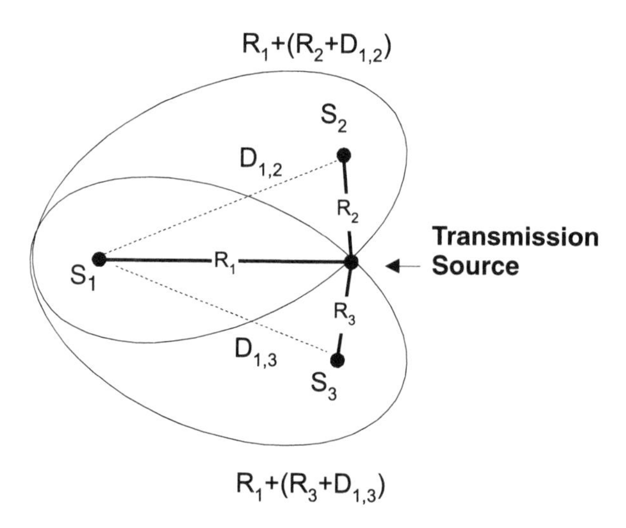

Figure 10–3 2-D Elliptical Position Location Solution.

In a practical implementation of this approach, a base station may transmit a test probe to the mobile and have the mobile re-transmit the probe to other base station receivers. If the turn-around time of the mobile re-transmission is known exactly, and if each base station is synchronized to GPS-time, an elliptical PL system can be implemented.

10.4 Hyperbolic PL Systems

Hyperbolic position location systems estimate the location of a source by the intersection of hyperboloids, which are the set of range-difference measurements between three or more base stations. The range-difference between two receivers is determined by measuring the difference in Time-Of-Arrival of a signal between them. The relationship between range-difference and the TDOA between receivers is given by

$$R_{i,j} = c\tau_{i,j} = c(\tau_i - \tau_j) = R_i - R_j, \tag{10.5}$$

where $\tau_{i,j}$ is the TDOA between receiver i and j. The TDOA estimate, in the absence of noise and interference, restricts the possible source locations to a hyperboloid of revolution with the receiver as the foci. Figure 10–4 illustrates a 2-D hyperbolic position location solution. In a 3-D system, the hyperboloids that describe the range-difference, $R_{i,j}$, between receivers are given by

$$\begin{aligned} R_{i,j} = & \sqrt{(X_i - x)^2 + (Y_i - y)^2 + (Z_i - z)^2} \\ & - \sqrt{(X_j - x)^2 + (Y_j - y)^2 + (Z_j - z)^2} \end{aligned} \tag{10.6}$$

where (X_i, Y_i, Z_i) and (X_j, Y_j, Z_j) define the locations of base station receivers i and j, respectively, $R_{i,j}$ is the range-difference measurement between base stations i and j, and (x, y, z) are the unknown coordinates of the source. If the number of unknowns, or coordinates of the source to be determined, is equal to the number of equations, or range-difference measurements, then the system is consistent and a unique solution exists. However, if redundant range-difference measurements are made, then the system may be inconsistent and a unique solution may or may not exist. As in the ranging PL approach, techniques such as Least Squares can be used to find a solution for such an overdetermined problem.

If the source and receivers are coplanar, the two-dimensional (2-D) source location can be estimated from the intersection of two or more hyperboloids produced from three or more TDOA measurements, resulting in a hyperbolic trilateration solution. Three-dimensional (3-D) source location estimation is produced by the intersection of three or more independently generated hyperboloids generated from four or more TDOA measurements, resulting in a hyperbolic multilateration solution.

If the hyperbola determined from multiple receivers intersect at more than one point, then there is ambiguity in the estimated position of the source. This location ambiguity may be resolved by using *a priori* information about the approximate source location (such as which cell site is serving the mobile), Lines-of-Bearing measurements at one or more of the base station receivers, or redundant range-difference measurements at additional base stations to generate additional hyperbolas.

A major advantage of the TDOA, or Hyperbolic PL, method is that it does not require knowledge of the transmit time from the source, as do TOA methods. Consequently, strict clock synchronization between the source and receiver is not required. As a result, hyperbolic position location techniques may not require additional hardware or software implementation within the mobile unit. However, clock synchronization is required of all receivers used for the PL esti-

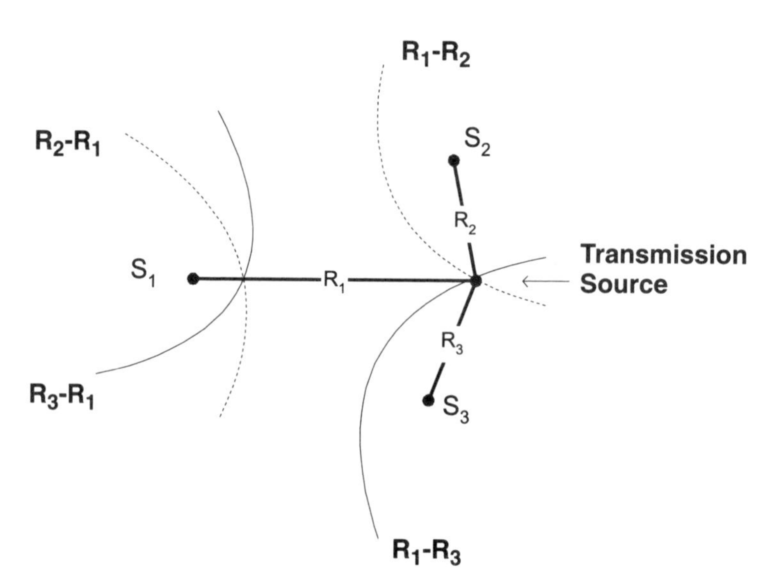

Figure 10–4 2-D Hyperbolic Position Location Solution.

mate. Furthermore, unlike TOA methods, the hyperbolic position location method is able to reduce or eliminate common time delay errors experienced at all receivers. This method lends itself extremely well to IS-95 CDMA systems, since all base stations rely on CDMA system time which is synchronized to GPS-time [Ans95b][Rap96b][Ree98].

10.5 Hyperbolic versus DF PL Systems

The two most commonly used PL techniques are Direction Finding (DF) and hyperbolic methods. While DF systems exploit adaptive arrays for beam/null steering to estimate the Direction-Of-Arrival (DOA) of the signal of interest, hyperbolic methods exploit the relative time differences of a signal arriving at different receivers.

Accuracy and spatial resolution capabilities of array-based DF methods may become worse as the distance between the mobile source and the receiving platform increases. However, spatial and temporal resolution capabilities of Time Difference Of Arrival (TDOA)-based ranging methods become *better* as the relative distance between base stations and the source increases, causing the hyperbolas to intersect closer to right angles to each another. However, when the mobile is close to either base station, TDOA degrades rapidly in a CDMA system [Aat97a]. Thus, a powerful position location strategy for CDMA is to combine DF with TDOA. This technique is known as an Direction-Of-Arrival (DOA) with TDOA approach. Various DOA estimation techniques were presented in Chapter 9. The remainder of this chapter focuses on the hyperbolic position location methed. *Data fusion* techniques can be used to develop linearized PL estimates using a combination of TDOA and DOA measurments.

10.6 TDOA Estimation Techniques

Hyperbolic position location (PL) estimation is accomplished in two stages. The first stage involves estimation of the Time Difference Of Arrival (TDOA) between base station receivers through the use of time delay estimation techniques. The estimated TDOAs are then transformed into range-difference measurements between base stations, resulting in a set of nonlinear hyperbolic range-difference equations. The second stage utilizes efficient algorithms to produce an unambiguous solution to these nonlinear hyperbolic equations. The solution produced by these algorithms results in the estimated position location of the source.

The Time Difference of Arrival (TDOA) of a signal can be estimated by two general methods: subtracting TOA measurements from two base stations to produce a relative TDOA, or through the use of cross-correlation techniques, in which the received signal at one base station is correlated with the received signal at another base station. The former method requires knowledge of the transmit timing, and thus, strict clock synchronization between the base stations which measure the signal received from the mobile unit. As shown in Chapter 2, this is available in the IS-95 standard. To eliminate the need for knowledge of the source transmit timing, differencing of arrival times at the receivers is commonly employed. Differencing the observed Time-of-Arrival eliminates some of the errors in TOA estimates common to all receivers and reduces other errors because of spatial and temporal coherence. While determining the TDOA from TOA estimates is a feasible method for some air interfaces, cross-correlation techniques dominate the field of TDOA estimation techniques. Therefore, we limit the discussion of TDOA estimation to cross-correlation estimation techniques. In the following section, a general model for TDOA estimation is developed and the techniques for TDOA estimation are presented.

10.6.1 General Model for TDOA Estimation

For a signal, $s(t)$, radiating from a remote source through a channel with interference and noise, a model for the time delay estimation between received signals at two base stations, $x_1(t)$ and $x_2(t)$, is given by

$$\begin{aligned} x_1(t) &= A_1 s(t-d_1) + n_1(t) \\ x_2(t) &= A_2 s(t-d_2) + n_2(t) \end{aligned} \tag{10.7}$$

where A_1 and A_2 scale the amplitude of each signal, $n_1(t)$ and $n_2(t)$ consist of noise and interfering signals, and d_1 and d_2 are the signal delay times, or arrival times. This model assumes that $s(t)$, $n_1(t)$ and $n_2(t)$ are real and jointly stationary, zero-mean (time average) random processes and that $s(t)$ is uncorrelated with noise $n_1(t)$ and $n_2(t)$. Referring the delay time and scaling amplitudes to the receiver with the shortest Time-of-Arrival, assuming $d_1 < d_2$, the model of (10.7) can be rewritten as

$$\begin{aligned} x_1(t) &= s(t) + n_1(t) \\ x_2(t) &= As(t-D) + n_2(t) \end{aligned} \tag{10.8}$$

where A is the amplitude ratio between the two versions of $s(t)$, and $D = d_2 - d_1$. It is desired to estimate D, the Time Difference of Arrival (TDOA) of $s(t)$ between the two receivers. It may also be desirable to estimate the scaling amplitude A, to augment TDOA estimates with path loss considerations. If the transmitted signal has some cyclical properties, it is possible to use the cyclostationary properties to estimate the time difference D and the amplitude ratio A. It follows that the limit cyclic cross-correlation and autocorrelations are given by

$$R^{\alpha}_{x_2x_1}(\tau) = AR^{\alpha}_{s}(\tau - D)e^{-j\pi\alpha D} + R^{\alpha}_{n_2n_1}(\tau) \quad (10.9)$$

where

$$R^{\alpha}_{x_1}(\tau) = R^{\alpha}_{s}(\tau) + R^{\alpha}_{n_1}(\tau) \quad (10.10)$$

$$R^{\alpha}_{x_2}(\tau) = |A|^2 R^{\alpha}_{s}(\tau)e^{-j\pi\alpha D} + R^{\alpha}_{n_2}(\tau) \quad (10.11)$$

where the parameter α is called the cycle frequency [Gar92a]. If $\alpha = 0$, the above equations are the conventional limit cross-correlation and autocorrelations.

If $s(t)$ exhibits a cycle frequency α not shared by $n_1(t)$ and $n_2(t)$, then by using this value of α in the measurements in (10.10) and (10.11), we obtain through infinite time averaging

$$R^{\alpha}_{n_1}(\tau) = R^{\alpha}_{n_2}(\tau) = R^{\alpha}_{n_2n_1}(\tau) = 0 \quad (10.12)$$

and the general model for time delay estimation between base stations becomes

$$R^{\alpha}_{x_2x_1}(\tau) = AR^{\alpha}_{s}(\tau - D)e^{-j\pi\alpha D} \quad (10.13)$$

$$R^{\alpha}_{x_1}(\tau) = R^{\alpha}_{s}(\tau) \quad (10.14)$$

$$R^{\alpha}_{x_2}(\tau) = |A|^2 R^{\alpha}_{s}(\tau)e^{-j\pi\alpha D} \quad (10.15)$$

Accurate estimation of D requires the use of estimation techniques that provide resistance to noise and interference and the ability to resolve multipath signal components. Many techniques have been developed with varying degrees of accuracy and robustness. These include the generalized cross-correlation (GCC) and cyclostationarity-exploiting cross-correlation methods. Cyclostationarity-exploiting methods include the Cyclic Cross-Correlation (CYCCOR), the Spectral-Coherence Alignment (SPECCOA) method, the Band-Limited Spectral Correlation Ratio (BL-SPECCORR) method and the Cyclic Prony method [Gar94]. While signal selective cyclostationarity-exploiting methods have been shown in [Gar94] and [Gar92a] to outperform GCC methods in the presence of noise and interference, they do so only when spectrally overlapping noise and interference exhibit a cycle frequency different than the Signal-of-Interest. When spectrally overlapping signals exhibit the same cycle frequency, as is encountered in multi-user CDMA systems, these methods do not offer an advantage over GCC methods, so that generalized cross-correlation methods for TDOA estimation must be used.

10.6.2 Generalized Cross-Correlation Methods

Conventional correlation techniques used to solve the problem of TDOA estimation are referred to as generalized cross-correlation (GCC) methods. These methods have been explored in [Gar92a], [Gar92b], [Kna76], [Car87], [Rot71], [Hah73] and [Hah75]. GCC methods cross-correlate pre-filtered versions of the received signals at two receiving stations, then estimate the TDOA between them as the location of the peak of the cross-correlation estimate. Pre-filtering is intended to accentuate frequencies for which signal-to-noise (SNR) is highest and to attenuate the noise before the signal is passed to the correlator.

Generalized cross-correlation methods for TDOA estimation are based on (10.13) with $\alpha = 0$ [Gar92a]. Thus (10.13) is rewritten as

$$R^{0}_{x_2x_1}(\tau) = AR^{0}_{s}(\tau - D). \tag{10.16}$$

The argument τ that maximizes (10.16) provides an estimate of the TDOA, D. Equivalently, (10.16) can be written as

$$R_{x_2x_1}(\tau) = R^{0}_{x_2x_1}(\tau) = \int_{-\infty}^{\infty} x_1(t)x_2(t-\tau)dt. \tag{10.17}$$

However, $R_{x_2x_1}(\tau)$ can be estimated only from a finite observation time. Thus, an estimate of the cross-correlation is given by

$$\hat{R}_{x_2x_1}(\tau) = \frac{1}{T}\int_{0}^{T} x_1(t)x_2(t-\tau)dt \tag{10.18}$$

where T represents the observation interval. Equation (10.18) is based on the use of an analog correlator. An integrate and dump correlation receiver of this form is one realization of a matched filter receiver [Zie85]. The correlation process can also be implemented digitally if sufficient sampling of the waveform is used. The output of a discrete correlation process using N digital samples of the signal is given by

$$\hat{R}_{x_2x_1}(m) = \frac{1}{N}\sum_{n=0}^{N-|m|-1} x_1(n)x_2(n+m)dt \tag{10.19}$$

The cross-power spectral density function, $G_{x_2x_1}(f)$, related to the cross-correlation of $x_1(t)$ and $x_2(t)$ in (10.18) is

$$R_{x_2x_1}(\tau) = \int_{-\infty}^{\infty} G_{x_2x_1}(f)e^{j\pi f\tau}df \tag{10.20}$$

or

$$G_{x_2x_1}(f) = \int_{-\infty}^{\infty} R_{x_2x_1}(\tau)e^{-j\pi f\tau}df \tag{10.21}$$

Because only a finite observation time of $x_1(t)$ and $x_2(t)$ is possible, only the estimate $\hat{G}_{x_2x_1}(f)$ of $G_{x_2x_1}(f)$ can be obtained.

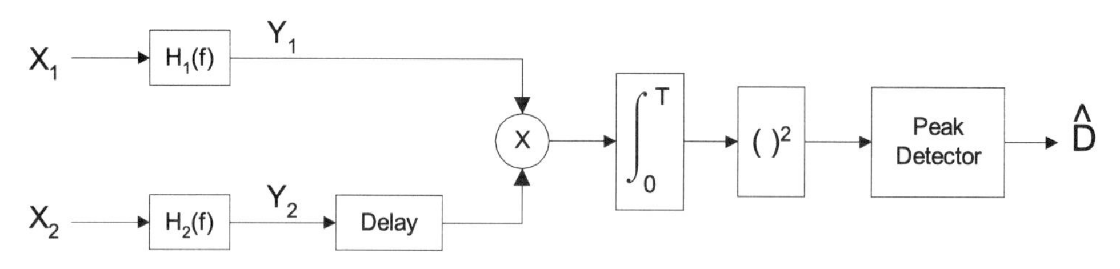

Figure 10–5 Generalized Cross-Correlation Method for TDOA Estimation.

To improve the accuracy of the delay estimate, filtering of the two signals is performed before integrating in (10.18). As shown in Figure 10–5, the received signals $x_1(t)$ and $x_2(t)$ are filtered, then correlated, integrated and squared. This is performed for a range of time shifts, τ, until a peak correlation is obtained. The time delay causing the cross-correlation peak is $\hat{D}$, an estimate of the TDOA having value *D*. If the correlator is to provide an unbiased estimate of *D*, the two filters must exhibit the same phase characteristics and hence are usually taken to be identical filters [Hah73].

When $x_1(t)$ and $x_2(t)$ are filtered, the cross-power spectrum between the filtered outputs is given by

$$G_{y_2y_1}(f) = H_1(f)H_2^*(f)G_{x_2x_1}(f) \tag{10.22}$$

where * denotes the complex conjugate. Therefore, the generalized cross-correlation, specified by superscript *G*, between $x_1(t)$ and $x_2(t)$ is

$$R^G_{y_2y_1}(\tau) = \int_{-\infty}^{\infty} \Psi_G(f)G_{x_2x_1}(f)e^{j\pi t\tau}df \tag{10.23}$$

where

$$\Psi_G(f) = H_1(f)H_2^*(F) \tag{10.24}$$

and denotes the general frequency weighting, or filter function. Because only an estimate of $R^G_{y_2y_1}(\tau)$ can be obtained, (10.23) is rewritten as

$$\hat{\boldsymbol{R}}^G_{y_2y_1}(\tau) = \int \Psi_G(f)\hat{G}_{x_2x_1}(f)e^{j\pi f\tau}df, \tag{10.25}$$

which is used to estimate *D*. The GCC methods use filter functions $\Psi_G(f)$ to minimize the effect of noise and interference.

The choice of the frequency function, $\Psi_G(f)$, is very important, especially when the received signals have delays resulting from a multipath environment. Several frequency functions, or processors, have been proposed to facilitate the estimate of D. When the filters $H_1(f) = H_2(f) = 1$, $\forall f$, then $\Psi_G(f) = 1$, and the estimate $\hat{G}$ is simply the delay abscissa at which the cross-correlation reaches a peak. This is considered cross-correlation processing. Other processors include the Roth Impulse Response processor [Rot71], the Smoothed Coherence Transform (SCOT) [Car73], the Eckart filter [Kna76], [Hah73] and the Hannan-Thomson

Table 10–1 GCC Frequency Functions.

Processor Name	Frequency Function $\Psi_G(f)$
Cross-correlation	1
Roth Impulse Response	$1/G_{x_1x_1}(f)$ or $1/G_{x_2x_2}(f)$
Smoothed Coherence Transform	$1/\sqrt{G_{x_1x_1}(f)G_{x_2x_2}(f)}$
Eckart	$G_{s_1s_1}(f)/(G_{n_1n_1}(f)[G_{x_2x_2}(f)])$
Hannon-Thomson or Maximum Likelihood	$\dfrac{\lvert\gamma x_1x_2(f)\rvert^2}{\lvert G_{x_1x_1}(f)\rvert[1-\lvert\lvert\gamma x_1x_2(f)\rvert^2\rvert]}$

(HT) processor or Maximum Likelihood (ML) estimator [Hah73]. A list of GCC frequency functions is provided in Table 10–1.

GCC methods require the differences in the TOAs for each signal to be greater than the widths of the cross-correlation functions so that the peaks can be resolved. Consequently, if the TOAs are not sufficiently separated, the overlapping of cross-correlations can introduce significant errors in the TDOA estimate. Also, if $s(t)$, $n_1(t)$ and $n_2(t)$ result in resolvable peaks, conventional GCC methods must still determine which peaks correspond to the Signal-of-Interest and which peaks are due to interference. These problems arise because GCC methods are not signal selective and produce TDOA peaks for all signals in the received data unless they are spectrally disjoint and can be filtered out [Gar94]. For IS-95 and wideband CDMA, knowledge of the delays of particular multipath components, determined from the Rake receiver, can ensure proper TDOA estimation.

10.6.3 Hyperbolic Position Location Estimation Technologies

Accurate position location (PL) estimation of a source requires an efficient hyperbolic position location estimation algorithm. Once the TDOA information has been acquired, a hyperbolic PL algorithm is responsible for producing an accurate and unambiguous solution to the position location problem. Many processing algorithms, with different complexities and restrictions, have been proposed for position location estimation based on TDOA estimates.

When base station receivers are placed in a linear fashion relative to the mobile source, the estimation of the PL is simplified. Carter's beamforming method provides an exact solution for the source range and bearing [Car81]. However, this is a very special case and requires an extensive search over a set of possible source locations, which can become computationally intensive. Hahn's method estimates the source range and bearing from the weighted sum of ranges and bearings obtained from the TDOAs of every possible combination of sensors. This method is very sensitive to the choice of weights, which can be complicated to obtain and is valid only for distant sources [Hah73][Hah75]. Abel and Smith provide an explicit solution that can achieve the Cramér-Rao Lower Bound (CRLB) in the small error region [Abe89]. The CRLB, as discussed in Section 10.7.1, is a Minimum Mean Square Error limit on the position accuracy one may expect in a PL system, relative to variations in the TDOA measurement system.

When the base stations are arbitrarily placed relative to the source (which is the typical scenario of a mobile unit within the infrastructure of a wireless system), position location becomes more complex. In this situation, the position location of a source is determined from the intersection of hyperbolic curves produced from the TDOA estimates. These equations, given by (10.6), are non-linear and are not easily solved. If the set of nonlinear hyperbolic equations equals the number of unknown coordinates of the source, then the system is consistent and a unique solution can be determined from iterative techniques. For an inconsistent system in which redundant range-difference measurements are made, the problem of solving for the position location of the source becomes more difficult because no unique solution exists.

While direct nonlinear solutions to the inconsistent system can provide accurate results, they tend to be very computationally intensive. Consequently, linearized approximations of these equations are commonly used to simplify the computation of the position location solution. One method used to approximate (10.6) is to employ Taylor Series expansion, retaining the first two terms. For most situations, linearization of the nonlinear equations of a ranging PL system does not introduce large position errors. However, linearization can introduce significant errors when determining a PL solution in certain geometries, particularly when the mobile source is very close to at least one base station. As discussed in Section 10.7.3, the Geometric Dilution Of Precision (GDOP) provides a means of quantifying sensitivity of position location errors to errors in measurements, as a function of the geometric layout of the source and receivers. It has been shown by Bancroft [Ban85] that eliminating the second order terms in a linearization can lead to significant GDOP in some situations. The effect of linearization of hyperbolic equations on the position location solution is also explored by Nicholson in [Nic73] and [Nic76].

For an inconsistent system of equations, some error criteria must be determined for selecting an optimum solution. Classical techniques for solving these equations include the Least Squares (LS) and Weighted Least-Squares (WLS) methods. These techniques can achieve the Maximum Likelihood (ML) estimate which maximizes the probability that a particular position estimate is the true position location. If the range-difference errors are uncorrelated and Gaussian distributed with zero mean and equal variances, then the LS solution provides the ML estimate [Sta94]. If the variances are unequal, then the WLS solution is the ML estimate. The WLS utilizes weighting coefficients inversely proportional to the variances of the range-difference estimates. However, a problem exists because the variances are either not known *a priori* or are difficult to estimate.

For arbitrarily placed base stations and a consistent system of equations, Fang provides an exact solution to the nonlinear equations [Fan90]. For arbitrarily distributed base stations and redundant TDOA estimates, the Spherical-Intersection (SX) [Sch87], Spherical-Interpolation (SI) [Fri87], [Smi87a], [Smi87b], [Abe87], Divide and Conquer (DAC) [Abe90], Chan's method [Cha94] and the Taylor Series [Foy76], [Tor84] methods can be used. The Taylor Series estimation method provides a more accurate solution, even at reasonable TDOA noise levels, than the other methods, but it is also more computationally intensive. Consequently, a trade-off exists between position location accuracy and computational requirements.

10.6.4 Methods for Hyperbolic PL Estimation

Solving for the two-dimensional location of a source using M base station receivers is now discussed. First, we refer all TDOAs to the first base station, which is assumed to be the first to receive the transmitted signal. Then, let (x, y) be the transmission source location and (X_i, Y_i) be the known location of the ith base station receiver. Following (10.2), the distance between the source and the ith receiver is

$$\begin{aligned} R_i(x, y) &= \sqrt{(X_i - x)^2 + (Y_i - y)^2} \\ &= \sqrt{X_i^2 + Y_i^2 - 2X_i x - 2Y_i y + x^2 + y^2} \end{aligned} \tag{10.26}$$

The range-difference between base stations with respect to the first arriving base station, from (10.6), is

$$\begin{aligned} R_{i,1}(x, y) &= c\tau_{i,1} = R_i(x, y) - R_1(x, y) \\ &= \sqrt{(X_i - x)^2 + (Y_i - y)^2} - \sqrt{(X_1 - x)^2 + (Y_1 - y)^2} \end{aligned} \tag{10.27}$$

where c is the speed of light, $R_{i,1}$ is the range-difference distance between the first base station and the ith base station, R_1 is the unknown distance between the first base station and the source, and $\tau_{i,1}$ is the estimated TDOA between the first base station and the ith base station. This defines the set of nonlinear hyperbolic equations whose solution gives the 2-D coordinates of the source. Note that the arguments of (10.26) and (10.27) may be the exact known coordinates of the mobile user (x, y), or estimates of the position of the mobile.

As mentioned earlier, solving the set of nonlinear equations for (x, y) is difficult. Consequently, linearizing this set of equations is commonly performed. One way of linearizing these equations is through the use of Taylor Series expansion and retaining the first two terms [Foy76] [Tor84]. The iterative Taylor Series solution is computed by writing the first order approximation of the TDOA in (10.27) as

$$R_{i,1}(x, y) = R_{i,1}(x_v + \delta x, y_v + \delta y) \tag{10.28}$$

$$\approx R_{i,1}(x_v, y_v) - \left(\delta x\frac{\partial}{\partial x} + \delta y\frac{\partial}{\partial y}\right)R_{i,1}(x, y)\Big|_{\substack{x = x_v \\ y = y_v}}$$

$$\approx R_{i,1}(x_v, y_v) + \delta x\left[\frac{(X_i - x_v)}{R_i(x_v, y_v)} - \frac{(X_1 - x_v)}{R_1(x_v, y_v)}\right] + \delta x\left[\frac{(Y_i - y_v)}{R_i(x_v, y_v)} - \frac{(Y_1 - y_v)}{R_1(x_v, y_v)}\right]$$

where the true location of the mobile is (x, y). The Taylor Series solution is estimated in an iterative manner, where the current estimate of the mobile's location is (x_v, y_v). The true mobile position is related to the estimated location by

$$\begin{aligned} x &= x_v + \delta x \\ y &= y_v + \delta y \end{aligned} \tag{10.29}$$

The measured TDOA also contains inherent equipment-induced measurement error, $e_{i,1}$. Then the error terms, δx and δy, in the current estimated position can be expressed as a linear function of measured variables, variables calculated from the estimated position, and an error term:

$$\left[\frac{(X_i - x_v)}{R_i(x_v, y_v)} - \frac{(X_1 - x_v)}{R_1(x_v, y_v)}\right]\delta x + \left[\frac{(Y_i - y_v)}{R_i(x_v, y_v)} - \frac{(Y_1 - y_v)}{R_1(x_v, y_v)}\right]\delta y \tag{10.30}$$

$$\approx R_{i,1}(x, y) - (R_i(x_v, y_v) - R_1(x_v, y_v)) - e_{i,1}$$

Given a set of TDOA measurements between two or more pairs of base stations, along with a previous estimate of the mobile's location, (x_v, y_v), and an estimate of the error terms, $\{e_{i,1}\}$, it is possible to determine values of δx and δy, to update the estimated location (x_v, y_v) to more closely approximate the actual position, (x, y). This process is repeated until the values of δx and δy become smaller than a desired threshold, indicating convergence [Foy76].

A Divide and Conquer (DAC) method, proposed by Abel [Abe90], consists of dividing the TDOA measurements into groups, each group having a size equal to the number of unknowns. A solution for the unknowns is calculated for each group, then combined to provide a final solution. Although this method can achieve optimum performance, the solution uses a stochastic approximation and requires that the Fisher information be sufficiently large. The Fisher Information Matrix (FIM) is the inverse of the Cramér-Rao Matrix Bound (CRMB) (i.e., $(\text{FIM})=(\text{CRMB})^{-1}$) [Hah75], where the CRMB is discussed in the Section 10.7.1. The estimator provides optimum performance when the errors are small, thus implying a low noise threshold in which the method deviates from the CRLB. This method requires an equal number of range-difference measurements in each group, and as a result, the TDOA estimates from the remaining sensors cannot be used to improve accuracy.

A non-iterative solution to the hyperbolic position estimation problem, which is capable of achieving optimum performance for arbitrarily placed sensors, was proposed by Chan [Cha94]. The solution is in closed-form and is valid for both distant and close sources. When TDOA estimation errors are small, this method is an approximation to the maximum likelihood (ML) estimator.

The Spherical-Intersection (SX) and Spherical-Interpolation (SI) methods were introduced in [Sch87][Smi87b][Abe87], and these provide rapid convergence using a LS approach. While these techniques are suboptimal, they offer insight into methods for finding non-linear solutions.

Chan's method performs significantly better than the SI method and has a higher noise threshold than the DAC method before the performance deviates from the Cramér-Rao Lower

Bound. Furthermore, it provides an explicit solution form that is not available in the Taylor Series method. A detailed approach to Chan's solution method is given in [Miz96b].

The hyperbolic PL estimation algorithms presented above offer different accuracies and complexities. The Taylor Series LS method offers accurate position location estimation at reasonable noise levels and is applicable to any number of range-difference measurements, but it can be computationally intensive. Fang's method provides an optimal solution when the system of equations is consistent, but it does not make use of redundant measurements. Another approach developed by Freidlander [Fri87] reduces computation, but, like the SI and SX methods, it is suboptimal since it eliminates a fundamental relationship in the solution. Chan's method offers a closed form solution, thus eliminating the need for an iteration approach, but it requires *a priori* information to eliminate ambiguities. The optimal PL algorithm for a given situation depends on the geometric configuration of the base stations, the range of geographic coordinates of the source to be solved, the range of TDOA measurements utilized, computational requirements and complexity, assumptions on the statistical nature of the channel, and desired accuracy.

10.7 Measures of Position Location Accuracy

A set of benchmarks is required to evaluate the accuracy of the hyperbolic position location technique. A commonly used measure of PL accuracy is the comparison of the mean square error (MSE) of the position location solution to the theoretical MSE based on the Cramér-Rao Lower Bound (CRLB). Another useful measure of PL accuracy is the circular error probability (CEP). The effect of the geometric configuration of the base stations on the accuracy of the position location estimate is measured by the geometric dilution of precision (GDOP). A simple relationship exists between GDOP and CEP. In [Lee75a] and [Lee75b], H. B. Lee provides a novel procedure for assessing the accuracy of hyperbolic multilateration PL systems and accuracy limitations. In [McG89], MMSE errors, as a function of base station placement, are demonstrated for a practical DOA position location system. Hepsaydir and Yates provide a performance analysis of position location systems using existing CDMA networks in [Hep94].

10.7.1 MSE and the Cramér-Rao Lower Bound

A commonly used measure of accuracy of a PL estimator is the comparison of the mean square error (MSE) of the PL solution (x, y) to the theoretical MSE based on the Cramér-Rao Lower Bound on the variance of unbiased estimators for ranging systems. The classical method for computing the MSE of a 2-D position location estimate is

$$MSE = \varepsilon = E[(x - x_v)^2 + (y - y_v)^2] \tag{10.31}$$

where (x, y) is the position of the source, (x_v, y_v) is the estimated position of the source, and $E[\bullet]$ denotes the ensemble average over all channel conditions and hardware anomalies for a user at a particular position location. The root-mean square (RMS) position location error, which can also be used as a measure of PL accuracy, is calculated as the square root of the MSE, as follows:

$$RMS \text{ Error} = \sqrt{\varepsilon} = \sqrt{E[(x - x_v)^2 + (y - y_v)^2]} \tag{10.32}$$

To gauge the best achievable accuracy of the PL estimator, the calculated MSE or RMS PL is compared to the theoretical minimum MSE based on the Cramér-Rao Lower Bound (CRLB). The conventional CRLB sets a lower (i.e., MMSE) bound for the variance of any unbiased parameter estimator and is typically used for a stationary Gaussian signal in the presence of stationary Gaussian noise [Gar92b]. For non-Gaussian and nonstationary (cyclostationary) signals and noise, alternate methods have been used to evaluate the performance of the estimators [Gar92b]. The derivation of the CRLB for Gaussian noise is provided in [Cha94], [Kna76], [Hah73], [Hah75], [Sto89].

A simple form of the CRLB is given by [Sti95]

$$TDOA_{rms} = (2\pi f \sqrt{2b\tau} RSNR)^{-1} \tag{10.33}$$

where $TDOA_{rms}$ represents the *rms* error in the time difference measurement between two sensors, and where RSNR represents the *Raw-Signal-to-Noise-Ratio* prior to any processing gain. The RSNR is given by

$$RSNR = \frac{\sqrt{SNR_1 SNR_2}}{\sqrt{(1 + SNR_1 + SNR_2)}} \tag{10.34}$$

where SNR_i represents the linear (not dB) value of the signal-to-noise ratio received by the sensor at position 1 or 2. The value τ in (10.33) is the coherent integration time in seconds, b is the baseband signal bandwidth of the source in Hz, and f is the *rms* RF bandwidth of the signal in Hz.

Using (10.33), it is possible to estimate the error in TDOA estimates for a pair of base stations. This perturbation is used in the hyperbolic position location algorithm to determine geographic errors in the PL solution. By mapping the PL solution over all possible source locations, it is possible to determine where a particular PL algorithm performs well, and where a PL algorithm performs poorly. This provides the basis of the Geometric Dilution of Precision.

10.7.2 Circular Error Probability

A crude but simple measure of the accuracy of position location estimates is the circular error probability (CEP) [Tor84] [Foy76]. The CEP is a measure of the uncertainty in the location estimator relative to its mean. For a 2-D system, the CEP is defined as the radius of a circle which contains half of the realizations of the random vector with the mean as its center. If the position location estimator is unbiased, the CEP is a measure of the uncertainty relative to the true transmitter position. If the estimator is biased and bound by bias B, then with a probability of one-half, a particular estimate is within a distance B + CEP from the true transmitter position. Figure 10–6 illustrates the 2-D geometrical relations.

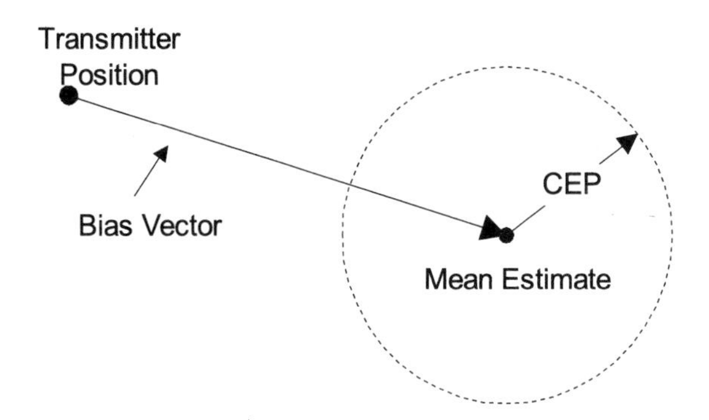

Figure 10–6 Circle of Error Probability.

The CEP is a complicated function and is usually approximated. Details of its computation are found in [Foy76] and [Tor84]. For hyperbolic position location estimator, the CEP is approximated with an accuracy within 10% as

$$CEP \approx 0.75\sqrt{\sigma_x^2 + \sigma_y^2} \tag{10.35}$$

where σ_x^2 and σ_y^2 are the variances in the estimated position on the x and y axes [Tor84].

10.7.3 Geometric Dilution of Precision

The accuracy of Range-Based PL systems depends to a large extent on the geometric relationship between the locations of the base stations and the location of the source. This is because errors due to the channel or hardware perturbations in the TDOA estimate are transformed into geographic position errors in the hyperbolic solution. One measure that quantifies the position accuracy based on this geometric configuration is called the *geometric dilution of precision (GDOP)* [Wel86][Tor84][Ban85][Lee75a]. The GDOP is defined as the ratio of the RMS position error to the RMS ranging error. The GDOP for an unbiased estimator and a ranging system is given by [Jor84] as

$$GDOP = \sqrt{tr[(A^T A)^{-1}]} \tag{10.36}$$

where A is expressed in equation (8.12) and $tr(\bullet)$ indicates the trace of the matrix. The GDOP for an unbiased estimator and a 2-D hyperbolic system is given by [Tor84] and [Lee75a] as

$$\begin{aligned} GDOP &= (\sqrt{(c\sigma_x)^2 + (c\sigma_y)^2}) / \sqrt{(c\sigma_s)^2} \\ &= (\sqrt{\sigma_x^2 + \sigma_y^2}) / \sigma_s \end{aligned} \tag{10.37}$$

where $(c\sigma_s)^2$ is the mean square ranging error and $(c\sigma_x)^2$ and $(c\sigma_y)^2$ are the mean square position errors in the x and y directions. The GDOP may be computed for all locations of sources in a coverage area, based on the TDOA errors and the particular hyperbolic algorithm used. The GDOP is related to the CEP by

$$CEP \approx (0.75\sigma_s)GDOP \tag{10.38}$$

Finding the smallest GDOP is often used as a criterion for selecting a set of base station sensors from a larger set of base station measurements, in order to produce minimum PL estimation error for a particular zone from which moble users are to operate. The GDOP may also be used for selecting base station receiver locations within a new system in order to provide sufficient position location accuracy for specific TDOA hardware or desired service regions.

10.8 Summary

This chapter introduced basic ranging and TDOA estimation techniques and hyperbolic PL algorithms used in the hyperbolic position location method. When used with DOA techniques, it is possible to provide accurate position location over wireless channels [Ken96]. A general model for the TDOA estimation problem was developed and Generalized Cross-Correlation (GCC) techniques commonly used for time delay estimation were presented. The effect of the frequency functions on the TDOA estimation and the importance of the choice of frequency function were discussed. Although GCC methods do facilitate the estimation of the TDOA, they encounter problems which are critical to the position location problem. First, GCC methods require the differences in the TDOA for each signal to be greater than the widths of the cross-correlation functions so that the correlation peaks can be resolved. This is generally not a problem in IS-95 based systems where a consecutive series of several chips would still offer distinct correlation peaks. If not separated sufficiently, overlapping cross-correlation functions corrupt the TDOA estimate. Furthermore, because GCC methods are not signal-selective, they produce correlation peaks for all signals and are faced with the problem of identifying the TDOA estimate of interest. Signal-selective TDOA estimation techniques outperform GCC methods; however, they do not offer any advantages over GCC methods when the spectrally overlapping noise and interference exhibit the same cycle frequency as the Signal-of-Interest, which is encountered in multi-user CDMA systems.

For the problem of geolocating mobile units within a cellular infrastructure, a number of hyperbolic position location algorithms applicable to arbitrarily placed mobile and base stations were presented. Finally, the measures of position location accuracy commonly used to evaluate hyperbolic position location algorithms, in conjunction with TDOA uncertainties, were reviewed.

As wireless service providers throughout the world are mandated to provide emergency location capabilities, the techniques described in Chapters 9 and 10 will be fundamental in offering solutions that will be implemented with smart antennas and CDMA communication systems.

APPENDIX A

Multiple Access Interference and the Gaussian Approximation

In analyzing the impact of Multiple Access Interference in DS-CDMA systems, as described in Chapter 1, we must consider the detailed structure of the contribution of each interfering signal to the decision statistic, Z_0. This is given by I_k in (1.32).

The relationship between $b_k(t-\tau_k)$, $a_k(t-\tau_k)$, and $a_0(t)$ is illustrated in Figure A–1. The quantities γ_k and Δ_k in Figure A–1 are defined from the delay of user k relative to user 0, τ_k, such that [Pur77]:

$$\tau_k = \gamma_k T_c + \Delta_k \qquad 0 \le \Delta_k < T_c \tag{A.1}$$

Based on Figure A–1, we rearrange (1.32)

$$\begin{aligned} I_k &= \sqrt{\frac{P_k}{2}}\cos\varphi_k \\ &\times\left\{\left(b_{k,-1}\sum_{l=j-\gamma_k}^{j-1} a_{k,l}a_{0,l+i-j+\gamma_k} + b_{k,0}\sum_{l=j}^{j+N-\gamma_k-1} a_{k,l}a_{0,l-j+i+\gamma_k}\right)(T_c-\Delta_k)\right. \\ &+\left.\left(b_{k,-1}\sum_{l=j-\gamma_k-1}^{j-1} a_{k,l}a_{0,l+i-j+\gamma_k+1} + b_{k,0}\sum_{l=j}^{j+N-\gamma_k-2} a_{k,l}a_{0,l-j+i+\gamma_k+1}\right)\Delta_k\right\} \end{aligned} \tag{A.2}$$

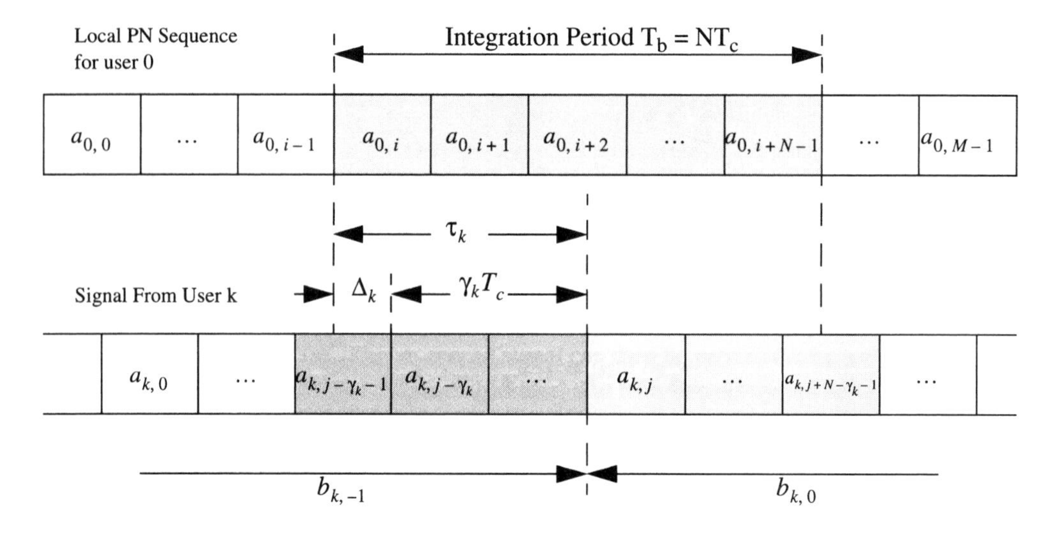

Figure A–1 Timing of the local PN sequence for user 0, $a_o(t)$ and the received signal from user k, $s_k(t-\tau_k)$.

It is convenient to rewrite this as

$$I_k = T_c\sqrt{\frac{P_k}{2}}\cos\varphi_k$$

$$\times\left\{\left(b_{k,-1}\sum_{l=0}^{\gamma_k-1}a_{k,l+j-\gamma_k}a_{0,l+i}+b_{k,0}\sum_{l=\gamma_k}^{N-1}a_{k,l+j-\gamma_k}a_{0,l+i}\right)\left(1-\frac{\Delta_k}{T_c}\right)\right.$$

$$\left.+\left(b_{k,-1}\sum_{l=-1}^{\gamma_k-1}a_{k,l+j-\gamma_k}a_{0,l+i+1}+b_{k,0}\sum_{l=\gamma_k}^{N-2}a_{k,l+j-\gamma_k}a_{0,l+i+1}\right)\left(\frac{\Delta_k}{T_c}\right)\right\} \tag{A.3}$$

This may be rearranged to give

$$I_k = T_c\sqrt{\frac{P_k}{2}}\cos\varphi_k$$

$$\times\left\{\left(\sum_{l=0}^{\gamma_k-1} b_{k,-1}a_{k,l+j-\gamma_k}a_{0,l+i}+\sum_{l=\gamma_k}^{N-2} b_{k,0}a_{k,l+j-\gamma_k}a_{0,l+i}+b_0a_{k,N-1+j-\gamma_k}a_{0,N-1+i}\right)\left(1-\frac{\Delta_k}{T_c}\right)\right.$$

$$\left.+\left(\sum_{l=0}^{\gamma_k-1} b_{k,-1}a_{k,l+j-\gamma_k}a_{0,l+i+1}+\sum_{l=\gamma_k}^{N-2} b_{k,0}a_{k,l+j-\gamma_k}a_{0,l+i+1}+b_{-1}a_{k,j-\gamma_k-1}a_{0,i}\right)\left(\frac{\Delta_k}{T_c}\right)\right\} \tag{A.4}$$

It is useful to define $Z_{k,l}$ as [Leh87a]

$$Z_{k,l}=\begin{cases} b_{k,-1}a_{k,l+j-\gamma_k}a_{0,l+i} & l=0,\gamma_k-1 \\ b_{k,0}a_{k,l+j-\gamma_k}a_{0,l+i} & l=\gamma_k,N-2 \\ b_{k,0}a_{k,N-1+j-\gamma_k}a_{0,N-1+i} & l=N-1 \\ b_{k,-1}a_{k,j-\gamma_k-1}a_{0,i} & l=N \end{cases} \tag{A.5}$$

where each of the $Z_{k,j}$ are independent Bernoulli trials, equally distributed on {-1,+1}. Then, using (A.5) and the fact that $a_{0,l}a_{0,l}=1$, (A.4) may be expressed as

$$I_k = T_c\sqrt{\frac{P_k}{2}}\cos\varphi_k\times\left\{\left(\sum_{l=0}^{N-2} Z_{k,l}\left(\left(1-\frac{\Delta_k}{T_c}\right)+a_{0,l+i}a_{0,l+i+1}\left(\frac{\Delta_k}{T_c}\right)\right)\right)\right.$$

$$\left.+Z_{k,N-1}\left(1-\frac{\Delta_k}{T_c}\right)+Z_{k,N}\left(\frac{\Delta_k}{T_c}\right)\right\} \tag{A.6}$$

We define the set $\mathcal{A}$ as the set of all integers in $[0,N-2]$ for which $a_{0,l+i}a_{0,l+i+1}=1$. Similarly, the set $\mathcal{B}$ is defined as the set of all integers in $[0,N-2]$ for which $a_{0,l+i}a_{0,l+i+1}=-1$. Then (A.6) is

$$I_k = T_c\sqrt{\frac{P_k}{2}}\cos\varphi_k\left\{\left(\sum_{l\in\mathcal{A}} Z_{k,l}+(1-2\Delta_k/T_c)\sum_{l\in\mathcal{B}} Z_{k,l}\right)\right.$$

$$\left.+Z_{k,N-1}\left(1-\frac{\Delta_k}{T_c}\right)+Z_{k,N}\left(\frac{\Delta_k}{T_c}\right)\right\} \tag{A.7}$$

We define

$$X_k = \sum_{l \in \mathcal{A}} Z_{k,l} \tag{A.8}$$

$$Y_k = \sum_{l \in \mathcal{B}} Z_{k,l} \tag{A.9}$$

$$U_k = Z_{k,N-1} \tag{A.10}$$

$$V_k = Z_{k,N} \tag{A.11}$$

Note that $\{Z_{k,l}\}$ is a set of independent Bernoulli trials [Coo86], each with each outcome identically distributed on {-1,+1}.

As described in [Leh87a] and [Mor89]. the contribution from interfering user k to the decision statistic is given by

$$I_k = T_c\sqrt{\frac{P_k}{2}}\cos\varphi_k\left\{\left(X_k + \left(1-\frac{2\Delta_k}{T_c}\right)Y_k\right) + \left(1-\frac{\Delta_k}{T_c}\right)U_k + \left(\frac{\Delta_k}{T_c}\right)V_k\right\} \tag{A.12}$$

The term A is the number of integers i such that $a_{0,i}a_{0,i+1} = 1$ for $0 \le i \le N-2$. Similarly, B is defined as the number of integers i such that $a_{0,i}a_{0,i+1} = -1$ for $0 \le i \le N-2$. The quantities, X_k, Y_k, U_k, and V_k in (A.12) have distributions, conditioned on A and B which are given by

$$p_{X_k}(l) = \binom{A}{\frac{l+A}{2}} 2^{-A} \qquad l = -A, -A+2, \ldots, A-2, A \tag{A.13}$$

$$p_{Y_k}(l) = \binom{B}{\frac{l+B}{2}} 2^{-B} \qquad l = -B, -B+2, \ldots, B-2, B \tag{A.14}$$

$$p_{U_k}(l) = \frac{1}{2} \qquad l = -1, 1 \tag{A.15}$$

$$p_{V_k}(l) = \frac{1}{2} \qquad l = -1, 1 \tag{A.16}$$

Thus the interference contributed to the decision statistic by a particular multiple access interferer is entirely defined by the quantities P_k, φ_k, Δ_k, A and B. Note that A and B are solely dependent on the sequence of user 0. Furthermore, $A + B = N - 1$ since sets $\mathcal{A}$ and $\mathcal{B}$ are disjoint and span the set of total possible signature sequences of length N in which there are a total of $N - 1$ possible locations for a chip level transition.

A.1 The Gaussian Approximation

The use of the Gaussian Approximation to determine the bit error rate in CDMA multiple access communication systems is based on the argument that the decision statistic, Z_0, given by (1.26), may be modeled as a Gaussian random variable [Pur77], [Mor89]. The first component in (1.26), I_0, is deterministic and its value is given by (1.27). The other two components of Z_0, ζ and η are assumed to be zero mean Gaussian random variables. If the additive receiver noise, $n(t)$, is a Gaussian random process, then η, which is given by (1.28), is a zero-mean Gaussian random variable. In this section, the standard expression for the bit error rate is derived based on the assumption that the multiple access interference term, ζ, may be approximated by a Gaussian random variable.

We will first define a combined noise and interference term ξ, given by

$$\xi = \zeta + \eta \tag{A.17}$$

such that the decision statistic, Z_0, is given by

$$Z_0 = I_0 + \xi \tag{A.18}$$

where Z_0 is a Gaussian random variable with mean I_0 and a variance which is equal to the variance, σ_ξ^2, of ξ.

The probability of error in determining the value of a received bit is equal to the probability that $\xi < -I_0$ when I_0 is positive and $\xi > I_0$ when I_0 is negative. Due to its structure, ξ is symmetrically distributed such that these two conditions occur with equal probability; therefore, we may say that the probability of error is equal to the probability that $\xi > |I_0|$. If we may assume that ξ is a zero mean Gaussian random variable with variance, σ_ξ^2, the probability of a bit error is given by

$$\begin{aligned} P_e &= \int_{|I_0|}^{\infty} f_\xi(x)dx \\ &= \int_{|I_0|}^{\infty} \frac{1}{\sqrt{2\pi}\sigma_\xi} \exp\left(\frac{-x^2}{2\sigma_\xi^2}\right)dx \\ &= \frac{1}{\sqrt{2\pi}} \int_{|I_0|/\sigma_\xi}^{\infty} \exp\left(\frac{-u^2}{2}\right)du \\ &= Q\left(\frac{|I_0|}{\sigma_\xi}\right) \end{aligned} \tag{A.19}$$

Using (1.27), this may be rewritten as

$$P_e = Q\left(\sqrt{\frac{P_0 T_b^2}{2\sigma_\xi^2}}\right) \tag{A.20}$$

Assuming that the multiple access interference contribution to the decision statistic, ζ, may be modeled as a zero mean Gaussian random variable with variance σ_ζ^2, and that the noise contribution to the decision statistic, η, may be modeled as a zero mean Gaussian noise process with variance $\sigma_\eta^2 = N_o T_b/4$, then, since the noise and the multiple access interference are independent, the variance of ξ is given by

$$\sigma_\xi^2 = \sigma_\zeta^2 + \sigma_\eta^2 \tag{A.21}$$

All that remains is to justify the Gaussian Approximation for ζ, and to determine the value of σ_ζ^2.

The total multiple access interference contribution to the decision statistic for user 0, from (1.31), is

$$\zeta = \sum_{k=1}^{K-1} I_k \tag{A.22}$$

where I_k is the contribution from user k to the decision statistic for user 0. Using (A.12), we write I_k, as

$$I_k = T_c W_k \sqrt{\frac{P_k}{2}} \cos(\varphi_k) \tag{A.23}$$

and W_k is given by

$$W_k = X_k + \left(1 - \frac{2\Delta_k}{T_c}\right) Y_k + \left(1 - \frac{\Delta_k}{T_c}\right) U_k + \left(\frac{\Delta_k}{T_c}\right) V_k \tag{A.24}$$

The distributions for X_k and Y_k are given by (A.13) and (A.14) and the distributions of U_k and V_k are given in (A.15) and (A.16) as derived by [Mor89] and shown in Appendix [Leh87a]. Note that W_k may take on only discrete values (in fact a maximum of $(N-1)B - B^2 + 6$ values) given Δ_k.

The Central Limit Theorem (C.L.T.), described in [Sta86] and [Coo86] is used to justify the approximation of ζ as a Gaussian random variable. According to the more general statement of the C.L.T. [Sta86], the sum, y,

$$y = \sum_{i=0}^{M-1} x_i \tag{A.25}$$

of M independent random variables, x_i (which are not necessarily identically distributed), each with mean μ_{x_i} and variance $\sigma_{x_i}^2$, has a distribution which approaches a Gaussian distribution as M gets large, provided that

$$\sigma_{x_j}^2 \ll \sigma_y^2 = \sum_{i=0}^{M-1} \sigma_{x_i}^2 \qquad j = 0 \ldots M-1 \tag{A.26}$$

Furthermore, the mean and variance of the sum, y, are given by

$$\mu_y = \sum_{i=0}^{M-1} \mu_{x_i} \tag{A.27}$$

$$\sigma_y^2 = \sum_{i=0}^{M-1} \sigma_{x_i}^2 \tag{A.28}$$

In this application, the condition given in (A.26) is equivalent to specifying that no single user dominates the total multiple access interference.

The terms I_k in (A.22) are not actually independent. Since the distributions of X_k and Y_k given in (A.13) and (A.14) are both dependent on B, which is itself a random variable, the random variables X_k and Y_k are independent only when conditioned on B. Since the terms I_k are not independent, one of the conditions of the C.L.T. is violated. To continue development of the Gaussian Approximation, we will assume that the C.L.T. may be applied; however, this issue is addressed at length in [Lib95].

The variance of the multiple access interference is computed from

$$\begin{aligned}\sigma_\zeta^2 &= E[\zeta^2] - (E[\zeta])^2 = E\left[\left(\sum_{k=1}^{K-1} I_k\right)^2\right] - \left(E\left[\sum_{k=1}^{K-1} I_k\right]\right)^2 \\ &= E\left[\left(\sum_{k=1}^{K-1} I_k\right)^2\right] - \left(\sum_{k=1}^{K-1} E[I_k]\right)^2\end{aligned} \tag{A.29}$$

where the expected value is taken over all values of $\{\phi_k\}$, $\{\Delta_k\}$, and B. The phases, $\{\phi_k\}$, are independent and uniformly distributed on $[0, 2\pi]$. Similarly, the fractional chip delays, $\{\Delta_k\}$ are independent and distributed on $[0, T_c]$. For random PN sequences, B is distributed as

$$p_B(i) = \binom{N-1}{i} 2^{1-N} \qquad i = 0, \ldots, N-1 \tag{A.30}$$

The expected value of I_k is

$$E[I_k] = E\left[T_c W_k \sqrt{\frac{P_k}{2}} \cos\varphi_k\right] = T_c \sqrt{\frac{P_k}{2}} (E[W_k] E[\cos\varphi_k]) = 0 \tag{A.31}$$

Therefore (A.29) reduces to

$$\sigma_\zeta^2 = E\left[\left(\sum_{k=1}^{K-1} I_k\right)^2\right] = E\left[\sum_{k=1}^{K-1}\sum_{l=1}^{K-1} I_k I_l\right] = \sum_{k=1}^{K-1}\sum_{l=1}^{K-1} E[I_k I_l] \tag{A.32}$$

To proceed with the derivation of the standard Gaussian Approximation, it is necessary to assume that the terms I_k are independent, which, as already stated, is not a strictly valid assumption. Assuming that the terms I_k are independent, and using (A.31), we may write

$$\sigma_\zeta^2 = \sum_{k=1}^{K-1} E[I_k^2] = \sum_{k=1}^{K-1} \sigma_{I_k}^2 \tag{A.33}$$

The variance of I_k is

$$\sigma_{I_k}^2 = E\left[\left(T_c W_k \sqrt{\frac{P_k}{2}}\cos(\varphi_k)\right)^2\right] = \frac{T_c^2 P_k}{2} E[W_k^2] E[\cos^2(\phi_k)] \tag{A.34}$$

where

$$E[\cos^2(\phi_k)] = \int_0^{2\pi} \frac{1}{2\pi}\cos^2(\phi)d\phi = \frac{1}{2} \tag{A.35}$$

and

$$E[W_k^2] = E\left[\left(X_k + \left(1 - \frac{2\Delta_k}{T_c}\right)Y_k + \left(1 - \frac{\Delta_k}{T_c}\right)U_k + \left(\frac{\Delta_k}{T_c}\right)V_k\right)^2\right] \tag{A.36}$$

As noted earlier, since distributions of random variables X_k and Y_k both depend on random variable B, they are not strictly independent. In the standard derivation of the Gaussian Approximation, this fact is not taken into account. To proceed, we assume that X_k, Y_k, U_k, and V_k are all independent with zero means. Then we may write

$$\begin{aligned} E[W_k^2] &= E[X_k^2] + E\left[\left(1 - \frac{2\Delta_k}{T_c}\right)^2\right]E[Y_k^2] \\ &+ E\left[\left(1 - \frac{\Delta_k}{T_c}\right)^2\right]E[U_k^2] + E\left[\left(\frac{\Delta_k}{T_c}\right)^2\right]E[V_k^2] \end{aligned} \tag{A.37}$$

The term X_k is the sum of A independent and identically distributed Bernoulli trials, x_i, each with equally likely outcomes from *{-1,+1}*. As described in [Pur77], A is the number of integers i such that first lag of the autocorrelation function for the chip sequence for the desired user is $a_{0,i}a_{0,i+1} = 1$ for $i \in [0, N-2]$. Note that $A + B = N - 1$ or $A = N - B - 1$. Therefore, when conditioned on A, the mean square value of X_k is

$$E[X_k^2|A] = E\left[\left(\sum_{i=0}^{A-1} x_i\right)^2\right] = \sum_{i=0}^{A-1} E[x_i^2] = A \tag{A.38}$$

Then

$$E[X_k^2] = E[A] \tag{A.39}$$

Similarly, Y_k, is the summation of the B independent Bernoulli trials y_i, where B is the number of integers i such that $a_{0,i}a_{0,i+1} = -1$ for $i \in [0, N-2]$. Note that $A + B = N - 1$ or $A = N - B - 1$.

$$E[Y_k^2|B] = E\left[\left(\sum_{i=0}^{B-1} y_i\right)^2\right] = \sum_{i=0}^{B-1} E[y_i^2] = B \tag{A.40}$$

Then

$$E[Y_k^2] = E[B] \tag{A.41}$$

The random variables U_k and V_k take on values of {-1,+1} with equal probability, therefore $E[U_k^2] = 1$ and $E[V_k^2] = 1$, Then we may express (A.37) as

$$E[W_k^2] = E[A] + E\left[1 - 4\frac{\Delta_k}{T_c} + 4\left(\frac{\Delta_k}{T_c}\right)^2\right]E[B] + E\left[1 - 2\frac{\Delta_k}{T_c} + 2\left(\frac{\Delta_k}{T_c}\right)^2\right] \quad (A.42)$$

The statistics of Δ_k are

$$E[\Delta_k] = \frac{1}{T_c}\int_0^{T_c} \Delta d\Delta = \frac{T_c}{2} \quad (A.43)$$

and

$$E[\Delta_k^2] = \frac{1}{T_c}\int_0^{T_c} \Delta^2 d\Delta = \frac{T_c^2}{3} \quad (A.44)$$

Using these values and $A = N - 1 - B$ in (A.42) we obtain

$$E[W_k^2] = E[A] + \frac{1}{3}E[B] + \frac{2}{3} = N - \frac{1}{3}(2E[B] + 1) \quad (A.45)$$

Finally, we find the expected value of B using the distribution given by (A.30),

$$E[B] = 2^{1-N}\sum_{i=0}^{N-1}\binom{N-1}{i}i = 2^{1-N}\sum_{j=0}^{N-1}\binom{N-1}{j}(N-1-j)$$

$$= (N-1)2^{1-N}\sum_{j=0}^{N-1}\binom{N-1}{j} - 2^{1-N}\sum_{j=0}^{N-1}\binom{N-1}{j}j$$

$$= \frac{(N-1)}{2}2^{1-N}\sum_{j=0}^{N-1}\binom{N-1}{j} = \frac{N-1}{2} \quad (A.46)$$

Therefore,

$$E[W_k^2] = \frac{2N}{3} \quad (A.47)$$

Substituting (A.47) and (A.35) into (A.34) we obtain

$$\sigma_{I_k}^2 = \frac{T_c^2 P_k N}{6} \quad (A.48)$$

If the $K-1$ values of $\sigma_{I_k}^2$ satisfy (A.26), we use the C.L.T. to model ζ as a zero mean Gaussian random variable as $K-1$ gets large. The variance of ζ is given by

$$\sigma_\zeta^2 = \frac{NT_c^2}{6}\sum_{k=1}^{K-1} P_k \quad (A.49)$$

Therefore, we model the decision statistic Z_0 as a Gaussian random variable with mean given by (1.27) and variance given by (A.21)

$$\sigma_{\xi}^2 = \sigma_{\zeta}^2 + \sigma_{\eta}^2$$

$$= \frac{NT_c^2}{6} \sum_{k=1}^{K-1} P_k + \frac{N_0 T_b}{4} \tag{A.50}$$

Using (A.20), the bit error rate is given by

$$P_e = Q\left(\frac{\sqrt{\frac{P_0}{2}} T_b}{\sqrt{\frac{NT_c^2}{6} \sum_{k=1}^{K-1} P_k + \frac{N_0 T_b}{4}}} \right)$$

$$= Q\left(\sqrt{\frac{P_0}{\frac{1}{3N} \sum_{k=1}^{K-1} P_k + \frac{N_0}{2T_b}}} \right) \tag{A.51}$$

This is the bit error rate for BPSK-DS-CDMA, using random, asynchronous, sequences and the Gaussian Approximation.

APPENDIX B

Q, *erf*, & *erfc* Functions[1]

B.1 The Q-Function

Computation of probabilities that involve a Gaussian process require finding the area under the tail of the Gaussian (normal) probability density function as shown in Figure B–1.

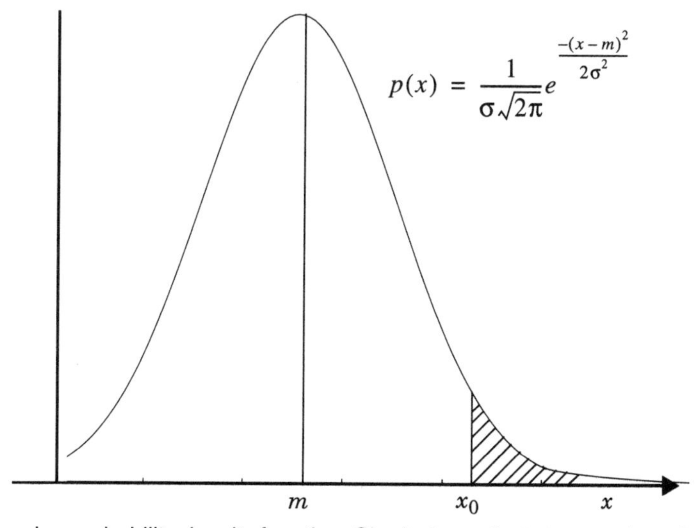

Figure B–1 Gaussian probability density function. Shaded area is $Pr(x \geq x_0)$ for a Gaussian random variable.

[1] Adapted from T. S. Rappaport, *Wireless Communications: Principles and Practice*, Prentice Hall, NJ, ©1996. Reprinted by permission of Prentice Hall.

Figure B–1 illustrates the probability that a Gaussian random variable x exceeds x_0, $Pr(x \geq x_0)$, which is evaluated as

$$Pr(x \geq x_0) = \int_{x_0}^{\infty} \frac{1}{\sigma\sqrt{2\pi}} e^{-(x-m)^2/(2\sigma^2)} dx \tag{B.1}$$

The Gaussian probability density function in (B.1) cannot be integrated in closed form.

Any Gaussian probability density function may be rewritten through use of the substitution

$$y = \frac{x-m}{\sigma} \tag{B.2}$$

to yield

$$Pr\left(y > \frac{x_0 - m}{\sigma}\right) = \int_{\left(\frac{x_0 - m}{\sigma}\right)}^{\infty} \frac{1}{\sqrt{2\pi}} e^{-y^2/2} dy \tag{B.3}$$

where the kernel of the integral on the right hand side of (B.3) is the normalized Gaussian probability density function with mean of 0 and standard deviation of 1. Evaluation of the integral in equation (B.3) is designated as the Q-function, which is defined as

$$Q(z) = \int_{z}^{\infty} \frac{1}{\sqrt{2\pi}} e^{-y^2/2} dy \tag{B.4}$$

Hence equations (B.1) or (B.3) can be evaluated as

$$P\left(y > \frac{x_0 - m}{\sigma}\right) = Q\left(\frac{x_0 - m}{\sigma}\right) = Q(z) \tag{B.5}$$

The Q-function is bounded by two analytical expressions as follows:

$$\left(1 - \frac{1}{z^2}\right)\frac{1}{z\sqrt{2\pi}} e^{-z^2/2} \leq Q(z) \leq \frac{1}{z\sqrt{2\pi}} e^{-z^2/2} \tag{B.6}$$

For values of z greater 3.0, both of these bounds closely approximate $Q(z)$.

Two important properties of $Q(z)$ are

$$Q(-z) = 1 - Q(z) \tag{B.7}$$

$$Q(0) = \frac{1}{2} \tag{B.8}$$

$Q(z)$ is tabulated in Table B–1, and a graph of $Q(z)$ versus z is given in Figure B–2.

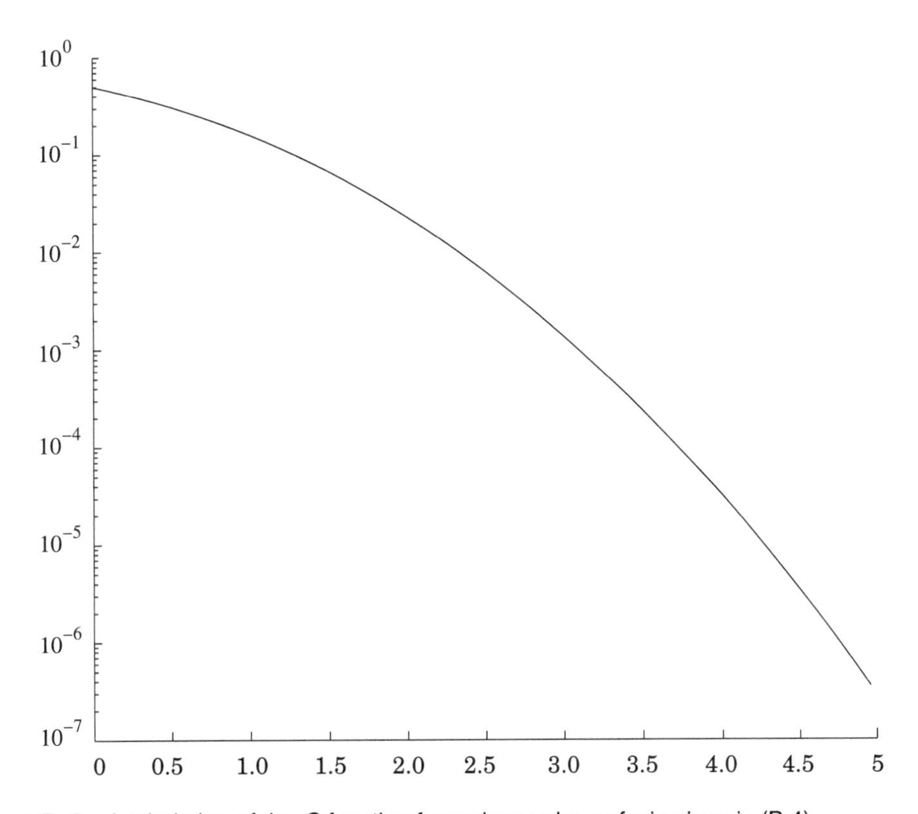

Figure B–2 A tabulation of the *Q*-function for various values of *z* is given in (B.4).

B.2 The *erf* and *erfc* functions

The error function (*erf*) is defined as

$$erf(z) = \frac{2}{\sqrt{\pi}} \int_{0}^{z} e^{-x^2} dx \tag{B.9}$$

and the complementary error function (*erfc*) is defined as

$$erfc(z) = \frac{2}{\sqrt{\pi}} \int_{z}^{\infty} e^{-x^2} dx \tag{B.10}$$

The *erfc* function is related to the *erf* function by

$$erfc(z) = 1 - erf(z) \tag{B.11}$$

The Q-function is related to the erf and $erfc$ functions by

Table B–1 Tabulation of the *Q*-function.

z	$Q(z)$	z	$Q(z)$
0.0	0.50000	2.0	0.02275
0.1	0.46017	2.1	0.01786
0.2	0.42074	2.2	0.01390
0.3	0.38209	2.3	0.01072
0.4	0.34458	2.4	0.00820
0.5	0.30854	2.5	0.00621
0.6	0.27425	2.6	0.00466
0.7	0.24196	2.7	0.00347
0.8	0.21186	2.8	0.00256
0.9	0.18406	2.9	0.00187
1.0	0.15866	3.0	0.00135
1.1	0.13567	3.1	0.00097
1.2	0.11507	3.2	0.00069
1.3	0.09680	3.3	0.00048
1.4	0.08076	3.4	0.00034
1.5	0.06681	3.5	0.00023
1.6	0.05480	3.6	0.00016
1.7	0.04457	3.7	0.00011
1.8	0.03593	3.8	0.00007
1.9	0.02872	3.9	0.00005

$$Q(z) = \frac{1}{2}\left[1 - erf\left(\frac{z}{\sqrt{2}}\right)\right] = \frac{1}{2}erfc\left(\frac{z}{\sqrt{2}}\right) \tag{B.12}$$

$$erfc(z) = 2Q(\sqrt{2}z) \tag{B.13}$$

$$erf(z) = 1 - 2Q(\sqrt{2}z) \tag{B.14}$$

The relationships in equations B.12 - B.14 are widely used in error probability computations. Table B–2 displays values for the *erf* function.

Table B–2 Tabulation of the Error Function *erf(z)*.

z	$erf(z)$	z	$erf(z)$
0.1	0.11246	1.6	0.97635
0.2	0.22270	1.7	0.98379
0.3	0.32863	1.8	0.98909
0.4	0.42839	1.9	0.99279
0.5	0.52049	2.0	0.99532
0.6	0.60385	2.1	0.99702
0.7	0.67780	2.2	0.99814
0.8	0.74210	2.3	0.99885
0.9	0.79691	2.4	0.99931
1.0	0.84270	2.5	0.99959
1.1	0.88021	2.6	0.99976
1.2	0.91031	2.7	0.99987
1.3	0.93401	2.8	0.99993
1.4	0.95228	2.9	0.99996
1.5	0.96611	3.0	0.99998

APPENDIX C

Mathematical Tables[1]

Table C–1 Trigonometric Identities

$\sin(A \pm B) = \sin A \cos B \pm \cos A \sin B$
$\cos(A \pm B) = \cos A \cos B \mp \sin A \sin B$
$\cos A \cos B = (1/2)[\cos(A+B) + \cos(A-B)]$
$\sin A \sin B = (1/2)[\cos(A-B) - \cos(A+B)]$
$\sin A \cos B = (1/2)[\sin(A+B) + \sin(A-B)]$
$\sin A + \sin B = 2\sin\left(\frac{A+B}{2}\right)\cos\left(\frac{A-B}{2}\right)$
$\sin A - \sin B = 2\sin\left(\frac{A-B}{2}\right)\cos\left(\frac{A+B}{2}\right)$
$\cos A + \cos B = 2\cos\left(\frac{A+B}{2}\right)\cos\left(\frac{A-B}{2}\right)$
$\cos A - \cos B = -2\sin\left(\frac{A+B}{2}\right)\sin\left(\frac{A-B}{2}\right)$

Continued on next page

[1] Adapted from T. S. Rappaport, *Wireless Communications: Principles and Practice*, Prentice Hall, NJ, ©1996. Reprinted by permission of Prentice Hall.

Table C–1 Trigonometric Identities (Continued)

$\sin 2A = 2\sin A\cos A$
$\cos 2A = \cos^2 A - 1 = 1 - 2\sin^2 A = \cos^2 A - \sin^2 A$
$\sin A/2 = \sqrt{(1-\cos A)/2}$ $\quad \cos A/2 = \sqrt{(1+\cos A)/2}$
$\sin^2 A = (1-\cos 2A)/2$ $\quad \cos^2 A = (1+\cos 2A)/2$
$\sin x = \dfrac{e^{jx}-e^{-jx}}{2j}$ $\quad \cos x = \dfrac{e^{jx}+e^{-jx}}{2}$ $\quad e^{jx} = \cos x + j\sin x$
$A\cos(\omega t+\phi_1) + B\cos(\omega t+\phi_2) = C\cos(\omega t+\phi_3)$ where $C = \sqrt{A^2+B^2-2AB\cos(\phi_2-\phi_1)}$ $\phi_3 = \tan^{-1}\left[\dfrac{A\sin\phi_1 + B\sin\phi_2}{A\cos\phi_1 + B\cos\phi_2}\right]$
$\sin(\omega t+\phi) = \cos(\omega t+\phi-90^0)$
$\cos(\omega t+\phi) = \sin(\omega t+\phi+90)$

Table C–2 Approximations

Taylor's series	$f(x) = f(a) + \dot{f}(a)\dfrac{(x-a)}{1!} + \ddot{f}(a)\dfrac{(x-a)^2}{2!} + \ \ldots$
Maclaurin's series	$f(0) = f(0) + \dot{f}(0)\dfrac{x}{1!} + \ddot{f}(0)\dfrac{x^2}{2!} + \ \ldots$
For small values of x (x<< 1)	$\dfrac{1}{1+x} \cong 1 - x$
	$(1+x)^n \cong 1 + nx \quad n \geq 1$
	$e^x \cong 1 + x$
	$\ln(1+x) \cong x$
	$\sin(x) \cong x$
	$\cos(x) \cong 1 - \dfrac{x^2}{2}$
	$\tan(x) \cong x$

Table C–3 Indefinite Integrals

$$\int \sin(ax)dx = -(1/a)\cos ax \qquad \int \cos(ax)dx = (1/a)\sin ax$$

$$\int \sin^2(ax)dx = \frac{x}{2} - \frac{\sin 2ax}{4a}$$

$$\int x\sin(ax)dx = (1/a^2)(\sin ax - ax\cos ax)$$

$$\int x^2\sin(ax)dx = (1/a^3)(2ax\cos ax + 2\cos ax - a^2x^2\cos ax)$$

$$\int \sin(ax)\sin(bx) = \frac{\sin(a-b)x}{2(a-b)} - \frac{\sin(a+b)x}{2(a+b)} \qquad (a^2 \neq b^2)$$

$$\int \sin(ax)\cos(bx) = -\left[\frac{\cos(a-b)x}{2(a-b)} + \frac{\cos(a+b)x}{2(a+b)}\right] \qquad (a^2 \neq b^2)$$

$$\int \cos(ax)\cos(bx) = \frac{\sin(a-b)x}{2(a-b)} + \frac{\sin(a+b)x}{2(a+b)} \qquad (a^2 \neq b^2)$$

$$\int e^{ax}dx = \frac{e^{ax}}{a}$$

$$\int xe^{ax}dx = \frac{e^{ax}}{a^2}(ax-1)$$

$$\int x^2e^{ax}dx = \frac{e^{ax}}{a^3}(a^2x^2 - 2ax + 2)$$

$$\int e^{ax}\sin(bx)dx = \frac{e^{ax}}{a^2+b^2}(a\sin(bx) - b\cos(bx))$$

$$\int e^{ax}\cos(bx)dx = \frac{e^{ax}}{a^2+b^2}(a\cos(bx) + b\sin(bx))$$

$$\int \cos^2 ax dx = \frac{x}{2} + \frac{\sin 2ax}{4a}$$

$$\int x\cos(ax)dx = (1/a^2)(\cos(ax) + ax\sin(ax))$$

$$\int x^2\cos(ax)dx = (1/a^3)(2ax\cos ax - 2\sin ax + a^2x^2\sin ax)$$

Table C–4 Definite Integrals

$$\int_0^\infty x^n e^{-ax} dx = \frac{n!}{a^{n+1}}$$

$$\int_0^\infty e^{-r^2x^2} dx = \frac{\sqrt{\pi}}{2r}$$

$$\int_0^\infty x e^{-r^2x^2} dx = \frac{1}{2r^2}$$

$$\int_0^\infty x^2 e^{-r^2x^2} dx = \frac{\sqrt{\pi}}{4r^3}$$

$$\int_0^\infty x^n e^{-r^2x^2} dx = \frac{\Gamma[(n+1)/2]}{2r^{n+1}}$$

$$\int_0^\infty \frac{\sin ax}{x} dx = \frac{\pi}{2}, 0, -\frac{\pi}{2} \qquad \text{for } a > 0, a = 0, a < 0$$

$$\int_0^\infty \frac{\sin^2 x}{x} dx = \frac{\pi}{2}$$

$$\int_0^\infty \frac{\sin^2 x}{x^2} dx = \frac{\pi}{2}$$

$$\int_0^\infty \frac{\sin^2 ax}{x^2} dx = |a|\frac{\pi}{2}$$

For m and n integers

$$\int_0^\pi \sin^2(mx)dx = \int_0^\pi \sin^2(x)dx = \int_0^\pi \cos^2(mx)dx = \int_0^\pi \cos^2(x)dx = \frac{\pi}{2}$$

Continued on next page

Table C–4 Definite Integrals (Continued)

$$\int_0^{\pi} \sin(mx)\cos(nx)dx = \begin{cases} \dfrac{2m}{m^2 - n^2} & \text{if } (m+n) \text{ odd} \\ 0 & \text{if } (m+n) \text{ even} \end{cases}$$	
$$\frac{1}{2\pi}\int_0^{2\pi} e^{jr\cos x}dx = J_o(r)$$	
$$\frac{j^{-n}}{2\pi}\int_0^{2\pi} e^{jr\cos x}\cos(nx)dx = J_n(r)$$	

Table C–5 Probability Functions

Discrete Distribution

Binomial

$$Pr(k) = \binom{n}{k}p^k q^{n-k} \qquad k = 0, 1, 2, \ldots\ldots\ldots n$$

$$= 0 \qquad \text{otherwise}$$

$$0 < p < 1, \quad q = 1 - p$$

$$p(x) = \sum_{k=0}^{n} \binom{n}{k} p^k q^{n-k} \delta(x - k)$$

$$\bar{x} = np$$

$$\sigma_x^2 = npq$$

Poisson

$$Pr(k) = \frac{\lambda^k e^{-\lambda}}{k!} \qquad k = 0, 1, 2, \ldots\ldots.$$

$$p(x) = \sum_{k=0}^{n} \frac{\lambda^k e^{-\lambda}}{k!}\delta(x - k)$$

$$\bar{x} = \lambda$$

$$\sigma_x^2 = \lambda$$

Continued on next page

Table C–5 Probability Functions (Continued)

Continuous Distribution

Exponential

$p(x) = ae^{-ax} \qquad x > 0$

$\quad = 0 \qquad$ otherwise

$\bar{x} = a^{-1}$

$\sigma_x^2 = a^{-2}$

Gaussian (normal)

$$p(x) = \frac{1}{\sigma_x\sqrt{2\pi}}\exp\left[-\frac{(x-\bar{x})^2}{2\sigma_x^2}\right] \qquad -\infty \le x \le \infty$$

$E\{x\} = \bar{x}$

$E\{(x-\bar{x})^2\} = \sigma_x^2$

Bivariate Gaussian (normal)

$$p(x, y) = \frac{1}{2\pi\sigma_x\sigma_y\sqrt{1-\rho^2}}\exp\left\{-\frac{1}{2(1-\rho^2)}\left[\left(\frac{x-\bar{x}}{\sigma_x}\right)^2 + \left(\frac{y-\bar{y}}{\sigma_y}\right)^2 - \frac{2\rho}{\sigma_x\sigma_y}(x-\bar{x})(y-\bar{y})\right]\right\}$$

$E\{x\} = \bar{x}$

$E\{y\} = \bar{y}$

$E\{(x-\bar{x})^2\} = \sigma_x^2$

$E\{(y-\bar{y})^2\} = \sigma_y^2$

$\rho = \frac{E[xy] - \mu_x\mu_y}{\sigma_x\sigma_y}$ is the correlation coefficient

$E\{(x-\bar{x})(y-\bar{y})\} = \sigma_x\sigma_y\rho$

Continued on next page

Table C–5 Probability Functions (Continued)

Continuous distribution

Rayleigh

The *pdf* of the <u>envelope</u> of Gaussian random noise having zero mean and variance σ_n^2

$$p(r) = \frac{r}{\sigma_n^2}\exp[-r^2/\sigma_n^2] \qquad r \geq 0$$

$$E\{r\} = \bar{r} = \sigma_n\sqrt{\pi/2}$$

$$E\{(r-\bar{r})^2\} = \sigma_r^2 = \left(2-\frac{\pi}{2}\right)\sigma_n^2$$

Rician

The *pdf* of the <u>envelope</u> of a sinusoid with amplitude A plus zero mean Gaussian noise with variance σ^2

$$p(r) = \frac{r}{\sigma^2}\exp\left[-\frac{(r^2+A^2)}{2\sigma^2}\right]I_0\left(\frac{Ar}{\sigma^2}\right) \qquad r \geq 0$$

For $A/\sigma \gg 1$, this is closely approximated by the following Gaussian *PDF*:

$$p(r) \cong \frac{1}{\sigma\sqrt{2\pi}}\exp\left[-\frac{(r-A)^2}{2\sigma^2}\right]$$

Uniform

$$p(x) = \begin{cases} \dfrac{1}{b-a} & a < x < b \\ 0 & \text{elsewhere} \end{cases}$$

$$\bar{x} = \frac{a+b}{2}$$

$$\sigma_x^2 = \frac{(b-a)^2}{12}$$

Table C–6 Functions

Function	Definition
Rectangular	$\Pi(t/T) = \begin{cases} 1 & \lvert t\rvert \le T/2 \\ 0 & \lvert t\rvert > T/2 \end{cases}$
Unit Pulse	$\psi(t/T) = \begin{cases} 1 & 0 \le t \le T \\ 0 & \text{otherwise} \end{cases}$
Triangular	$\Lambda(t/T) = \begin{cases} 1 - \dfrac{\lvert t\rvert}{T} & \lvert t\rvert \le T \\ 0 & \lvert t\rvert > T \end{cases}$
Unit Step	$u(t) = \begin{cases} 1 & t > 0 \\ 0 & t < 0 \end{cases}$
Signum	$\operatorname{sgn}(t) = \begin{cases} 1 & t > 0 \\ -1 & t < 0 \end{cases}$
Impulse	$\delta(t) = \begin{cases} 1 & t = 0 \\ 0 & t \ne 0 \end{cases}$
Bessel	$J_n(\beta) = \dfrac{1}{2\pi}\displaystyle\int_{-\pi}^{\pi} e^{j(\beta\sin\theta - n\theta)}\,d\theta$
n th moment of a random variable X	$E[X^n] = \displaystyle\int_{-\infty}^{\infty} x^n f_X(x)\,dx$ where $n = 0,1,2,\ldots$ and $f_X(x)$ is the *pdf* of X
n th central moment of X	$E[(X-\mu)^n] = \displaystyle\int_{-\infty}^{\infty} (x-\mu)^n f_X(x)\,dx$ where $\mu = E[X]$
Variance of X	$\sigma^2 = \displaystyle\int_{-\infty}^{\infty} (x-\mu)^2 f_X(x)\,dx$ where $\mu = E[X]$

APPENDIX D

Abbreviations and Acronyms

A

AC	Access Channel
ACA	Adaptive Channel Allocation
ACF	Autocorrelation Function
ACI	Adjacent Channel Interference
ACK	Acknowledge
ADM	Adaptive Delta Modulation
ADPCM	Adaptive Digital Pulse Code Modulation
AIN	Advanced Intelligent Network
AIC	Akaike Information theoretic Criteria
AM	Amplitude Modulation
AMPS	Advanced Mobile Phone System
ANSI	American National Standards Institute
AOA	Angle Of Arrival
APC	Adaptive Predictive Coding
ARFCN	Absolute Radio Frequency Channel Numbers
ARQ	Automatic Repeat Request
ASIC	Application Specific Integrated Circuit
ATC	Adaptive Transform Coding
ATM	Asynchronous Transfer Mode

AUC Authentication Center
AWGN Additive White Gaussian Noise

B

BER Bit Error Rate
BCH Bose-Chaudhuri-Hocquenghem, also Broadcast Channel
BCCH BroadCast Control Channel
BFN BeamForming Network
BFSK Binary Frequency Shift Keying
BISDN Broadband Integrated Services Digital Network
BIU Base Station Interface Unit
BL-SPECCORR Band-Limited Spectral Correlation Ratio
BOC Bell Operating Company
BPSK Binary Phase Shift Keying
BRI Basic Rate Interface
BSC Base Station Controller
BSIC Base Station Identity Code
BS Base Station
BSS Base Station Subsystem
BT 3 dB Bandwidth-bit-duration product for GMSK
BTA Basic Trading Area
BTS Base Transceiver Station
BU Bad Urban

C

CAI Common Air Interface
CCCH Common Control Channel
CCH Control Channel
CCI Co-channel Interference
CCIR Consultative Committee for International Radiocommunications
CCITT International Telegraph and Telephone Consultative Committee

CCS	Common Channel Signaling
CD	Collision Detection
CDF	Cumulative Distribution Function
CDG	The CDMA Development Group (now cdmaOne)
CDMA	Code Division Multiple Access
cdmaOne	The CDMA development body (formerly CDG) and umbrella term for EIA IS-95A, ANSI J-STD-008, and their successors.
CDPD	Cellular Digital Packet Data
CDVCC	Coded Digital Verification Color Code
CELP	Code Excited Linear Predictor
CFP	Cordless Fixed Part
C/I	Carrier-to-Interference Ratio
CIC	Circuit Identification Code
CINR	Carrier-to-Interference-and-Noise-Ratio
CIR	Carrier-to-Interference-Ratio
CIU	Cellular Controller Interface Unit
CLEC	Competitive Local Exchange Carier
CLNP	Connectionless Protocol (Open System Interconnect)
CMA	Constant Modulus Algorithm
CO	Central Office
CODEC	Coder/decoder
CPE	Customer Premises Equipment
CPFSK	Continuous Phase Frequency Shift Keying
CPP	Cordless Portable Part
CRC	Cyclic Redundancy Code
CRLB	Cramér-Rao Lower Bound
CRMB	Cramér-Rao Matrix Bound
CSMA	Carrier Sense Multiple Access
CT2	Cordless Telephone -2
CVSDM	Continuously Variable Slope Delta Modulation
CW	Continuous Wave
CYCCOR	Cyclic Cross-Correlation

D

DAM	Diagnostic Acceptability Measure
DAC	Divide And Conquer
DBAS	Database Service Management System
DCA	Dynamic Channel Allocation
DCCH	Dedicated Control Channel
DCE	Data Circuit Terminating Equipment
DCS	Digital Communication System
DCS1800	Digital Communication System — 1800
DCT	Discrete Cosine Transform
DD	Decision-Directed Algorithm
DDC	Digital DownConversion
DECT	Digital European Cordless Telephone
DEM	Digital Elevation Models
DF	Direction Finding
DFE	Decision Feedback Equalization
DFT	Discrete Fourier Transform
DLC	Data Link Control
DLL	Delay Locked Loop
DM	Delta Modulation
DOA	Direction-Of-Arrival
DPCM	Differential Pulse Code Modulation
DQPSK	Differential Quadrature Phase Shift Keying
DRT	Diagnostic Rhyme Test
DS	Direct Sequence
DSAT	Digital Supervisory Audio Tone
DS-CDMA	Direct Sequence CDMA
DSE	Data Switching Exchange
DS/FHMA	Hybrid Direct Sequence/Frequency Hopped Multiple Access
DSP	Digital Signal Processing
DS-SS	Direct Sequence Spread Spectrum
DST	Digital Signaling Tone
DTC	Digital Traffic Channel
DTE	Data Terminal Equipment

DTMF	**Dual Tone Multiple Frequency**
DTX	**Discontinuous Transmission Mode**
DUP	**Data User Part**

E

EbNo	**Energy per Bit to Noise Spectral Density Ratio**
E_b/N_0	**Bit Energy-to-Noise Density**
EIA	**Electronic Industry Association**
EIR	**Equipment Identity Register**
EIRP	**Effective Isotropic Radiated Power**
EOC	**Embedded Operations Channel**
erf	**Error Function**
erfc	**Complementary Error Function**
E-SMR	**Extended — Specialized Mobile Radio**
ESN	**Electronic Serial Number**
ESPRIT	**Estimation of Signal Parameters via Rotational Invariance Technique**
E-TACS	**Extended Total Access Communication System**
ETSI	**European Telecommunications Standards Institute**
EVRC	**Enhanced Variable Rate Coder**

F

FACCH	**Fast Associated Control Channel**
FAF	**Floor Attenuation Factor**
FBF	**Feedback Filter**
FC	**Fast Channel**
FCC	**Federal Communications Commission, Inc.,** also **Forward Control Channel**
FCCH	**Frequency Correction Channel**
FCDMA	**Hybrid FDMA/CDMA**
FDD	**Frequency Division Duplex**
FDMA	**Frequency Division Multiple Access**
FDTC	**Forward Data Traffic Channel**

FEC	Forward Error Correction
FER	Frame Error Rate
FFF	Feedforward Filter
FFSR	Feedforward Signal Regeneration
FH	Frequency Hop
FHMA	Frequency Hop Multiple Access
FH-SS	Frequency Hop Spread Spectrum
FIM	Fisher Information Matrix
FM	Frequency Modulation
FN	Frame Number
FPLMTS	Future Public Land Mobile Telephone System
FQI	Frame Quality Indicator
FSE	Fractionally Spaced Equalizer
FSK	Frequency Shift Keying
FTF	Fast Transversal Filter
FVC	Forward Voice Channel

G

GA	Gaussian Approximation
GAA	Gaussian Angle-of-Arrival
GBSB	Geometrically Based Single Bounce
GBSBCM	GBSB Circular Model
GBSBEM	GBSB Elliptical Model
GCC	Generalized Cross-Correlation
GDOP	Geometric Dilution Of Precision
GIS	Geographical Information System
GIU	Gateway Interface Unit
GLE	Gerschgorin Likelihood Estimator
GMSK	Gaussian Minimum Shift Keying
GOS	Grade of Service
GPS	Global Positioning System
GSM	Global System for Mobile Communications
GSO	Gram-Schmidt Orthogonalization
GWSSUS	Gaussian Wide Sense Stationary Uncorrelated Scattering Model

H

HDB	Home Database
HDC	Hybrid Downconverter
HLR	Home Location Register
HT	Hannan-Thomson

I

IDCT	Inverse Discrete Cosine Transform
IDFT	Inverse Discrete Fourier Transform
IEEE	Institute of Electrical and Electronics Engineers
IF	Intermediate Frequency
IFFT	Inverse Fast Fourier Transform
IGA	Improved Gaussian Approximation
IIR	Infinite Impulse Response
IM	Intermodulation
IMSI	International Mobile Subscriber Identity
IMT-2000	International Mobile Telecommunications 2000
IMTS	Improved Mobile Telephone Service
IP	Internet Protocol
IS-54	EIA Interim Standard for U.S. Digital Cellular (USDC)
IS-95	EIA Interim Standard for U.S. Code Division Multiple Access
IS-96A	EIA Interim Standard for the 8.6 kbps vocoder
IS-136	EIA Interim Standard 136 — USDC with Digital Control Channels
ISDN	Integrated Services Digital Network
ISI	InterSymbol Interference
ISM	Industrial, Scientific, and Medical
ISLP-CMA	Iterative Least Squares Projection based CMA
ISUP	ISDN User Part
ITFS	Intructional Television Fixed Service
ITU	International Telecommunications Union
IXC	Interexchange Carrier

J

JDC	Japanese Digital Cellular (later called Pacific Digital Cellular)
JRC	Joint Radio Committee
JTACS	Japanese Total Access Communication System
JTC	Joint Technical Committee

L

LAN	Local Area Network
LAR	Log-Area Ratio
LATA	Local Access and Transport Area
LBT	Listen-Before-Talk
LCC	Lost Call Cleared
LCMV	Linearly Constrained Minimum Variance
LCD	Lost Call Delayed
LCR	Level Crossing Rate
LEC	Local Exchange Carrier
LEO	Low Earth Orbit
LES	Linear Equally Spaced
LMDS	Local Multipoint Distribution Service
LMS	Least Mean Square
LOB	Lines-Of-Bearing
LOS	Line-Of-Sight
LPC	Linear Predictive Coding
LPD	Low Probability Detection
LPI	Low Probability of Interception
LS	Least Squares
LS-DD	Least Squares Decision-Directed
LS-DRMTA	Least Squares Despread-Respread Multitarget Array
LS-DRMTCMA	Least Squares Despread-Respread Multitarget Constant Modulus Algorithm
LSSB	Lower Single Side Band
LTE	Linear Transversal Equalizer
LTP	Long Term Prediction

M

MAC	Medium Access Control
MAI	Multiple Access Interference
MAHO	Mobile Assisted Handoff
MAN	Metropolitan Area Network
M-ary	Multiple Level Modulation
MDBS	Mobile Database Stations
MDLP	Mobile Data Link Protocol
MDS	Multipoint Distribution Service
MDL	Minimum Descriptive Length
MFJ	Modified Final Judgment
MFSK	Minimum Frequency Shift Keying
MIN	Mobile Identification Number
ML	Maximum Likelihood
MLSE	Maximum Likelihood Sequence Estimation
MMDS	Multichannel Multipoint Distribution Service
MMSE	Minimum Mean Square Error
MOS	Mean Opinion Score
MoU	Memorandum of Understanding
MPE	Multi-pulse Excited
MPSK	Minimum Phase Shift Keying
MRC	Maximal Ratio Combiner
MSB	Most Significant Bit
MSC	Mobile Switching Center
MSCID	MSC Identification
MSE	Mean Square Error
MSK	Minimum Shift Keying
MSU	Message Signal Unit
MTA	Major Trading Area
MT-DD	Multitarget Decision Directed
MT-LSCMA	Multitarget Least Squares Constant Modulus Algorithm
MT-LSDD	Multitarget Least Squares Decision-Directed
MT-SDDD	Multitarget Steepest-Descent Decision Directed
MTP	Message Transfer Part

MTSO Mobile Telephone Switching Office
MUX MUltipleXer
MUSIC MUltiple SIgnal Classification
MVDR Minimum Variance Distortionless Response

N

NACK Negative Acknowledge
NAMPS Narrowband Advanced Mobile Phone System
NBFM Narrowband Frequency Modulation
N-ISDN Narrowband Integrated Service Digital Network
NMT-450 Nordic Mobile Telephone — 450
NRZ Non-Return to Zero
NSP Network Service Part
NSS Network and Switching Subsystem
NTACS Narrowband Total Access Communication System
NTT Nippon Telephone & Telegraph

O

OBS Obstructed
OFDM Orthogonal Frequency Division Multiplexing
OMAP Operations Maintenance and Administration Part
OMC Operation Maintenance Center
OQPSK Offset Quadrature Phase Shift Keying
OSI Open System Interconnect
OSS Operation Support Subsystem

P

PABX Private Automatic Branch Exchange
PACS Personal Access Communications System
PAD Packet Assembler Disassembler
PBX Private Branch Exchange

PCH	**Paging Channel**
PCM	**Pulse Code Modulation**
PCN	**Personal Communication Network**
PCS	**Personal Communication System**
pda-gram	**Power-Delay-Angle Profiles**
PDC	**Pacific Digital Cellular**
pdf	**probability density function**
PG	**Processing Gain**
PH	**Portable Handset**
PHP	**Personal Handyphone**
PHS	**Personal Handyphone System**
PIC	**Parallel Interference Canceler**
PL	**Position Location**
PLL	**Phase Locked Loop**
PLMR	**Public Land Mobile Radio**
PMRM	**Power Measurement Report Message**
PN	**Pseudo-Noise**
PR	**Packet Radio**
PRI	**Primary Rate Interface**
PSD	**Power Spectral Density**
PSK	**Phase Shift Keying**
PSMM	**Pilot Strength Measurement Message**
PSTN	**Public Switched Telephone Network**
PTI	**Permanent Terminal Identifier**

Q

QAM	**Quadrature Amplitude Modulation**
QCELP	**Qualcomm Code Excited Linear Predictive Coder**
QMF	**Quadrature Mirror Filter**
QPSK	**Quadrature Phase Shift Keying**

R

RACE	Research on Advanced Communications in Europe
RACH	Random Access Channel
RCC	Reverse Control Channel
RCS	Radar Cross Section
RD-LAP	Radio Data Link Access Protocol
RDTC	Reverse Data Traffic Channel
RELP	Residual Excited Linear Predictor
RF	Radio Frequency
RFP	Radio Fixed Part
RLC	Resistor Inductor Capacitor
RLS	Recursive Least-Squares
rms	Root Mean Square
RPCU	Radio Port Control Unit
RRMS	Radio Resource Management Protocol
RS	Reed Solomon
RSSI	Received Signal Strength Indication
RVC	Reverse Voice Channel
Rx	Receiver
RZ	Return to Zero

S

SACCH	Slow Associated Control Channel
SAT	Supervisory Audio Tone
SBC	System Broadcasting Channel
	also Sub-Band Coding
SC	Slow Channel
SCCP	Signaling Connection Control Part
SCH	Synchronization Channel
SCM	Station Class Mark
SCORE	Self COherence REstoral
SCOT	Smoothed COherence Transform
SCP	Service Control Point

SDCCH	**Stand-alone Dedicated Control Channel**
SD	**Steepest-Descent**
SD-DD	**Steepest-Descent Decision-Directed**
SDMA	**Space Division Multiple Access**
SEIGA	**Simplified Expression for the Improved Gaussian Approximation**
SELP	**Stochastically Excited Linear Predictive coder**
SEP	**Switching End Points**
SFM	**Spectral Flatness Measure**
SH	**Sequential Hypothesis**
SI	**Spherical-Interpolation**
S/I	see **SIR**
SID	**Station Identity**
SIM	**Subscriber Identity Module**
SINR	**Signal-to-Interference-and-Noise-Ratio**
SIR	**Signal-to-Interference-Ratio**
SIRCIM	**Simulation of Indoor Radio Channel Impulse response Models**
SISP	**Site Specific Propagation**
SMR	**Specialized Mobile Radio**
SMRCIM	**Simulation of Mobile Radio Channel Impulse response Models**
SMS	**Short Messaging Service**
	also **Service Management System**
S/N	see **SNR**
SOI	**Signals-Of-Interest**
SNOI	**Signals-Not-Of-Interest**
SNR	**Signal-to-Noise Ratio**
SP	**Signaling Point**
SPECCOA	**Spectral-Coherence Alignment**
SQNR	**Signal-to-Quantization Noise Ratio**
SS	**Spread Spectrum**
SSB	**Single Side Band**
SSMA	**Spread Spectrum Multiple Access**
SS7	**Signaling System No. 7**
ST	**Signaling Tone**
STP	**Short Term Prediction,**
	also **Signaling Transfer Point**

SX Spherical-Intersection
SYN Synchronization Channel

T

T_DROP: power Threshold to determine when a pilot channel is DROPed by a mobile unit
T_TDROP: Threshold Timer to determine when pilot channels are DROPed by a mobile unit
TACS Total Access Communications System
TCDMA Time Division CDMA
TCH Traffic Channel
TCM Trellis Coded Modulation
TDD Time Division Duplex
TDFH Time Division Frequency Hop
TDMA Time Division Multiple Access
TDN Temporary Directory Number
TDOA Time Difference Of Arrival
TIA Telecommunications Industry Association
TIU Trunk Interface Unit
TLS Total Least Squares
TOA Time-Of-Arrival
TSOA Time-Sum-Of-Arrival
TTIB Transparent Tone-In-Band
TU Typical Urban
TUP Telephone User Part
Tx Transmitter

U

UF Urban Factor
UHF Ultra High Frequency
UMTS Universal Mobile Telecommunications System
U.S. United States of America
USD Uniform Sectored Distribution

USGS	United States Geological Survey
USSB	Upper Single Side Band
UTRA	UMTS Terrestrial Radio Access

V

VAD	Voice Activity Detector
VCI	Virtual Circuit Identifier
VCIR	Vector Channel Impulse Response
VCO	Voltage Controlled Oscillator
VDB	Visitor Database
VGA	Variable Gain Amplifier
VIU	Visitor Interface Unit
VLSI	Very Large-Scale Integration
VMAC	Voice Mobile Attenuation Code
VPD	Variable Phase Device
VQ	Vector Quantization
VSELP	Vector Sum Excited Linear Predictor

W

WACS	Wireless Access Communication System (later called PACS)
WAN	Wide Area Network
WARC	World Administrative Radio Conference
W-CDMA	Wideband CDMA
WIN	Wireless Information Network
WIU	Wireless Interface Unit
WLAN	Wireless Local Area Network
WLL	Wireless Local Loop
WLS	Weighted Least Squares
WUPE	Wireless User Premises Equipment

X

XOR **eXclusive-OR**

Z

ZF **Zero Forcing**

References

[Aat97a] M. Aatique, G. Mizusawa, and B. D. Woerner, "Performance of Hyperbolic Position Location Techniques for Code Division Multiple Access," *Wireless 97*, Boston, MA, July 1997.

[Aat97b] M. Aatique and B. D. Woerner, "Evaluation of TDOA Techniques for Position Location in CDMA Systems," *Masters Thesis MPRG-TR-97-15*, Mobile & Portable Radio Research Group, Virginia Tech, Blacksburg, VA, 1997.

[Abe87] J. S. Abel and J. O. Smith, "The Sphereical Interpolation Method for Closed Form Passive Localization Using Range Difference Measurements," *Proc. of the IEEE Int'l. Conf. on Acoustics, Speech, and Signal Processing*, pp. 471–474, 1987.

[Abe89] J. S. Abel and J. O. Smith, "Source Range and Depth Estimation from Multipath Range Difference Measurements," *IEEE Trans. on Acoustics, Speech, and Signal Processing*, Vol. 37, No. 8, pp. 1157–1165, Aug. 1989.

[Abe90] J. S. Abel and J. O. Smith, "A Divide-and-Conquer Approach to Least-Squares Estimation," *IEEE Trans. on Aerospace and Electronic Systems*, Vol. 26, pp. 423–427, Mar. 1990.

[Ada86] F. Adachi, M. T. Feeney, A. G. Williamson, and J. D. Parsons, "Crosscorrelation between the Envelopes of 900 MHz Signals Received at a Mobile Radio Base Station Site," *IEE Proceedings*, Vol. 133, Pt. F, No 6, pp. 506–512, Oct. 1986.

[Age86] B. G. Agee, "The Least-Squares CMA: A New Technique for Rapid Correction of Constant Modulus Signals," *Proc. of the IEEE Int'l. Conf. on Acoustics, Speech, and Signal Processing*, pp. 19.2.1–19.2.4, 1986.

[Age89] B. Agee, "Blind Separation and Capture of Communications Signals Using a Multi-target Constant Modulus Beamformer," *1989 IEEE Military Communications Conf.*, Boston, MA, Oct. 1989.

[Age90] B. G. Agee, S. V. Schell, and W. A. Gardner, "Spectral Self-Coherence Restoral: A New Approach to Blind Adaptive Signal Extraction Using Antenna Arrays," *Proc. of the IEEE*, Vol. 78, No. 4, pp. 753–767, 1990.

[Age93] B. Agee, "Solving the Near-Far Problem: Exploitation of Spatial and Spectral Diversity in Wireless Personal Communication Networks," *Proc. Third Va. Tech Symposium on Wireless Personal Comm.*, June 1993.

[Alt56] F. J. Altman and W. Sichak, "A Simplified Diversity Communciation System for Behind-the-Horizon Links," *IRE Trans. Comm. Sys.*, CS-4, Mar. 1956.

[Amo91] F. Amoroso and W. W. Jones, "Modelling Direct Sequence Pseudonoise (PN) Satellite Reception with Directional Antennas in the Dense Scatterer Mobile Environment" *Int. Journ. of Satellite Comm.*, Vol. 9, No. 2, Mar.-Apr.1991.

[Amo93] F. Amoroso and W. W. Jones, "Geometric Model for DSPN Satellite Reception in the Dense Scatterer Mobile Environment," *IEEE Trans. on Comm.*, Vol. 41, No. 3, Mar. 1993.

[Ana95] Anaren, *Microwave Components*, Anaren Microwave, Inc., Syracuse, NY 13057, 1995.

[And63] T. W. Anderson, "Asymptotic Theory for Principle Component Analysis," *Ann. Math. Stat.*, Vol. 34, pp. 122–148, 1963.

[And91] S. Anderson, M. Millnert, M. Viberg, and B. Wahlberg, "An Adaptive Array for Mobile Communication Systems," *IEEE Trans. on Vehicular Technology,* Vol. 40, No. 1, Feb. 1991.

[Ans95a] ANSI, "Personal Communications Services Air Interface Specification," *J-STD-007*, 1995.

[Ans95b] ANSI, "Personal Station-Base Station Compatibility Requirements for 1.8 to 2.0 GHz Code Division Multiple Access (CDMA) Personal Communcations Systems," *J-STD-008*, (EIA SP-3384), 1995.

[App76] S. P. Applebaum, "Adaptive Arrays," *IEEE Trans. on Antennas & Propagation*, Vol. AP-24, No. 5, Sept. 1976.

[Asz96] D. Aszetly, "On Antenna Arrays in Mobile Communication Systems: Fast Fading and GSM Base Station Receiver Algorithms," *Ph.D. Dissertation*, Royal Institute of Technology, Stockholm, Sweden, Mar. 1996.

[Ban85] S. Bancroft, "An Algebraic Solution of the GPS Equations," *IEEE Trans. on Aerospace and Electronic Systems*, Vol. AES-21, No. 7, Jan. 1985.

[Bar83] A. J. Barabell, "Improving the Resolution Performance of Eigenstructure-based Direction Finding Algorithms," *Proc. of the IEEE Int'l. Conf. on Acoustics, Speech, and Signal Processing-83,* pp. 336–339, 1983.

[Bar84] A. J. Barabell, J. Capon, D. F. Delong, J. R. Johnson, and K. Senne, "Performance Comparison of Superresolution Array Processing Algorithms," *Technical Report TST-72,* Lincoln Laboratory, M.I.T., 1984.

[Bei79] C. S. Beightler, D. T. Phillips, and D. J. Wilde, *Foundations of Optimizations*, Prentice Hall, NJ, 1979.

[Ber87] R. C. Bernhardt, "Macroscopic Diverity in Frequency Reuse Systems," *IEEE Journal on Selected Areas in Comm.*, Vol-SAC 5, pp. 862–878, June 1987.

[Bie79] Bienvenu, W. J., Kopp, "Principle de la Goniometrie Passive Adaptive," *Proc. 7'eme Colloque GRESIT.* Nice, France, 106/1-106/10, 1979.

[Bie93] T. E. Biedka, "Subspace Constrained SCORE Algorithms," *Proc. 27th Asilomar Conf. on Signals, Systems*, pp. 716–720, Nov. 1993.

[Bie95] T. E. Biedka, "A Method for Reducing Computations in Cyclostationarity-Exploiting Beamforming," *Proc. of the IEEE Int'l. Conf. on Acoustics, Speech, and Signal Processing*, pp. 1828–1831, 1995.

[Big95] L. Bigler, H. P. Lin, S. S. Jeng, and G. Xu, "Experimental Direction-of-Arrival and Spatial Signature Measurements at 900 MHz for Smart Antenna Systems," *1995 IEEE 45th Vehicular Technology Conf.*, Chicago, IL, 1995.

[Bla96] J. J. Blanz, A. Klein, and W. Mohr, "Measurement-Based Parameter Adaptation of Wideband Spatial Mobile Radio Channel Models," *IEEE Fourth Int'l. Symposium on Spread Spectrum Techniques & Applications*, pp. 91–97, 1996.

[But61] J. Butler and R. Lowe, "Beamforming Matrix Simplifies Design of Electronically Scanned Antennas," *Electronic Design*, Apr. 1961.

[Cal88] G. Calhoun, *Digital Cellular Radio*, Artech House, Norwood, MA, 1988.

[Cam92] R. Cameron and B. D. Woerner, "An Analysis of CDMA with Imperfect Power Control," *Proc. of the IEEE Vehicular Technology Conf.*, May 1992.

[Cap69] J. Capon, "High Resolution Frequency-Wavenumber Spectral Analysis," *Proc. of the IEEE*, Vol. 57, No. 8, pp. 1408–1418, Aug. 1969.

[Cap79] J. Capon, "Maximum Likelihood Spectral Estimation," *Nonlinear Methods of Spectral Analysis*. Ed. by S. Haykin, Springler, NY, pp. 155–179, 1979.

[Car73] G. C. Carter, A. H. Nuttall, and P. C. Cable, "The Smoothed Coherence Transform," *Proc. of the IEEE*, Vol. 61, pp. 1497–1498, Oct. 1973.

[Car81] G. C. Carter, "Time Delay Estimation for Passive Sonar Signal Processing," *IEEE Trans. on Acoustics, Speech, and Signal Processing*, Vol. ASSP-29, No. 3, pp. 463–470, June 1982.

[Car87] G. C. Carter, "Coherence and Time Delay Estimation," *Proc. of the IEEE*, Vol. 75, pp.236–255, Feb. 1991.

[Cas92a] L. Castedo, C. Y. Tseng, and L. J. Griffiths, "A New Cost Function for Adaptive Beamforming Using Cyclostationary Signal Properties," Vol. 4, pp. 284–287, Apr. 1993.

[Cas92b] L. Castedo and C. Y. Tseng, "Behavior of Adaptive Beamformers Based on Cyclostationary Signal Properties in Multipath Environments," *Proc. 27th Asilomar Conf. on Signals, Systems*, Vol. 1, pp. 653–657, Nov. 1993.

[Cha92] G. K. Chan, "Effects of Sectorization on the Spectrum Efficiency of Cellular Radio Systems," *IEEE Trans. on Vehicular Technology,* Vol. 41, No. 3, Aug. 1992.

[Cha94] Y. T. Chan and K. C. Ho, "A Simple and Efficient Estimator for Hyperbolic Location," *IEEE Trans. on Signal Processing*, Vol. 42, No. 8, pp. 1905–1915, Aug. 1994.

[Che84] C. T. Chen, *Linear System Theory and Design*, Holt, Rinehart, and Winston, Inc., NY, 1984.

[Chi94a] I. Chiba, T. Takahashi, and Y. Karasawa, "Transmitting Null Beam Forming with Beam Space Adaptive Array Antennas," *Proc. of the IEEE Vehicular Technology Conf.*, pp. 1498–1502, 1994.

[Chi95] I. Chiba, W. Chujo, and M. Fujise, "Beam-Space CMA Adaptive Array Antennas," *Electronics and Communications in Japan*, Part 1, Vol. 78, No. 2, pp. 85–95, 1995.

[Chu87] J. C. I. Chuang, "The Effects of Time Delay Spread on Portable Radio Communications Channels with Digital Modulation," *IEEE Journal on Selected Areas in Comm.*, Vol. SAC-5, No. 5, June 1987.

[Cla68] R. H. Clarke, "A Statistical Theory of Mobile-Radio Reception," *Bell Systems Technical Journal*, Vol. 47, pp. 957–1000, 1968.

[Col85] R. E. Collin, *Antennas and Radiowave Propagation*, McGraw-Hill, NY, 1985.

[Com78] R. T. Compton, "An Adaptive Array in a Spread-Spectrum Communication System," *Proc. of the IEEE*, Vol. 66, No. 3, Mar. 1978.

[Com88] R. T. Compton, *Adaptive Antennas*, Prentice Hall, NJ, 1988.

[Coo78] G. R. Cooper and R. W. Nettleton, "A Spread-Spectrum Technique for High-Capacity Mobile Communications," *IEEE Trans. on Vehicular Technology,* Vol. VT-27, No. 4, Nov. 1978.

[Coo86] G. R. Cooper and C.D. McGillem, *Probabilistic Methods of Signal and System Analysis*, Holt, Rinehart, and Winston, NY, 1986.

[Cou83] L. W. Couch, *Digital and Analog Communication Systems*, Second Edition, Maxmillan Publishing Co., NY, 1983.

[Cox72] D. C. Cox, "Delay Doppler Characteristics of Multipath Propagation at 910 MHz in a Suburban Mobile Radio Environment," *IEEE Trans. Antennas & Propagation*, Vol. AP-20, No. 5, Sept. 1972.

[Cox84] D. C. Cox, R. Murray, and A. Norris, "800 MHz Attenuation Measured In and Around Suburban Houses," *AT&T Bell Laboratory Technical Journal*, Vol. 673, No. 6, July-Aug. 1984.

[Dar91] M. J. Darcy, "Digital Down-Conversion," *IEE Int'l. Conf. on Analogue to Digital and Digital to Analogue Conversion*, Swansea,Wales, Sept. 1991.

[Dev87a] D. M. J. Devasirvatham, "Multipath Time Delay Spread in the Digital Portable Radio Environment," *IEEE Comm. Magazine,* Vol. 25, No. 6, June 1987.

[Dev87b] D. M. J. Devasirvatham, "A Comparison of Time Delay Spread and signal level Measurements with Two Dissimilar Office Buildings," *IEEE Trans. on Antennas & Propagation*, Vol. AP-35, No. 3, pp. 319–324, Mar. 1994.

[Dix84] R. C. Dixon, *Spread Spectrum Systems*, John Wiley and Sons, NY, 1984.

[EIA90] Telecommunications Industry Association, *EIA/TIA Interim Standard, Cellular System Dual-Mode Mobile Station - Base Station Compatibility Standard, IS-54*, Telecommunications Industry Association, Washington, D.C., May, 1990.

[EIA93] Telecommunications Industry Association, *TIA/EIA Interim Standard, Mobile Station - Base Station Compatibility Standard for Dual-Mode Wideband Spread Spectrum Cellular System, TIA/EIA/IS-95*, Telecommunications Industry Association, Washington, D.C., July 1993.

[Ert98] R. Ertel, P. Cardieri, K. W. Sowerby, T. S. Rappaport, J. H. Reed, "Overview of Spatial Channel Models for Antenna Array Communication Systems," *IEEE Personal Communications*, Vol. 5, No. 1, pp. 10-22, February 1998.

[Eva82] J. E. Evans, J. R. Johnson, and D. F. Sun, "High Resolution Angular Spectrum Estimation Techniques for Terrain Scattering Analysis and Angle of Arrival Estimation in ATC Navigation and Surveillance System," M.I.T. Lincoln Lab., Lexington, MA, Rep. 582, 1982.

[Fan90] B. T. Fang, "Simple Solutions for Hyperbolic and Related Position Fixes," *IEEE Trans. on Aerospace and Electronic Systems*, Vol. 26, No. 5, pp. 748–753, Sept. 1990.

[Fed88] M. Feder and E. Weinstein, "Parameter Estimation of Superimposed Signals Using the EM Algorithm," *IEEE Trans. on Acoustics, Speech, and Signal Processing*, Vol. 36, pp. 477–489, Apr. 1988.

[Feu94] M. J. Feuerstein, K. L. Blackard, T. S. Rappaport, S. Y. Seidel, and H. H. Xia, "Path Loss, Delay Spread, and Outage Models as Functions of Antenna Height for Microcellular System Design," *IEEE Trans. on Vehicular Technology*, Vol. 43, No. 3, Aug. 1994.

[Feu98] M. J. Feuerstein, "Controlling RF Coverage - Smart Antennas Know How to Optimize CDMA Networks," *America's Network*, Feb. 15, 1998.

[Foy76] W. H. Foy, "Position-Location Solutions by Taylor-series Estimation," *IEEE Trans. on Aerospace and Electronic Systems*, Vol. AES-12, pp.187–194, Mar. 1976.

[Fri87] B. Friedlander, "A Passive Localization Algorithm and Its Accuracy Analysis," *IEEE Journal of Oceanic Engineering*, Vol. OE-12, No. 1, pp. 234–244, Jan. 1987.

[Fro72] O. L. Frost, III, "An Algorithm for Linearly Constrained Adaptive Array Processing," *Proc. of the IEEE*, Vol. 60, No. 8, Aug. 1972.

[Fuh97] J. Fuhl, J-P Rossi, and E. Bonek, "High Resolution 3-D Direction-of-Arrival Determination for Urban Mobile Radio," *IEEE Trans. on Antennas & Propagation*, Vol. 45, No. 4, pp. 672–681, Apr. 1997.

[Fuj92] M. Fujimoto, N. Kikuma, and N. Inagaki, "Performance of CMA Adaptive Array Optimized by the Marquardt Method for Suppressing Multipath Waves, Electronics and Communications in Japan," Part 1, Vol. 75, No. 9, pp. 89–100, 1992.

[Gan72] M. J. Gans, "A Power Spectral Theory of Propagation in the Mobile Radio Environment," *IEEE Trans. on Vehicular Technology*, Vol.VT-21, pp. 27–38, Feb. 1972.

[Gar88] W. A. Gardner, "Simplification of MUSIC and ESPRIT by Exploitation of Cyclostationarity," *Proc. of the IEEE*, Vol. 76, pp. 845–847, July 1988.

[Gar91] W. A. Gardner, "Exploitation of Spectral Redundancy in Cyclostationary Signals," *IEEE Acoustics, Speech, and Signal Processing Magazine*, pp. 14–36, Apr. 1991.

[Gar92a] W. A. Gardner and C. K. Chen, "Signal-Selective Time-Difference-of-Arrival Estimation for Passive Location of Man-Made Signal Sources in Highly Corruptive Environments, Part I: Theory and Method," *IEEE Trans. on Signal Processing*, Vol. 40, No. 5, pp. 1168–1184, May 1992.

[Gar92b] W. A. Gardner, S. V. Schell, and P. A. Murphy, "Multiplication of Cellular Radio Capacity by Blind Adaptive Spatial Filtering," *IEEE Int'l. Conf. on Selected Topics in Wireless Comm. Mobile,* Vancouver, B.C., Canada, June 1992.

[Gar94] W. A. Gardner, Ed., *Cyclostationarity in Communications and Signal Processing*, IEEE Press, 1994.

[Gar97] V. K. Garg, K. Smolik, and J. E. Wilkes, *Applications of CDMA in Wireless Personal Communications*, Prentice Hall, NJ, 1997

[Ger82a] E. A. Geraniotis and M. B. Pursley, "Error Probability for Direct-Sequence spread-Spectrum Multiple-Access Communications - Part II: Approximations," *IEEE Trans. on Comm.*, Vol. Com-30, No. 5, May 1982.

[Ger82b] E. A. Geraniotis and M. B. Pursley, "Performance of Coherent Direct-Sequence Spread-Spectrum Communications Over Specular Multipath Fading Channels," *IEEE Trans. on Comm.*, Vol. Com-33, No. 6, June 1985.

[Ger93] D. Gerlach and A. Paulraj, "Base Station Transmitter Antenna Arrays with Mobile to Base Feedback," *Proc 27th Asilomar Conf. on Signals, Systems, and Computers,* Nov. 1993.

[Gil91] K. S. Gilhousen, *et al*, "On the Capacity of a Cellular CDMA System," *IEEE Trans. on Vehicular Technology*, Vol. 40, No. 2, May 1991.

[Gol73] G. H. Golub and V. Pereyra, "The Differentiation of Pseudo-inverses and Nonlinear Least-Squares Problems Whose Variables Separate," *SIAM Journal of Numerical Analysis*, Vol. 10, pp. 413–432, 1973.

[Goo86] R. Gooch and J. Lundell, "The CM Array: An Adaptive Beamformer for Constant Modulus Signals," *IEEE Int'l. Conf. on Acoustics, Speech, and Signal Processing*, Tokyo, Japan, 1986.

[Gro90] U. Grob, A. L. Welti, E. Zollinger, R. Kung, and H. Kaufmann, "Micro-Cellular Direct-Sequence Spread-Spectrum Radio System Using N-path RAKE Receiver," *IEEE Journal on Selected Areas in Comm.*, Vol. 8, No. 5, pp. 772–780, June 1990.

[Hab86] F. Haber and M. Zoltowski, "Spatial Spectrum Estimation in a Coherent Signal Environment Using an Array in Motion," *IEEE Trans. on Antennas & Propagation*, Vol. AP-34, pp. 301–310, Mar. 1986.

[Hah73] W. R. Hahn and S. A Tretter, "Optimum Processing for Delay-Vector Estimation in Passive Signal Arrays," *IEEE Trans. on Information Theory*, Vol. IT-19, pp. 608–614, Sept. 1973.

[Hah75] W. R. Hahn, "Optimum Signal Processing for Passive Sonar Range and Bearing Estimation," *Journal of the Acoustical Society of America*, Vol. 58, No. 1, pp. 201–207, July 1975.

[Han73] E. J. Hannan and P. J. Thomson, "Estimating Group Delay," *Biometrika*, Vol. 60, pp. 241–253, 1973.

[Han92] S. A. Hanna, M. S. El-Tanany, S. A. Mahmoud, and S. R. Todd, "Applications of Adaptive Antenna Combining to Digital Radio Communications within Buildings," *IEEE Int'l. Conf. on Selected Topics in Wireless Comm. Mobile,* Vancouver, B.C., Canada, June 1992.

[Hay91] S. Haykin, *Adaptive Filter Theory*, Prentice Hall, NJ, 1991.

[Hay95] S. Haykin, "Advances in Spectrum Analysis and Array Processing," Vol. 3, Prentice Hall, NJ, 1995.

[Hep94] E. Hepsaydir and W. Yates, "Performance Analysis of Mobile Positioning using Existing CDMA Network," *IEEE Position Location and Navigation Systems*, pp. 190–192, 1994.

[Hig93] A. Higashi and T. Matsumoto, "Combined Adaptive RAKE Diversity (ARD) and Coding for DPSK DS/CDMA Mobile Radio," *IEEE Journal on Selected Areas in Comm.*, Vol. 11, No. 7, pp. 1076–1084, Sept. 1993.

[Hil93] K. Hilal and P. Duhamel, "A Blind Equalizer Allowing Soft Transition Between the Constant Modulus and Decision-Directed Algorithm for PSK Modulated Signals," *Proc. of the IEEE Int'l. Conf. on Comm.*, Vol. 2, pp. 1144–1148, 1993.

[Hol92] J. M. Holtzman, "A Simple, Accurate Method to Calculate Spread-Spectrum Multiple Access Error Probabilities," *IEEE Trans. on Comm.*, Vol. 40, No. 3, Mar. 1992.

[ITU98] http://www.itu.int/imt/2-radio-dev/proposals/index.html, Nov. 1998.

[Jak74] W. C. Jakes, Ed., *Microwave Mobile Communications*, IEEE Press, Piscataway, NJ, 1974.

[Jen95] S. S. Jeng, H. P. Lin, G. Xu, and W. J. Vogel, "Measurements of Spatial Signature of an Antenna Array," *IEEE 6th Int'l. Symposium on Personal, Indoors and Mobile Radio Comm.*, Vol. 2, pp. 669–672, Sept. 1995.

[Jer92] M. C. Jeruchim, P. Balaban, and K. S. Shanmugan, *Simulation of Communcation Systems*, Plenum Press, NY, 1992.

[Joh86] R. L. Johnson, "Eigenvector Matrix Partition and Radio Direction Finding Performance," *IEEE Trans. on Antennas and Propagation*, Vol. AP-34, No. 8, pp. 985–991, Aug. 1986.

[Jon95] M. A. Jones and M. A. Wickert, "Direct-Sequence Spread-Spectrum Using Directionally Constrained Adaptive Beam Forming to Null Interference," *IEEE Journal on Selected Areas in Comm.*, Vol. 13, No. 1, pp. 71–79, Jan. 1995.

[Jor84] P. S. Jorgenson, "Navstar/Global Positioning System 18 Satellite Constellations," *Global Positioning System: Papers Published NAVIGATION*, Washington: The Institute of Navigation, Vol. II, 1984.

[Jun93] P. Jung, P. W. Baier, and A. Steil, "Advantages of CDMA and Spread Spectrum Techniques over FDMA and TDMA in Cellular Radio Applications," *IEEE Trans. Vehicular Technology*, Vol. 42, No. 3, Aug. 1993.

[Ken96] J. P. Kennedy and S. W. Ellingson, "Smart Antenna Testbed for Mobile Wireless Systems," *Proc. Sixth Va. Tech Symposium on Wireless Personal Comm.*, Blacksburg, VA, 1996.

[Ket96] J. Ketchum, M. Wallace, and R. Walton, "CDMA Network Deployment of 8 kbps and 13 kbps Voice Services," *1996 5th IEEE Int'l. Conf. on Universal Personal Communications Record*, Cambridge, MA, Oct. 1996.

[Kha93] B. H. Khalaj, A. Paulraj, and T. Kailath, "Antenna Arrays for CDMA Systems with Multipath," *1993 IEEE Military Comm. Conf.*, Boston, MA, Oct. 1993.

[Kha94] B. H. Khalaj, A. Paulraj, and T. Kailath, "2D RAKE Receivers for CDMA Cellular Systems," *Proc. IEEE Globecom*, pp. 400–404, 1994.

[Kle96a] A. Kein, W. Mohr, R. Thomas, P. Weber, and B. Wirth, "Direction-of-Arrival of Partial Waves in Wideband Mobile Radio Channels for Intelligent Antenna Concepts," *IEEE Vehicular Technology Conf.*, pp. 849–853, 1996.

[Kle96b] A. Kein and W. Mohr, "A Statistical Wideband Mobile Radio Channel Model Including the Direction of Arrival," *IEEE Fourth Int'l. Symposium on Spread Spectrum Techniques & Applications*, pp. 102–106, 1996.

[Kna76] C. H. Knapp and G. C. Carter, "The Generalized Correlation Method for Estimation of Time Delay," *IEEE Trans. on Acoustics, Speech, and Signal Processing*, Vol. ASSP-24, No. 4, pp. 320–327, Aug. 1976.

[Koh90] R. Kohno, H. Imai, M. Hatori, and S. Pasupathy, "Combination of an Adaptive Array Antenna and a Canceller of Interference for Direct-Sequence Spread-Spectrum Multiple-Access System," *IEEE Journal on Selected Areas in Comm.*, Vol. 8, No. 4, May 1990.

[Kri97] K. J. Krizman, T. E. Biedka, and T. S. Rappaport, "Wireless Position Location: Fundamentals, Implementation Strategies, and Sources of Error," *IEEE Vehicular Technology Conf.*, Phoenix, AZ, May 1997.

[Kum83] R. Kumaresan and D. W. Tufts, "Estimating the Angles of Arrival of Multiple Plane Waves," *IEEE Trans. on Aerospace and Electronic Systems*, Vol. AES-19, No. 1, pp. 134–139, Jan. 1983

[Kur97] R. Kuruppillai, M. Dontamsetti, and F. J. Cosentino, *Wireless PCS*. McGraw Hill, NY, 1997.

[Lar83] M. G. Larimore and J. R. Treichler, "Convergence Behavior of the Constant Modulus Algorithm," *IEEE. Int. Conf. on Acoust., Speech, and Signal Processing*, Boston, MA, 1983.

[Lar85] M. G. Larimore and J. R. Treichler, "Noise Capture Properties of the Constant Modulus Algorithm," *Proc. IEEE Int'l. Conf. on Acoustics, Speech, and Signal Processing*, pp. 30.6.1–30.6.4, Apr. 1985.

[Lar86] M. G. Larimore and J. R. Treichler, "CMA-based Techniques for Adaptive Interference Rejection," *Proc. IEEE Military Communications Conf.*, Apr. 1986.

[Law76] Lawhead, "Position Location Systems Technology," *IEEE PLANS 76*, pp. 1–12, 1976.

[Lee75a] H. B. Lee, "A Novel Procedure for Accessing the Accuracy of Hyperbolic Multilateration Systems," *IEEE Trans. on Aerospace and Electronic Systems*, Vol. AES-11, No. 1, pp. 2–15, Jan. 1975.

[Lee75b] H. B. Lee, "Accuracy Limitations of Hyperbolic Multilateration Systems," *IEEE Trans. on Aerospace and Electronic Systems*, Vol. AES-11, No. 1, pp. 16–29, Jan. 1975.

[Lee82] W. C. Y. Lee, *Mobile Communications Engineering*. McGraw Hill Publications, NY, 1982.

[Lee89] W. C. Y. Lee, *Mobile Cellular Telecommunications Systems,* McGraw Hill, NY, 1989.

[Leh87a] J. S. Lehnert and M. B. Pursley, "Error Probabilities for Binary Direct-Sequence Spread-Spectrum Communications with Random Signature Sequences," *IEEE Trans. on Comm.*, Vol. Com-35, No. 1, Jan. 1987.

[Leh87b] J. S. Lehnert and M. B. Pursley, "Multipath Diversity Reception of Spread Spectrum Multiple Access Communications," *IEEE Trans. on Comm.,* Vol. Com-35, No. 11, Nov. 1987.

[Li93] J. Li and R.T. Compton, "Maximum Likelihood Angle Estimation for Signals with Known Waveforms," *IEEE Trans. on Signal Processing*, Vol. 41, No. 9, pp. 2850–2861, Sept. 1993.

[Li94] J. Li and R. T. Compton, "Angle-of-Arrival Estimation of Coherent Signals Using an Array Doublet in Motion," *IEEE Trans. on Aerospace and Electronic Systems*, Vol. 30, No. 1, pp. 126–133, Jan. 1994.

[Lia95] J.-W. Liang and A. Paulraj, "Forward Link Diversity Using Feedback for Indoor Communication Systems," *Proc. 1995 IEEE Int'l. Conf. on Acoustics, Speech, and Signal Processing,* Detroit, MI, May 1995.

[Lib93a] J. C. Liberti and T. S. Rappaport, "Reverse Channel Performance Improvements in CDMA Cellular Communication Systems Employing Adaptive Antennas," *4th WINLAB Workshop on Third Generation Wireless Information Networks*, East Brunswick, NJ, Oct. 1993.

[Lib93b] J. C. Liberti and T. S. Rappaport, "Performance Improvements in CDMA Cellular Communication Systems Employing Adaptive Antennas," *IEEE Globecom'93*, Houston, TX, Nov. 1993.

[Lib94] J. C. Liberti and T. S. Rappaport, "Analytical Results for Reverse Channel Performance Improvements in CDMA Cellular Communications Systems Employing Adaptive Antennas," *IEEE Trans. on Vehicular Technology*, Vol. 43, No. 3, Aug. 1994.

[Lib95] J. C. Liberti, "Analysis of CDMA Cellular Radio Systems Employing Adaptive Antennas," *Ph.D. Dissertation MPRG-TR-95-17*, Mobile & Portable Radio Research Group, Virginia Tech, Blacksburg, VA, Sept. 1995.

[Lib96a] J. C. Liberti and T. S. Rappaport, "A Geometrically Based Model for Line-of-Sight Multipath Radio Channels," *IEEE Vehicular Technology Conf.*, pp. 844–848, Apr. 1996.

[Lib96b] J. C. Liberti and T. S. Rappaport, "Reverse Channel Performance Improvements in CDMA Cellular Systems Employing Smart Antennas," *IEEE Vehicular Technology Conf.*, pp. 844–848, Apr. 1996.

[Lib96c] J. C. Liberti, *Smart Antennas for High Tier Cellular and PCS Systems*, Bellcore Internal Memo, Sept. 1996.

[Lib97] J. C. Liberti, "Propagation Measurements for Advanced Antenna Systems for Future Battlefield Radio Communications," *1997 U.S. Army Research Labs Advanced Telecommunications/Information Distribution Research Program (ATIRP) Annual Conf.*, College Park, MD, Jan. 1997.

[Lib98a] J. C. Liberti, "Array Measurements for Smart Antenna Applications on the Digital Battlefield," *1998 U.S. Army Research Labs Advanced Telecommunications/Information Distribution Research Program (ATIRP) Annual Conf.*, College Park, MD, Feb. 1998.

[Lib98b] J. C. Liberti, "Measuring and Modeling Spatial Radio Channels for Smart Antenna Systems," *Proc. 1998 IEEE Antennas and Propagation Society Int'l. Symposium*, Atlanta, GA, June 1998.

[Lin83] S. Lin, and D. J. Costello, *Error Control Coding: Fundamentals and Applications*, Prentice-Hall, NJ, 1983.

[Lin95] H. P. Lin, S. S. Jeng, I. Parra, G. Xu, W. J. Vogel, and G. W. Torrence, "Experimental Studies of SDMA Schemes for Wireless Communications," *Proc. 1995 IEEE Int'l. Conf. on Acoustics, Speech, and Signal Processing,* Detroit, MI, May 1995.

[Liu93] T-C. Liu, "The Modular Covariance Adjustment Adaptive Array for CDMA Wireless Communications," *Proc. 1993 IEEE Int'l. Conf. on Acoustics, Speech, and Signal Processing*, Minneapolis, MN, Apr. 1993.

[Lu97] M. Lu, T. Lo, and J. Litva, "A Physical Spatio-Temporal Model of Multipath Propagation Channels," *IEEE Vehicular Technology Conf.*, pp. 180–814, 1997.

[Lu98] M. Lu, T. Lo, and J. Litva, "A Decision-Directed Beamforming Algorithm for CDMA Systems with Noncoherent M-ary Orthogonal Modulation," Submitted to *IEEE Trans. on Comm.*, 1998.

[Luc68] R. W. Lucky, J. Salz, and E. J. Weldon, Jr., *Principles of Data Communication*, McGraw-Hill, NY, 1968.

[Mat95a] A. Mathur, A. V. Keerthi, and J. J. Shynk, "Estimation of Correlated Cochannel Signals Using the Constant Modulus Array," *Proc. IEEE Int'l. Conf. Comm.*, Vol. 3, pp. 1525–1529, June 1995.

[Mat95b] A. Mathur, A. V. Keerthi, and J. J. Shynk, "Cochannel Signal Recovery Using the MUSIC Algorithm and the Constant Modulus Array," *IEEE Signal Processing Letters*, Vol. 2, No. 10, pp. 191–194, Oct. 1995.

[May93] S. Mayrargue, "Spatial Equalization of a Radio-Mobile Channel Without Beamforming Using the Constant Modulus Algorithm (CMA)," *Proc. 1993 IEEE Int'l. Conf. on Acoustics, Speech, and Signal Processing*, Minneapolis, MN, Apr. 1993.

[McG89] C. D. McGillem and T. S. Rappaport, "A Beacon Navgation Method for Autonomous Vehicles," *IEEE Trans. on Vehicular Technology*, Vol. 27, No. 3, pp. 133–139, Aug. 1989.

[Mil92] L. B. Milstein, T. S. Rappaport, and R. Barghouti, "Performance Evaluation for Cellular CDMA," *IEEE Journal on Selected Areas in Comm.,* Vol. 10, May 1992.

[Miz96a] G. A. Mizusawa, "Radio Frequency Position Location Systems Review," *Technical Report MPRG-TR-96-16*, Mobile & Portable Radio Research Group, Virginia Tech, Blacksburg, VA, May 1996.

[Miz96b] G. A. Mizusawa, "Performance of Hyperbolic Position Location Techniques for Code Division Multiple Access," *Technical Report MPRG-TR-96-29*, Mobile & Portable Radio Research Group, Virginia Tech, Blacksburg, VA, Aug. 1996

[Mog96] P. Mogensen, P. Zetterberg, H. Dam, P. L. Espensen, S. L. Larsen, and K. Olesen, "Algorithms and Antenna Array Recommendations," Technical Report A020/AUC/A12/DR/P/1/xx-D2.1.2, Tsunami (II), Sept. 1996.

[Mol91] D. Molkdar, "Review on Radio Propagation into and within Buildings," *IEE Proc.*, Vol. 138, No. 1, Feb 1991.

[Mon80] R. Monzingo and T. Miller, *Introduction to Adaptive Arrays*, Wiley and Sons, NY, 1980.

[Mor89] R. K. Morrow, Jr. and J. S. Lehnert, "Bit-to-Bit Error Dependence in Slotted DS/SSMA Packet Systems with Random Signature Sequences," *IEEE Trans. on Comm.*, Vol. 37, No. 10, Oct. 1989.

[Mot89] Motorola, Inc., *DSP96002 IEEE Floating Point Dual-Port Processor User's Manual*, Motorola, Inc., 1989.

[Mou92] M. Mouly and M.-B. Pautet, *The GSM System for Mobile Communications*, Published by Authors, Palaiseau, France, 1992.

[Muh96a] R. Muhamed and T. S. Rappaport, "Direction of Arrival Estimation Using Antenna Arrays," *Technical Report MPRG-TR-96-03*, Mobile & Portable Radio Research Group, Virginia Tech, Blacksburg, VA, Jan. 1996.

[Muh96b] R. Muhamed and T. S. Rappaport, "Performance Comparison of Subspace-Based Algorithms and the Integrated Property-Restoral with Subspace-Based Algorithms for DOA Estimation," *1996 5th IEEE Int'l. Conf. on Universal Personal Communications Record*, Cambridge, MA, Oct. 1996.

[Nag94a] A. F. Naguib, A. Paulraj, and T. Kailath, "Capacity Improvement with Base Station Antenna Arrays in Cellular CDMA," *IEEE Trans. on Vehicular Technology,* Vol. 43, No. 3, Aug. 1994.

[Nag94b] A. F. Naguib, A. Paulraj, and T. Kailath, "Performance of CDMA Cellular Networks with Base-Station Arrays: The Downlink," *IEEE Int'l. Conf. on Comm.*, 1994.

[Nic73] D. L. Nicholson, "Multipath Sensitivity of a Linearized Algorithm used in Time Difference of Arrival Location Systems," *Digest of Int'l. Electrical and Electronics Conf. and Exposition*, Paper No. 73252, Oct. 1973.

[Nic76] D. L. Nicholson, "Multipath and Ducting Tolerant Location Techniques for Automatic Vehicle Location Systems," *Rec. of Papers Presented at IEEE Vehicular Technology Conf.*, Washington, D.C. , pp. 151–154, Mar. 24-26, 1976.

[Nor94] O. Norklit and J. B. Anderson, "Mobile Radio Environments and Adaptive Arrays," *IEEE Int'l. Symposium on Personal, Indoor and Mobile Radio Comm.*, pp. 725–728, 1994.

[Ohg91] T. Ohgane, "Characteristics of CMA Adaptive Array for Selective Fading Compensation in Digital Land Mobile Radio Communications," *Electronics and Communications in Japan*, Part 1, Vol. 74, No. 9, pp. 43–53, 1991.

[Ohg93] T. Ohgane, N. Matsuzawa, T. Shimura, M. Mizuno, and H. Sasaoka, "BER Performance of CMA Adaptive Array for High-Speed GMSK Mobile Communication - A Description of Measurements in Central Tokyo," *IEEE Trans. on Vehicular Technology*, Vol. 42, No. 4, Nov. 1993.

[Opp89] A. V. Oppenheim and R. W. Schafer, *Discrete-Time Signal Processing*, Prentice Hall, NJ, 1989.

[Ott95] B. Ottersten, "Spatial Division Multiple Access (SDMA) in Wireless Communications," *Proc. of Nordic Radio Symposium*, 1995.

[Par89] J. D. Parsons and J. G. Gardiner, *Mobile Communcation Systems*, Blackie and Son, Limited, Glasgow, Scotland, 1989.

[Par92] J. D. Parsons, *The Mobile Radio Propagation Channel*, John Wiley & Sons, NY, 1992.

[Par95] I. Parra, G. Xu, and H. Liu, "A Least Squares Projective Constant Modulus Approach," *Proc. Sixth IEEE Int'l. Symposium on*, Vol. 2, 1995, pp. 673-676.

[Par96] B. W. Parkinson, J. Spiker Jr., P. Axelrod, and P. Enge, *Global Positioning System: Theory and Applications Vol. I and II*, American Institute of Aeronautics and Astronautics, Inc., Washington, D.C., Vol. 136, 1996.

[Pau86a] A. Paulraj, R. Roy, and T. Kailath, "A Subspace Rotation Approach to Signal Parameter Estimation," *Proc. of the IEEE,* Vol. 74, July 1986.

[Pau86b] A. Paulraj, R. Roy, and T. Kailath, "Estimation of Signal Parameters vis Rotational Invariance Techniques - ESPRIT," *Proc. of 19th Asilomar Conf. on Circuits and Systems*, pp. 83–89, 1985.

[Pau93] A. Paulraj, *et al*, "Subspace Methods for Direction-of-Arrival Estimation," *Handbook of Statistics 10 - Signal Processing and its Applications*, Ed. by N. K. Bose and C. R. Rao, Chapter 16, pp. 693–640, 1993.

[Pet96] P. Petrus, J. H. Reed, and T. S. Rappaport, "Geometrically Based Statistical Channel Model for Macrocellular Mobile Environments," *Proc. IEEE Globecom*, pp. 1197–1201, 1996.

[Pet97a] P. Petrus, "Novel Adaptive Array Algorithms and Their Impact on Cellular System Capacity," *PhD Dissertation MPRG-TR-96-30*, Mobile & Portable Radio Research Group, Virginia Tech, Blacksburg, VA, Mar. 1997.

[Pet97b] P. Petrus, J. H. Reed, and T. S. Rappaport, "Effects of Directional Antennas at the Base Station on the Doppler Spectrum," *IEEE Comm. Letters*, Vol. 1, No. 2, Mar. 1997.

[Pic93] R. Pickholtz and K. Elbarbary, "The Recursive Constant Modulus Algorithm: A New Approach for Real-Time Array Processing," *Proc. 27th Asilomar Conf. on Signals, Systems,* pp. 627–632, 1993.

[Pil89a] S. U. Pillai and B. H. Kwon, "Forward/Backward Spatial Smoothing Techniques for Coherent Signal Identification," *IEEE Trans. on Acoustics, Speech, and Signal Processing*, Vol. 37, No. 1, pp. 8–15, Jan. 1989.

[Pil89b] S. U. Pillai, and B. H. Kwon, "Performance Analysis of MUSIC-Type High Resolution Estimators for Direction Finding in Correlated and Coherent Scenes," *IEEE Trans. on Acoustics, Speech, and Signal Processing*, Vol. 37, No. 8, pp. 1176–1189, Aug. 1989.

[Pron89] N. B. Pronios, "Performance Considerations for Slotted Spread-Spectrum Random-Access Networks with Directional Antennas," *Proc. IEEE Globecom '89*, May, 1989.

[Proa89] J. G. Proakis, *Digital Communications*, McGraw-Hill, NY, 1989.

[Pro95] J. G. Proakis, *Digital Communications*, McGraw-Hill, NY, 1995.

[Pur77] M. B. Pursley and D. V. Sarwate, "Performance Evaluation for Phase-Coded Spread-Spectrum Multiple-Access Communication - Part II: Code Sequence Analysis," *IEEE Trans. on Comm.*, Vol. Com-25, No. 8, Aug. 1977.

[Pur82] M. B. Pursley, D. V. Sarwate, and W. E. Stark, "Error Probability for Direct-Sequence Spread-Spectrum Multiple-Access Communications - Part I: Upper and Lower Bounds," *IEEE Trans. on Comm.*, Vol. Com-30, No. 5, May 1982.

[Pur87] M. B. Pursley and D. J. Taipale, "Error Probabilities for Spread-Spectrum Packet Radio with Convolutional Codes and Viterbi Decoding," *IEEE Trans. on Comm.*, Vol. Com-35, No. 1, Jan. 1987.

[Qua93] Qualcomm, *The CDMA Network Engineering Handbook,* Qualcomm, Inc., San Diego, CA, Mar. 1, 1993.

[Ral95a] G. G. Raleigh and A. Paulraj, "Time Varying Vector Channel Estimation for Adaptive Spatial Equalization," *Proc. IEEE Globecom*, pp 218–224, 1995.

[Rap89] T. S. Rappaport, "Characterization of UHF Multipath Radio Channels in Factory Buildings," *IEEE Trans. on Antennas and Propagation*, Vol. 37, No. 8, Aug. 1989.

[Rap91] T. S. Rappaport, "Wireless Personal Communications: Trends and Challenges," *IEEE Antennas and Propagation Magazine*, Vol. 33, No. 5, Oct. 1991.

[Rap92] T. S. Rappaport and L. B. Milstein, "Effects of Radio Propagation Path Loss on DS-CDMA Cellular Frequency Reuse Efficiency for the Reverse Channel," *IEEE Trans. on Vehicular Technology,* Vol. 41, No. 3, Aug. 1992.

[Rap94] T. S. Rappaport and S Sandhu, "Radio-wave Propagation for Emerging Wireless Personal Communication Systems", *IEEE Antennas and Propagation Magazine*, Vol. 36, No. 5, Oct. 1994.

[Rap96a] T. S. Rappaport, *Wireless Communications: Principles and Practice*, Prentice Hall, NJ, 1996.

[Rap96b] T. S. Rappaport, J. H. Reed, and B. D. Woerner, "Position Location Using Wireless Communications on Highways of the Future," *IEEE Comm. Magazine*, Vol. 34, No. 10, pp. 33–41, Oct. 1996.

[Rap98] T. S. Rappaport, *Smart Antennas: Adaptive Arrays, Algorithms, and Wireless Position Location: Selected Readings*, IEEE Press, 1998.

[Raz98] B. Razavi, *RF Microelectronics*, Prentice Hall, NJ, 1998.

[Ree98] J. H. Reed, K. J. Krizman, B. D. Woerner, and T. S. Rappaport, "An Overview of the Challenges and Progress in Meeting the E-911 Requirement for Location Service," *IEEE Personal Comm. Magazine*, Vol. 5, No. 3, pp. 30–37, Apr. 1998.

[Ris78] J. Rissanen, "Modeling by the Shortest Data Description," *Automatica*, Vol. 14, pp. 465–471, 1978.

[Ron96a] Z. Rong, "Simulation of Adaptive Array Algorithms for CDMA Systems," *Master's Thesis MPRG-TR-96-31*, Mobile & Portable Radio Research Group, Virginia Tech, Blacksburg, VA, Sept. 1996.

[Ron96b] Z. Rong, T. S. Rappaport, "Simulation of MultiTarget Adaptive Algorithms for Wireless CDMA Systems,"*Proc. IEEE Vehicular Technology Conf.*, Apr. 1996.

[Ros93] J. Rossi, and A. Levi, "A Ray Model for Decimetric Radiowave Propagation in an Urban Area." *Radio Science*, Vol. 27, No. 6, pp. 971–979, 1993.

[Ros96] A. H. M. Ross and K. S. Gilhousen, "CDMA Technology and the IS-95 North American Standard," *The Mobile Communications Handbook*, CRC Press, Inc., 1996.

[Rot71] P. R. Roth, "Effective Measurements Using Digital Signal Analysis," *IEEE Spectrum*, Vol. 8, pp. 62–70, Apr. 1971.

[Roy89] R. Roy and T. Kailath, "ESPRIT-Estimation of Signal Parameters via Rotational Invariance Techniques," *IEEE Trans. on Acoustics, Speech, and Signal Processing*, Vol. 37, pp. 984–995, July 1986.

[Roy90] R. Roy, and T. Kailath, "ESPRIT-Estimation of Signal Parameters Via Rotational Invariance Techniques," *Optical Engineering*, Vol. 29, No. 4, pp. 296–313, Apr. 1990.

[Rud89] M. J. Rude and L. J. Griffiths, "Incorporation of Linear Constraints into the Constant Modulus Algorithm," *Proc. IEEE Int'l. Conf. on Acoustics, Speech, and Signal Processing*, pp. 968–971, 1989.

[Sal87] A. A. M. Saleh and R. A. Valenzuela, "A Statistical Model for Indoor Multipath Propagation," *IEEE Journal. on Selected Areas in Comm.*, Vol. SAC-5, No. 2, Feb. 1987.

[Sal90] A. Salmasi, "An Overview of Advanced Wireless Telecommunication Systems Employing Code Division Multiple Access," *Conf. on Mobile, Portable & Personal Communications*, Kings College, England, Sept. 1990.

[San95] S. Sandhu, P. Koushik, and T. S. Rappaport, "Predicted Path Loss for Rosslyn, VA. - Second Set of Predictions," *Technical Report for ORD: MPRG-TR-95-03*, Mobile & Portable Radio Research Group, Virginia Tech, Blacksburg, VA, Mar. 1995.

[Sau90] G. J. Saul, *et al*, "A VLSI Demodulator for Digital RF Network Applications: Theory and Results," *IEEE Journal on Selected Areas in Comm.*, Vol. 8, No. 8, Oct. 1990.

[Sch79] R. O. Schmidt, "Multiple Emitter Location and Signal Parameter Estimation," *Proc. of RADC Spectrum Estimation Workshop*, Griffiss AFB, NY, pp. 243–258, 1979.

[Sch86a] R. O. Schmidt, "Multiple Emitter Location and Signal Parameter Estimation," *IEEE Trans. on Antennas and Propagation,* Vol. AP-34, No. 3, Mar. 1986.

[Sch86b] R. O. Schmidt, and R. E. Franks, "Multiple Source DF Signal Processing: An Experimental System," *IEEE Trans. on Antennas and Propagation,* Vol. AP-34, No. 3, pp. 281–290, Mar. 1986.

[Sch87] H. C. Schau, "Passive Source Localization Employing Intersection Spherical Surfaces for Time-of-Arrival Differences," *IEEE Trans. on Acoustics, Speech, and Signal Processing*, Vol. 29, No. 4, pp. 984–995, July 1989.

[Sch89] S. V. Schell, Calabretta, W. A. Gardner, B. G. Agee, "Cyclic MUSIC Algorithms for Signal Selective DOA Estimation," *Proc. of the Int'l. Conf. on Acoustics, Speech, and Signal Processing-89*, pp. 2278–2281, 1989.

[Sch92] K. R. Schaubach, "Microcellular Radio Channel Prediction Using Ray Tracing," *Master's Thesis MPRG-TR-92-15*, Mobile & Portable Radio Research Group, Virginia Tech, Blacksburg, VA, Aug. 1992.

[Sch93a] S. V. Schell, W. A. Gardner, and P. A. Murphy, "Blind Adaptive Antenna Arrays in Cellular Communcations for Increased Capacity," *Proc. Third Va. Tech Symposium on Wireless Personal Comm.*, June 1993.

[Sch93b] S. V. Schell and W. A. Gardner, "Maximum Likelihood and Common Factor Analysis-Based Blind Adaptive Spatial Filtering for Cyclostationary Signals," *Proc. 1993 IEEE Int. Conf. on Acoustics, Speech, and Signal Processing*, Minneapolis, MN, Apr. 1993.

[Sch93c] S. V. Schell and W. A. Gardner, "Blind Adaptive Spatiotemporal Filtering for Wide-Band Cyclostationary Signals," *IEEE Trans. on Signal Processing*, Vol. 41, No. 5, May 1993.

[Sch93d] S. V. Schell, W. A. Gardner, "High Resolution Direction Finding," Chapter 17, *K.* Bose and C. R. Rao, pp. 755–817, 1993.

[Sch94] S. V Schell, "Performance Analysis of the Cyclic MUSIC Method of Direction Estimation for Cyclostationary Signals," *IEEE Trans. on Signal Processing*, Vol. 42, No. 11, pp. 3043–3050, Nov. 1994.

[Sei89] S. Y. Seidel, "UHF Indoor Radio Channel Models for Manufacturing Environments," *Master's Thesis MPRG-TR-89-00*, Mobile & Portable Radio Research Group, Virginia Tech, Blacksburg, VA, Aug. 1989.

[Sei91] S. Y. Seidel, *et al*, "Path Loss, Scattering and Multipath Propagation Statistics for European Cities for Digital Cellular and Microcellular Radiotelephone," *IEEE Trans. on Vehicular Technology*, Vol. VT-40, No. 4, pp. 721–730, Nov. 1991

[Sei98] S. Y. Seidel, "Radio Propagation and Planning at 28 GHz for Local Multipoint Distribution Service (LMDS)," *Proc. 1998 IEEE Antennas and Propagation Society Int'l. Symposium*, Atlanta, GA, June 1998.

[Sha85] T. J. Shan, M. Wax, and T. Kailath, "On Spatial Smoothing for Estimation of Coherent Signals," *IEEE Trans. on Acoustics, Speech, and Signal Processing*, Vol. ASSP-33, pp. 802–811, Aug. 1985.

[Shy93] J. J. Shynk and R. P. Gooch, "Convergence Properties of the Multistage CMA Adaptive Beamformer," *Proc. 27th Asilomar Conf. on Signals, Systems*, pp. 622–626, 1993.

[Siw95] K. Siwiak, *Radiowave Propagation and Antennas for Personal Communications*, Artech House, Boston, MA, 1995.

[Sko64] M. I. Skolnik and D. D. King, "Self-Phasing Array Antennas," *IEEE Trans. on Antennas and Propagation*, Vol. AP-12, No. 2, Mar. 1964.

[Smi87a] J. O. Smith and J. S. Abel, "The Spherical Interpolation Method for Source Localization," *IEEE Journal of Oceanic Engineering*, Vol. OE-12, No. 1, pp. 246–252, Jan. 1987.

[Smi87b] J. O. Smith and J. S. Abel, "Closed-Form Least-Squares Source Location Estimation from Range-Difference Measurements," *IEEE Trans. on Acoustics, Speech, and Signal Processing,* Vol. ASSP-35, No. 12, pp. 1661–1669, Dec.1987.

[Sou94] E. S. Sousa, *et al*, "Delay Spread Measurements for the Digital Cellular Channel in Toronto," *IEEE Trans. on Vehicular Technology*, Vol. 43, No. 4, pp. 837–847, Nov. 1994

[Spe97a] Q. Spencer, M. Rice, B. Jeffs, and M. Jensen, "A Statistical Model For Angle-Of-Arrival In Indoor Multipath Propagation," *IEEE Vehicular Technology Conf.*, pp. 1415–1419, 1997.

[Spe97b] Q. Spencer, M. Rice, B. Jeffs, and M. Jensen, "Indoor Wideband Time/Angle of Arrival Multipath Propagation Results," *IEEE Vehicular Technology Conf.*, pp. 1410–1414, 1997.

[Sta86] H. Stark and J. W. Woods, *Probability, Random Variables, and Estimation Theory for Engineers*, Prentice-Hall, NJ, 1986.

[Sta94] S. P. Stapleton, X. Carbo, and T. McKeen, "Spatial Channel Simulator for Phased Arrays," *IEEE Vehicular Technology Conf.*, pp. 1789–1792, 1994.

[Sta96] S. P. Stapleton, X. Carbo, and T. Mckeen, "Tracking and Diversity for a Mobile Communications Base Station Array Antenna," *IEEE Vehicular Technology Conf.*, pp. 1695–1699, 1996.

[Ste92] R. Steele, *Mobile Radio Communications*, IEEE Press, NY, 1992.

[Sti95] L. A. Stilp, "Time Difference of Arrival Technology Locating Narrowband Cellular Signals," *SPIE Conf. on Voice, Data, and Video Communications*, pp. 134–144, Oct. 1995.

[Sto89] P. Stoica and A. Nehoral, "MUSIC, Maximum Likelihood, and Cramer-Rao Bound," *IEEE Trans. on Acoustics, Speech and Signal Processing*, Vol. 37, No. 5, pp. 720–741, May 1989.

[Stu81] W. L. Stutzman and G. A. Thiele, *Antenna Theory and Design*, John Wiley & Sons, NY, 1981.

[Sua93] B. Suard, A. F. Naguib, G. Xu, and A. Paulraj, "Performance of CDMA Mobile Communication Systems Using Antenna Arrays," *Proc. 1993 IEEE Int'l. Conf. on Acoustics, Speech, and Signal Processing*, Minneapolis, MN, Apr. 1993.

[Swa90] S. C. Swales, M. A. Beach, D. J. Edwards, and J. P. McGeehan, "The Performance Enhancement of Multibeam Adaptive Base-station Antennas for Cellular Land Mobile Radio Systems," *IEEE Trans. on Vehicular Technology,* Vol. 39, No. 1, Feb. 1990.

[Tak87] K. Takao and N. Kikuma, "An Adaptive Array Utilizing an Adaptive Spatial Averaging Technique for Multipath Environments," *IEEE Trans. on Antennas and Propagation*, Vol. AP-35, No. 12, pp. 1389–1396, Dec. 1987.

[Tal93] S. Talwar, M. Viberg, and A. Paulraj, "Blind Estimation of Multiple Co-Channel Digital Signals Arriving at an Antenna Array," *Proc. 27th Asilomar Conf. on Signals, Systems*, Vol. 1, pp. 349-353, Nov. 1993.

[Tal94] S. Talwar, M. Viberg, and A. Paulraj, "Blind Estimation of Multiple Co-Channel Digital Signals Using an Antenna Array," *IEEE Signal Processing Letters,* Vol. 1, No. 2, Feb. 1994.

[TIA98] Telecommunications Industry Association (TIA) TR45.5.4, "The cdma2000 ITU-R RTT Candidate Submission", available from http://www.itu.int/imt/2-radio-dev/proposals/cdma2000(0.18).pdf, July 27, 1998.

[Tei91] K. Teitelbaum, "A Flexible Processor for a Digital Adaptive Array Radar," *1991 IEEE Natl. Radar Conf.*, 1991.

[Tie98] E. Tiedemann, "The Evolution of CDMA," *Proc. Eighth Va. Tech Symposium on Wireless Personal Comm.*, June 1998.

[Tho92] H. J. Thomas, T. Ohgane, and M. Mizuno, "A Novel Dual Antenna Measurement of the Angular Distribution of Received Waves in the Mobile Radio Environment as a Function of Position and Delay Time," *IEEE Vehicular Technology Conf.*, Vol. 1, pp. 546–549, 1992.

[Tre83] J. R. Treichler and B. G. Agee, "A New Approach to Multipath Correction of Constant Modulus Signals," *IEEE Trans. on Acoustics, Speech, and Signal Processing*, Vol. ASSP-31, No. 2, pp. 459–471, Apr. 1983.

[Tre85] J. R. Treichler and M. G. Larimore, "New Processing Techniques Based on the Constant Modulus Adaptive Algorithm," *IEEE Trans. on Acoustics, Speech, and Signal Processing*, Vol. ASSP-33, No. 2, pp. 420–431, Apr. 1985.

[Ton93] L. Tong, G. Xu, and T. Kailath, “Fast Blind Equalization via Antenna Arrays,” *Proc. 1993 IEEE Int’l. Conf. on Acoustics, Speech, and Signal Processing*, Minneapolis, MN, Apr. 1993.

[Tor84] D. J. Torrieri, “Statistical Theory of Passive Location Systems,” *IEEE Trans. Aerospace and Electronic Systems*, Vol. AES-20, No. 2, Mar. 1992.

[Tur72] Turin, “A Statistical Model for Urban Multipath Propagation,” *IEEE Trans. on Vehicular Technology*, Vol. VT-21, No. 1, pp. 1–11, Feb. 1972.

[Vai92] P. P. Vaidyanathan, *Multirate Systems and Filter Banks*, Prentice Hall, NJ, 1992.

[Val93] R. A. Valenzuela, “A Ray Tracing Approach for Predicting Indoor Wireless Transmission,” *Proc. IEEE Vehicular Technology Conf.*, pp. 214–218, 1993.

[Van88] B. D. Van Veen and K. M. Buckley, “Beamforming: A Versatile Approach to Spatial Filtering,” *IEEE Acoustics, Speech, and Signal Processing Magazine*, Apr. 1988.

[Vau88] R. G. Vaughan, “On Optimum Combining at the Mobile,” *IEEE Trans. on Vehicular Technology*, Vol. 37, No. 4, Nov. 1988.

[Vib91] M. Viberg and B. Otterston, “Sensor Array Processing Based on Subspace Fitting,” *IEEE Trans. on Signal Processing*, Vol. 39, No. 5, pp. 1110–1120, May 1991.

[Vit95] A. J. Viterbi, *CDMA - Principles of Spread Spectrum Communications,* Addison-Wesley, NY, 1995.

[Wal94] M. Wallace and R. Walton, “CDMA Radio Network Planning,” *1994 3rd IEEE Int’l. Conf. on Universal Personal Communications Record*, 1994.

[Wan95] Y. Wang and J.R. Cruz, “Adaptive Antenna Arrays for Cellular CDMA Communication Systems,” *Proc. 1995 IEEE Int’l. Conf. on Acoustics, Speech, and Signal Processing,* Detroit, MI, May 1995.

[Wax85] M. Wax and T. Kailath, “Detection of Signals by Information Theoretic Criteria,” *IEEE Trans. on Acoustics, Speech, and Signal Processing*, Vol. ASSP-33, No. 2, pp. 387–392, Apr. 1985.

[Wel86] D. Wells, “Guide to GPS Positioning,” *Canadian GPS Associates*, 1986.

[Wid67] B. Widrow, P. E. Mantey, L. J. Griffiths, and B. B. Goode, “Adaptive Antenna Systems,” *Proc. of the IEEE,* Vol. 55, No. 12, Dec. 1967.

[Wid85] B. Widrow and S. Stearns, *Adaptive Signal Processing,* Prentice-Hall, NJ, 1985.

[Win82] J. H. Winters, “Spread Spectrum in a Four-Phase Communication System Employing Adaptive Antennas,” *IEEE Trans. on Comm.,* Vol. Com-30, No. 5, May 1982.

[Win84] J. H. Winters, “Optimum Combining in Digital Mobile Radio with Cochannel Interference,” *IEEE Trans. on Vehicular Technology*, Vol. VT-33, No. 3, Aug. 1984.

[Win84] J. H. Winters, J. Salz, and R. D. Gitlin, "Adaptive Antennas for Digital Mobile Radio," *Proc. IEEE Adaptive Antenna Systems Symposium*, Melville, NY, Nov. 1992.

[Win93] J. H. Winters, "Signal Acquisition and Tracking with Adaptive Arrays in the Digital Mobile Radio System IS-54 with Flat Fading," *IEEE Trans. on Vehicular Technology*, Vol. 42, No. 4, Nov. 1993.

[Wir98] Wireless Valley Communications, Inc., PO Box 10727, Blacksburg, VA, 24060, http://www.wvcomm.com

[Wu95] H. T. Wu, J. F. Yang, and F. K. Chen, "Source Number Estimators Using Transformed Gerschgorin Radii," *IEEE Trans. on Signal Processing*, Vol. 43, No. 6, pp. 1325–1333, June 1995.

[Xu94a] G. Xu, R. H. Roy, and T. Kailath, "Detection of Number of Sources via Exploitation of Centro-Symmetry Property," *IEEE Trans. on Signal Processing*, Vol. 42, No. 1, pp. 102–111, Jan. 1994.

[Xu94b] G. Xu and T. Kailath, "Fast Subspace Decomposition," *IEEE Trans. on Signal Processing*, Vol. 42, No. 3, pp. 539–550, Mar. 1994.

[Xu95] G. Xu and H. Liu, "An Effective Transmission Beamforming Scheme for Frequency-Division-Duplex Digital Wireless Communication Systems," *Proc. 1995 IEEE Int'l. Conf. on Acoustics, Speech, and Signal Processing,* Detroit, MI, May 1995.

[Yac93] M. D. Yacoub, *Foundations of Mobile Radio Engineering*, CRC Press, Boca Raton, FL, 1993.

[Yao77] K. Yao, "Error Probability of Asynchronous Spread Spectrum Multiple Access Communication Systems," *IEEE Trans. on Comm.*, Vol. Com-25, No. 8, Aug. 1977.

[Yeh82] Y-S. Yeh and D.O. Reudink, "Efficient Spectrum Utilization for Mobile Radio Systems Using Space Diversity," *IEEE Trans. on Comm.,* Vol. Com-30, No. 3, Mar. 1982.

[Zet94] P. Zetterberg and B. Ottersten, "The Spectrum Efficiency of a Basestation Antenna Array System for Spatially Selective Transmission," *IEEE Vehicular Technology Conf.*, 1994.

[Zet95] P. Zetterberg, Mobile Communication with Base Station Antenna Arrays: Propagation Modeling and System Capacity, *Thesis*, Royal Institute of Technology, Stolkholm, Sweden, Jan. 1995.

[Zet96a] P. Zetterberg and P. L. Espensen, "A Downlink Beam Steering Technique for GSM/DCS1800/PCS1900," *IEEE Int'l. Symposium on Personal, Indoor and Mobile Radio Communications*, Taipei, Taiwan, Oct. 1996.

[Zet96b] P. Zetterberg, P. L. Espensen, and P. Mogensen, "Propagation, Beamsteering and Uplink Combining Algorithms for Cellular Systems," *ACTS Mobile Communications Summit*, Granada, Spain, Nov. 1996.

[Zie85] R. E. Ziemer and R. L. Peterson, *Digital Communications and Spread Spectrum Systems*, Macmillan Publishing Company, NY, 1985.

[Zis88] I. Ziskind and M. Wax, "Maximum Likelihood Localization of Multiple Sources by Alternating Projection," *IEEE Trans. on Acoustics, Speech, and Signal Processing*, Vol. 36, No. 10, pp. 1553–1560, Oct. 1988.

[Zol86] M. D. Zoltowski and M. Haber, "A Vector Space Approach to Dirction Finding in a Coherent Multipath Environment," *IEEE Trans. on Antennas and Propagation*, Vol. AP-34, No. 9, pp. 1069–1079, Sept. 1986.

[Zol96] M. D. Zoltowski, M. Haardt, and C. P. Mathews, "Closed-form 2-D Angle Estimation with Rectangular Arrays in Element Space or Beamspace via Unitary ESPRIT," *IEEE Trans. on Signal Processing*, Vol. 44, pp. 316–328, Feb. 1996.

Index

A

B

C

D

N

O

P

Q

R

S

T

U

V

W

X

Z

The Authors

Joseph C. Liberti, Jr. joined Bellcore in Red Bank, New Jersey, after completing his Ph.D. degree from Virginia Tech in 1995. His areas of work include radio propagation measurements and modeling, development of space-time adaptive signal processing techniques for smart antennas, and system level techniques to exploit smart antennas and other emerging wireless technologies. He is the author of numerous technical papers related to radio propagation, smart antennas, and CDMA, and he holds three patents for inventions related to cellular system monitoring and signal processing techniques for combining signals in multi-antenna receivers.

Since joining Bellcore, Dr. Liberti has continued his work in smart antennas. He has led the development of several highly sensitive array-based systems to characterize the radio channel seen by smart antenna systems. This work has included measurements of static and dynamic spatial radio channels for a variety of signal types including frequency hopping and direct sequence spread spectrum emitters. This work has also demonstrated smart antenna applications for IS-95 CDMA and PACS. In addition to these areas, his efforts have also included direction finding and geolocation systems.

Beyond his smart antenna work, he has also contributed to several projects related to providing wireless broadband data communications and Internet access using the unlicensed spectrum in the United States, and he has participated in the development of GIS-based wireless planning and analysis tools.

Theodore S. Rappaport received BSEE, MSEE, and Ph.D. degrees from Purdue University in 1982, 1984, and 1987, respectively. Since 1988, he has been on the Virginia Tech electrical and computer engineering faculty, where he is the James S. Tucker Professor and the founding director of the Mobile & Portable Radio Research Group (MPRG), a university research and teaching center dedicated to the wireless communications field. In 1989, he founded TSR Technologies, Inc., a cellular radio/PCS manufacturing firm that he sold in 1993. He received the Marconi Young Scientist Award in 1990 and an NSF Presidential Faculty Fellowship in 1992. In 1996, he received Honorable Mention for the G. Holmes MacDonald Outstanding Teaching Award, an honor given by Eta Kappa Nu to electrical engineering professors age 35 or younger.

Dr. Rappaport has authored, co-authored, and co-edited 13 books in the wireless field, including the popular textbook *Wireless Communications: Principles & Practice* (Prentice Hall, 1996), and several compendia of papers including *Cellular Radio and Personal Communications: Selected Readings* (IEEE Press, 1995), *Cellular Radio and Personal Communications: Advanced Selected Readings* (IEEE Press, 1996), and *Smart Antennas; Selected Readings* (IEEE Press, 1998). He has co-authored more than 130 technical journal and conference papers and holds 3 patents. He serves on the editorial board of the *International Journal of Wireless Information Networks* (Plenum Press, NY), is a Fellow of the IEEE, and is active in the IEEE Communications and Vehicular Technology societies. Dr. Rappaport is also chairman of Wireless Valley Communications, Inc., a microcell and in-building computer-aided design (CAD) software development firm. He is a registered professional engineer in the state of Virginia and is a Fellow and past member of the board of directors of the Radio Club of America.